AF540276

MOLECULAR BIOLOGY OF ECOLOGY

ENCYCLOPAEDIA OF MOLECULAR BIOLOGY - IV

MOLECULAR BIOLOGY OF ECOLOGY

By

Dr. M.Prakash

Dept. of Zoology
M.M.H. Post Graduate College
Ghaziabad
(U.P.)

First Published – 2008

Reprinted – 2025

ISBN: 978-93-5056-572-8 (Set)
978-81-8356-269-0

Molecular Biology of Ecology

Published by:

DISCOVERY PUBLISHING HOUSE
4383/4B, Ansari Road, Darya Ganj
New Delhi-110 002 (India)
Phone: +91-11-23279245; 23253475; 43596065
Mobile: +91 9811179893 / +91 9871656464
E-mail: discoverybooksindia@gmail.com
orderdphbooks@gmail.com
namitwasan9@gmail.com
web: www.discoverypublishinggroup.com

Printed at:
Infinity Imaging Systems
Delhi

Preface

The Present title **Molecular Biology of Ecology** is a fast growing area of research from majors or careers in physics, chemistry, mathematics and engineering as well as animal, plant, cell biology and medicine. The overall objective of this publication is to provide a professional level reference work with comprehensive coverage of the molecular basis of life and the application of that knowledge in genetics, evolution, medicine, and agriculture. It deals with the life processes at a molecular level genetic disease diagnosis and genetic therapy; the theory and techniques for understanding manipulating, and synthesizing biological molecules and their aggregates; and the application of biological process to make or modify products to improve plants or animals or to develop microorganisms for specific uses.

Teachers and professors in schools and universities will use this publication for course preparation, and members of the press will find useful background information on new development in biotechnology and genetic medicine. Efforts have been made to prevent a concise treatment of their field of expertise at a level useful to both colleagues and researchers who are experts in related fields, as well as to university students requiring an introduction to a specific molecular biology discipline.

There can be no claim to originality except in the manner of treatment and much of the information has been obtained from the books and scientific journals available in the different libraries.

The author expresses his thanks to his friends and colleagues whose continue inspirations have initiated him to bring out this book.

The author expresses his gratitude to Mr. Wasan and staff of M/s Discovery Publishing House for their whole hearted co-operation in the publication of this book.

Author

CONTENTS

1

Historical Background

The basic information of ecology concerns the lives of organisms and the environments in which they live. In its raw form, fresh from the field, such information is called *natural history*. The observations of natural history provide a basis upon which scientific inquiry begins to establish patterns and determine the mechanisms responsible for those patterns. It should be obvious that the direction of scientific inquiry, and to some degree its ultimate success, depends on the quality of natural history observation. Many of ecology's greatest theoreticians, from Charles Darwin on, were great naturalists. Their insights sprang from close observation of organisms and habitats.

In this chapter we shall see how natural history colours our perception of ecological relationships by considering the lives of three species, each living in a different environment and having to find solutions to different problems. The first, the giant red velvet mite, lives in the harsh environment of the Mojave Desert of southern California; its activities are closely geared to physical factors in its environment, particularly to temperature and rainfall. The next species, the chestnut-headed oropendola, a large relative of North American blackbirds and grackles, lives in the lowland tropical forest of Panama. The oropendola is partner to a complicated arrangement among a diverse group of interacting species. Unlike the harsh desert environment, the climate in wet tropical areas is highly favourable for the development of life; therefore biological factors become more conspicuous parts of the environment, and physical factors are relegated to less important roles as influences on organism function and agents of natural selection. Our third example, that of the baboon on the savannas of Africa, is striking in its complex social life. As for all animals and plants, physical and biological aspects of the baboon's environment have laid the foundations of its adaptations and guide its daily life, but the rich social environment of the baboon has additionally shaped the behaviour of the species.

Environmental Conditions

The Mojave Desert of southern California is an area of little rain, searing summer heat, and chilling winter cold. The desert appears to be nearly devoid of life for most of the year. But the desert's silence and apparent sterility are occasionally broken during the milder days of winter by swarms of insects and other creatures that appear on the surface or fly above it for a few hours, and then disappear as mysteriously as they came. One of the more conspicuous of these creatures is the giant red velvet mite, *Dinothrombium pandorae*. The mite's generic name paints a vivid picture of this close relative of spiders. *Dinothrombium* is derived from the Greek *deinos*, meaning terrible, and *thrombos*, a lump or a clot, describes the mite's resemblance to a clot of blood.

Several decades ago, biologists Lloyd Tevis and Irwin Newell (1962) began a study of the behaviour of the mite in relation to the physical conditions of its environment. They found that the mites spend most of the year in burrows dug in the sand. The particular conditions that favour the emergence of mites occur infrequently in the Mohave Desert. During 4 years of observation, adults appear aboveground only 10 times, always during the cooler months of December, January, or February, when they can tolerate the temperatures on the desert's surface. Tevis and Newell could predict from their observations that an emergence would occur on the first sunny day after a rain of more than three-tenths of an inch, provided that air temperatures were moderate. An individual mite appeared only once each year.

On the day of a major emergence, the mites came out of their burrows between 9:00 and 10:00 A.M., and by late morning one could find thousands of mites scurrying across the desert sands in all directions. At midday, between 11:30 and 12:30, the mites dug back into the sand, waiting until the following year before emerging again.

These "terrible clots" do not leave their burrows to terrorize unsuspecting desert travelers. During their 2- or 3-hour stay aboveground each year, each mite must perform two important functions; feeding and mating. On the same day the mites emerge, large swarms of termites appear flying over the desert sand, their own emergence presumably triggered by the same physical factors that cause the mites to leave their burrows. It is upon these termites that the mites feed. A flying termite cannot, of course, be caught by the earthbound mites; a mite must locate its prey after the termite drops to the ground and sheds its wings, but before it burrows into the sand. All this happens very quickly, giving the mites about an hour to find their prey.

Because the mites are solitary in their burrows, they must mate during their brief period aboveground each year. Courtship of the giant red velvet mite is similar to that of spiders and their relatives. The males walk nervously around and over a feeding female, tapping and stroking her, and cover the sand around her with loosely spun webs. Males court *feeding* females for two reasons. First, the females have ravenous appetites (they have not eaten for a year, after all) and would just as likely devour a male mite as a termite. Second, and perhaps more important, females can produce eggs only if they have had a meal. Thus, by mating with a feeding female, the male guarantees that his efforts to reproduce will not be wasted.

About midday, after the mites have fed and mated, they congregate in troughs on the windward sides of sand dunes, where surface temperatures and the size of the sand particles are just right (less than one-half millimeter in diameter), and reenter the sand almost simultaneously, as if urged on by the same unseen hand. Mites continue to dig their new burrows on the first day until the coolness of the late winter afternoon slows their activity. Burrowing continues on subsequent days when the sand becomes warm enough, until the burrows are completed. During the rest of the year, the adult mite spends its time moving up and down in the burrow to follow the movement of its preferred temperature zone as the surface of the sand heats and cools each day.

Females lay eggs during the early spring. The eggs soon hatch, and the young mites crawl to the surface of the desert to search for hosts, usually grasshoppers, to which they attach themselves. While it is growing, the young mite remains with its host, obtaining its nourishment from the body fluids. When it is fully grown, the young mite drops off its unwilling host and seeks a suitable spot to dig its own burrow in the sand, thus renewing the life cycle of the giant red velvet mite.

Complex Interrelationship

The tropical surroundings of the oropendola in the Republic of Panama are a far cry from the rigorous environment of the giant red velvet mite. In the tropics, air temperatures vary little during the year, and abundant rainfall maintains the lush vegetation. Life abounds; diverse animals and plants are intricately interwoven into a rich fabric of biological interactions.

The situation we are going to examine involves two birds, the chestnut-headed oropendola and its brood parasite the giant cowbird. As we shall see, two insects—a bot fly and a wasp—also play an integral part in the interaction between them. This story was discovered

Fig. 1.1. Female chestnut-headed oropendola at her nest (above) and giant cowbird (below).

by Neal Smith (1968), a staff biologist at the Smithsonian Tropical Research Institute in the Republic of Panama.

Brood parasites (cowbirds and cuckoos are familiar ones) have been known for a long time. They are so-named because the female lays her eggs in the nest of another species, the host, and the young are raised by the foster parents. The presence of a brood parasite in a nest usually reduces survival of the host young because parasitic young compete for the food brought by the host parents. In many cases, the brood parasite removes one or more host eggs before laying her own. Many potential hosts can detect parasite eggs in their nests and eject them. To counter this defense, many species of brood parasite have evolved an elaborate egg mimicry to fool the host species into accepting the alien egg as one of its own.

Chestnut-headed oropendolas breed in colonies, which include anywhere from 10 to 100 of their long, hanging nests, placed along the branches of large, isolated trees. From the observations of other naturalists. Smith knew that the oropendolas were often parasitized by the giant cowbird. Eggshells found on the ground underneath the colonies, thrown out of nests by females after the young had hatched, included those of both oropendolas and cowbirds. But Smith noted that in some colonies the eggshells of the cowbirds were distinctly different from those of the oropendolas, whereas in other colonies the eggshells of the two species so closely resembled each other that they could be distinguished only on the basis of shell thickness. Thus it appeared that in some colonies the cowbirds had evolved to mimic the eggs of their host, while in others they had not.

On the basis of these observations. Smith set out to determine whether the presence or absence of egg mimicry in the cowbirds elicited different behaviour of the oropendolas toward foreign eggs in their nests. A number of objects—including mimetic cowbird eggs, nonmimetic cowbird eggs, other kinds of eggs, and a variety of objects only remotely, if at all, resembling eggs—were put into nests in both kinds of oropendola colonies. Smith found, as he had suspected, that in oropendola colonies where the cowbird eggs closely mimicked those of their hosts, the oropendolas removed virtually everything from their nests except their own eggs and very closely matching cowbird eggs. These oropendolas were discriminators; they tried, often in vain, to discover and eject the cowbird eggs. In oropendola colonies where cowbirds were poor egg mimics, the oropendolas were willing to accept all sorts of alien objects.

In what other ways did these two types of oropendola colonies differ? An extensive survey of the oropendolas in central Panama revealed that in nondiscriminator colonies, young oropendolas were often infested with the larvae of a species of bot fly. These parasites sometimes killed the nestling oropendolas, and frequently so weakened them that their chances of surviving to adulthood were slim. In discriminator colonies these parasites were rarely present. Here was a major difference between the two types of colonies. Was it possible that the role of the cowbird in the colonies was linked to bot-fly parasitism?

Smith examined oropendola young in nondiscriminator colonies (susceptible to bot-fly parasitism), and discovered that the incidence of bot flies was higher in nests that did not contain cowbirds than in nests that did, as shown by the following data:

	Number of nestling oropendolas in nests	
	With cowbirds	Without cowbirds
With bot-fly parasites	57	382
With no parasites	619	42

Further observations showed that the nestling cowbirds would snap at anything small that moved within the nest, including adult bot flies and, furthermore, they would remove bot-fly larvae from the skin of the nestling oropendolas. This behaviour on the part of the young cowbird benefited the oropendola and could account for the acceptance of the brood parasites in nondiscriminator colonies.

The cowbird young are well suited to groom their nest mates. They hatch 5 to 7 days before the oropendola young, and develop precociously. Their eyes are open within 48 hours after hatching, whereas the eyes of oropendola nestlings open 6 to 9 days after hatching. Also, cowbirds are born with a thick covering of down, absent in the oropendola young, which presumably deters bot flies from laying their eggs on the skin of the young cowbirds. By the time the oropendolas hatch, the cowbirds are sufficiently developed to groom them. In discriminator colonies, which are not troubled by bot-fly parasitism, cowbird young perform no such useful function for the oropendolas; and because they compete with the oropendola young for food, cowbirds reduce the productivity of the colony.

The role of the cowbird in this story is now evident, but we have not yet determined why bot flies are present in some colonies and absent from others. Smith noted that all the discriminator colonies, and none of the nondiscriminator colonies, were built near the nests of wasps or bees, whose occupants swarm in large numbers around their nests and virtually fill the air throughout the oropendola colony. Wasps and bees presumably prevent the bot flies from entering the colonies. At any rate, rolls of flypaper hung in the two types of colonies revealed that adult bot flies rarely enter the area around discriminator nests.

Oropendolas and cowbirds are not bothered by the same wasps and bees that viciously attack other intruders at their nests and hives. The birds could benefit the wasps inasmuch as their defenses against nest predators may also protect the wasp nests. In addition, the oropendolas have a characteristic, strong odor that may act as a chemical signal to the insects to suppress their normal defensive reactions. Clearly, the oropendola story may someday chronicle subtle mutual interdependences of which we now have only tantalizing hints.

The nesting success of the oropendolas in nondiscriminator colonies shows the advantage of having cowbirds as nest mates. Relatively few young are raised from the nest of nondiscriminating oropendolas—less than one-half of a young per nest on the average—owing to predation, starvation, and abandonment of nests. Nondiscriminator broods with cowbirds nonetheless produce more than twice as many oropendola young as nest without parasites. The difference in success between nondiscriminator nests and discriminator nests, in the absence of cowbirds, may be attributed to the detrimental effects of bot-fly parasitism in the nondiscriminator nests.

We might expect that in the presence of cowbirds, discriminator and nondiscriminator nests would produce similar numbers of oropendola young. Neither is beset by bot-fly parasitism, and both have the same number of mouths to feed. But, in fact, cowbirds reduce success more in discriminator colonies, where they are disadvantageous to the host, than in nondiscriminator colonies, where they confer some advantage. In discriminator colonies, a cowbird chick reduces the number of host young fledged by about one-half. In nondiscriminator colonies, however, a brood parasite more than doubles the nesting success of the host. These facts suggest that if there were no bot-fly parasitism, nondiscriminator colonies would be the more successful, perhaps because they can select colony sites without regard to the presence or absence of wasp nests. Furthermore, nondiscriminator colonies breed earlier in the year than discriminator colonies because they do not have to wait for wasp nests to become active. More food may be available earlier in the year.

For the oropendola, the presence or absence of wasps in the vicinity of the colony completely alters the role of the cowbird as a factor of the environment. Accordingly, the behaviour of the oropendolas toward cow- birds and their eggs also varies between the two types of colonies, and in turn affects the environment that molds the behaviour of the cowbird. In discriminator colonies, adult oropendolas not only eject the eggs of cowbirds from their nests when they can distinguish them, but they also chase adult cowbirds out of the colony. The oropendolas in nondiscriminating colonies are indifferent to cowbirds. The behaviour of the cowbird in the two types of colonies reflects this difference. In discriminator colonies, female cowbirds are cautious and always enter the colony singly. Behavioural adaptation has gone so far that the cowbirds mimic the behaviour of the discriminator oropendolas; they often gather nest-building materials and act as if they are beginning to build a nest, a most uncharacteristic behaviour of brood parasites. Conversely, cowbirds that parasitize nondiscriminator colonies are often gregarious and enter the colonies in small groups. They behave aggressively toward the oropendolas and sometimes even chase them from their nests.

It is not sufficient for a successful mimic merely to produce an egg that is indistinguishable from host eggs. It must also lay the egg at the proper time. A female oropendola normally lays her 2 eggs on consecutive days. She is sensitive to the appearance of eggs in the nest before or after her laying period, and will frequently desert her nest if it contains more than 3 eggs. Even when confronted by perfect egg mimicry, a discriminating oropendola can be fooled only if a single egg is laid by the cowbird soon after the first oropendola egg has been laid. If the cowbird lays its egg a day too early or too late, it reveals its presence and the oropendola will abandon the nest and start another. In nondiscriminator colonies, however, neither the number of eggs laid by the cowbird nor their appearance matters to the oropendola. Commonly, a female cowbird lays 2 to 5 eggs over several days in the nest of a single nondiscriminating oropendola. Smith referred to these birds as dumpers.

The interaction between the oropendola and the cowbird shows how the pattern of the environment determines the adaptations of the organism and how two or more kinds of organisms can be major factors in each other's environments. The oropendola, the cowbird, the bot fly, and the wasp are mutually important and affect each other's evolution. Such biotic relationships differ sharply from the relationship to the physical environment of an organism that neither evolves nor responds by adaptation to changes in the biotic environment. Whereas the physical environment is passive, the biotic environment is responsive. Each species continually readjusts to evolutionary changes in others with which it interacts. We might expect evolution in a biotically dominated environment to differ from evolution in a physically dominated environment. Different populations, like the oropendola and the cowbird, may coevolve with respect to each other to form a mutually beneficial relationship that could not be achieved between a population and its physical environment.

Behavioural interactions between individuals define both the social environment of an organism and the behavioural patterns that promote and cope with sociality. These patterns are clearly expressed in the social behaviour of baboons, which confers another layer of pattern to our observations of nature.

Baboon in Evolution

Social behaviour has developed to different degrees in various animals, but nowhere has its study excited so much interest as in the subhuman primates, the closest relatives of man. The social group consists of individuals of the same species, many of whom are closely related, upon whom the individual depends in part for its survival. The extent to which some animals have evolved in a social context is demonstrated by their utter dependence upon the group for survival. Such is the case among the baboons of East Africa. In groups, these animals are avoided by all predators except lions; individual baboons, despite their strength and fearsome teeth, have very poor prospects. As a consequence of living in a social group, the individual adapts to an environment determined largely by the behaviour of its group members. Whereas individual antagonism and avoidance are the general rules among nonsocial animals, group cohesiveness is absolutely necessary for the survival of such social animals as the baboon. This cohesiveness is promoted by the evolution of a high level of organization within the baboon troop.

Anthropologists S. L. Washburn and Irven DeVore (1961) studied baboons intensively at two localities in Kenya: a small park near Nairobi and the Amboseli Reserve at the foot of Mount Kilimanjaro. Months of careful observation revealed details of the social interactions among the baboons. For example, the troop, moving in an open savanna, has a spatial structure that reflects the roles of individuals. When resting in trees, baboons are normally safe from predators, but when they move across open grasslands without the immediate safety of trees, members of a troop assume an organized, defensive pattern. The less dominant adult males and some of the older juveniles occupy the periphery of the group. Females without infants and some of the older juveniles stay closer to the center. Females with infants and young juveniles, the most vulnerable members of the troop, move at the center with the dominant males. When the troop is threatened by a predator, the dominant males move into position between the center of the troop and the threat, while the rest of the baboons seek safety behind them. The importance of trees as potential escape routes is so great that baboons are not often found in areas lacking them, even if the habitat is otherwise perfectly suitable.

Interactions among individuals within the troop are many and diverse. The baboon has a strong tendency to stay with other baboons. Even within the troop, individuals form small, persistent friendship groups, within which there is much grooming and playing. Most baboons, particularly adult females, spend many hours each day grooming others that present themselves in the proper manner. To groom, one baboon parts the hair of the other with its hands and removes dirt, lice, ticks, and other objects with its hands and mouth. Grooming and play additionally strengthen the social bonds within the troop and promote its stability and cohesiveness.

Care of infant baboons is left to their mothers. The fathers' attentions are directed toward the well-being of the troop as a whole. After an infant has been weaned, its strong bond with its mother is replaced by looser associations within a play group of other juveniles. As young baboons reach adulthood, they enter the dominance hierarchy of the troop in which their position is determined by fighting, bluffing, and the relative rank of those with whom they associate most frequently. Several males may band together and enhance each other's position in the hierarchy, although individually they hold lower ranks. The position of an individual in the social hierarchy determines to some degree the amount of grooming it receives and its precedence over other baboons at a food and water source. Rank also extends privileges in mating with females.

The female is sexually receptive for about 1 week each month. At the beginning of her estrus cycle, a female often mates with the older juveniles and some of the subordinate males. But when she becomes fully fertile she mates only with the dominant males, who consequently sire most of the offspring in the troop. During the mating period, a male and female form what is called a consort pair for a few hours to several days; all other social functions are disrupted for them. Paired baboons generally move at the periphery of the troop, and the male can be extremely aggressive toward others. Baboons appear not to form long-lasting pair bonds, although lack of close association within the troop need not imply a weak social bond.

The baboon must cope with a wide range of behavioural interactions during its lifetime. The relationship between mother and infant, among individuals in a friendship or play group, among males in a dominance hierarchy or among those individuals that cooperate to defend their dominance rank, between the male and female in a consort pair—all produce a complex social environment.

Social adaptation does not occur independently of the physical and biotic environments. Baboon troops in different habitats exhibit considerable variation in their social organization. For example, in the open savannas, an average-sized baboon troop, perhaps 40 individuals near Nairobi and 80 individuals in the Amboseli, may occupy a home range commonly of 2 to 6 square miles, but sometimes as large as 15 square miles. In Uganda, the baboons inhabit forest or mixed forest-grassland where troops of 40 to 80 usually range over less than 2 square miles. In the Ugandan forests, the preferred food of the baboon, which is fruit, is much more plentiful than in the open savannas of Kenya. Thus the troops can satisfy their food needs in a smaller area.

The social organization within the troops of the two areas also differs in several important aspects. In the open savannas of Kenya, antagonistic interactions between individuals are more frequent than in Uganda. In the forest, an individual can go out of sight behind a bush until its antagonist "forgets" about the dispute between them. No such social escape exists on the open savannas. Forest baboons can fill their stomachs in an hour's feeding each day, whereas baboons of the open grasslands spend most of their time in search of food. Whereas grassland troops are often on the move, forest baboons are frequently engaged in grooming, playing, sitting, and drowsing. Because grassland baboons forage almost continuously, individuals interact less socially, and when they do they are more often antagonistic than are forest dwellers.

Hamadryas baboons, inhabiting the arid grassland at the southern edge of the Danakil Desert in Ethiopia, are organized at three social levels: (1) single males with several females and their offspring; (2) persistent bands of several of these single-male family groups; and (3) a larger troop composed of many bands. During the day, the hamadryas disperse to feed as small bands or single-male groups. At dusk, they congregate in large numbers to sleep on cliffs, where they are protected from nocturnal predators. Before breaking up in the morning, the troop congregates near the roosting cliffs for a morning social hour, filled with grooming and, among infant and juvenile males, play. The gelada baboon, a cliff-roosting species of the Ethiopian highlands, forages in large groups when food is abundant but disperses into single-male groups during seasons of scarcity. In contrast to hamadryas, geladas do not appear to have distinct bands within the troop. The patas monkey of the dry grasslands of Uganda has an entirely different social structure, based solely on single-male groups that seldom meet and are rarely on friendly terms. These studies, and others describing additional variations on the basic single-male groups, multi-male bands, and the larger troops of savanna-dwelling species, emphasize the role of the environment in determining social organization. The particular social structure itself influences adaptations of behaviour that strengthen social bonds and both organize and facilitate social interactions.

Natural history observations are not limited to details of the daily lives of organisms. They also include bacterial degradation of plant litter, population cycles of fur-bearing mammals, community development on bared patches of rocky shore, and variation in the diversity of biological communities over the surface of the earth. With this knowledge, ecologists begin to perceive patterns in the natural world and propose explanations for these patterns. These explanations are put to the test by experimentation and further observation as we seek to understand the basic processes responsible for the structure and function of ecological systems.

2

Environment and Its Inhabitants

The English word "ecology" is taken from the Greek *oikos*, meaning house, the immediate human environment. In 1870, the German zoologist Ernst Haeckel first gave the word its broader meaning, the study of the natural environment and of the relations of organisms to each other and to their surroundings. General use of the word came only in the late 1800s, when European and American scientists began to call themselves ecologists. The first societies and journals explicity devoted to ecology appeared in the early decades of this century. Since that time, ecology has undergone immense growth and diversification, so much so that persons devoting their professional lives to ecology now number in the tens of thousands. With the dual crises of rapid growth of human population and accelerating deterioration of the earth's environment, ecology has taken on the utmost importance to everyone. Management of biotic resources in a way that sustains a reasonable quality of human life depends upon wise ecological principles, not merely to solve or prevent environmental problems but to inform our economic, political, and social thought and practice.

Nitrogen cycles in forests, energy budgets of pond organisms, stability of marine food chains, pollination systems of plants, mating calls of toads, predator-prey oscillations—all fall within the realm of ecology. This diversity is well illustrated by the complexities that meet our effort toward control and management of populations, whether to maintain resources of food or fuel, to reduce the incidence of disease, or to preserve species of aesthetic or cultural worth. Management of populations requires us to understand the causes of population change and the consequences of population change in any natural system. Several years ago, the Nile perch was introduced into Lake Victoria in East Africa. This was done with the well-intentioned purpose of providing additional food for the peoples living in the area and additional income from export of the surplus catch. Because simple ecological principles were ignored, the result was virtual destruction of the lake's entire fishery. Until the introduction of the Nile perch, Lake Victoria supported endemic fishing for a variety of species of cichlids, which themselves fed primarily on detritus and plants. The large Nile perch feeds upon other fish, the smaller cichlids in this instance. Because energy is lost with each step in the food chain, a predatory fish cannot be harvested at as high a rate as its herbivorous prey. Furthermore, the perch was alien to Lake Victoria, and evolution of the local cichlids did not include behaviour to escape predation. Inevitably, the perch annihilated the cichlid populations, destroying the native fishery and all but eliminating its own food, thereby, in turn, hastening its own demise as an exploitable fish. To be sure, the native fishery was already precariously overexploited owing to growth of the local human population and recent application of less primitive fishing technology, but the appropriate solution to these problems would have been better management of the cichlids, not introduction of an efficient predator upon them.

Still other results of the introduction of the alien fish turn the story into a tragic comedy of errors. The flesh of the Nile perch is not particularly liked by the peoples living near the shores of Lake Victoria. They prefer the accustomed texture and flavor of the native fishes. Moreover, the oily meat of the Nile perch must be preserved by smoking rather than sun-drying, and local forests are being cut rapidly for firewood. Because the larger Nile perch requires larger and more elaborate nets, local subsistence fishermen cannot compete with more prosperous outsiders who are equipped for commercial fishing. The lesson is painful yet simple: Man is an integral part of the ecology of the Lake Victoria area. Traditional local fishing had been sustained for thousands of years until the pressure of population and the perception of an opportunity for an export fishery led to an ecologically unsound decision and to an economic and social disaster.

Half a world away, efforts to save the California sea otter illustrate the intricate intermingling of ecology and other human concerns. The sea otter was once widely distributed around the North Pacific Rim from Japan to Baja California, but in the 1700s and 1800s intense hunting for fine otter pelts reduced the population to near extinction. The fur hunting industry, of Course, collapsed from the overexploitation. Subsequent protection enabled the California sea otter population to rise to more than 1500 individuals by the 1980s. The sea otter's success, alas, has irked California fishermen, who claim that the otters —which do not need commercial fishing licenses—drastically reduce stocks of valuable abalone, sea urchins, and spiny lobsters. The matter has deteriorated to the marine equivalent of a range war between fishermen and conservationists, with the otter, often fatally, caught in the line of fire. Ironically, the otters benefit a different commercial marine enterprise, the harvesting of kelp, a large seaweed used in making fertilizer. The kelp grows in shallow waters in stands called kelp forests, which provide refuge and feeding grounds for several larval fish. The kelp is also grazed by sea urchins which, like goats on land, can, given the right conditions, denude an area of vegetation. One of those conditions is the absence of a principal sea urchin predator, the sea otter. When the expanding sea otter population has spread into new areas, urchin populations have been controlled, allowing kelp forests to regrow. Human involvement in the local ecosystem requires that management of otter populations and various commercial exploitation be balanced in an economically and socially acceptable manner, an impossibility without understanding the complex ecological role of the otter in the system.

Although the plight of endangered species may arouse us emotionally, there is a beginning realization that the only means of preserving and using natural resources is through the conservation of entire ecological systems and of broad-scale ecological processes. Individual species, including the ones that humans rely upon for food and other products, themselves depend upon the maintenance of a healthy ecological support system. We have already seen how such predators as the Nile perch and sea otter may assume key roles in maintaining the natural properties of a system or in dramatically altering its properties. On land, pollinators and seed-dispersers are necessary to the long-term maintenance of plant formations. Therefore, the management of forests without attention to maintaining populations of insects, birds, and mammals, upon which the long life of the forest depends, is ecologically naive. Natural systems are diverse and highly evolved in the sense that each species is adapted to its particular role within the system and the system depends to some degree on each player acting out its role. It turns out that any reduction in biodiversity can upset the balance of a system and alter its function.

Most natural systems sustain cycles of disturbance (fire, hurricanes, disease) and rejuvenation. Natural restoration depends on there being patches of intact habitat from which species may recolonize disturbed areas. But as habitats become more and more fragmented, as are forests by the spread of agriculture and urban development, disturbances can so thoroughly destroy an area as to leave little chance of complete recovery, even with substantial help from man. Disturbance cycles, whatever their spatial and temporal scale, are as much a part of the natural landscape as are key species within a local system and must be taken equally into account in any effort to maintain the system.

Many of the now rapid changes in the world bring a seeming improvement in the quality of human life, but those very changes increasingly strain the earth's ecological support systems and in reality bring decline in the quality of life. On any scale on which we view the surface of the earth, the deterioration of the environment is alarming. Partially enclosed systems, such as the estuaries of major rivers, allow us to isolate the causes of much of the change. In San Francisco Bay, for example, the pressures of population and agriculture have gradually but profoundly altered it to the point that it would be unrecognizable to nineteenth-century inhibitants of the region. Diking of wetlands to create farmland, filling to create land for urban construction, pollution from sewage and from agricultural and industrial waste, and introduction of exotic species have transformed a healthy, productive, and economically valuable system into an ecological disaster, whose greatly diminished value lies entirely in recreation, itself of ever less value, and in commercial shipping, not in its once generous biotic resources. The present human population of the San Francisco Bay area may well be incompatible with a pristine estuary, but much of the change came about through state subsidy of unprofitable agricultural ventures, absence of ecologically and economically sensible city planning, careless introduction of pest species, use of the bay as a cheap sewer, and uncontrolled harvesting of fish and shellfish.

The multiplication of technology and population has brought ecological crisis to far larger regions of the earth than San Francisco Bay. One of the most critical is the Sahel in Africa. The Sahel is a vast band of semiarid habitat lying at the southern edge of the Saharan Desert and extending from Gambia in the west to Ethiopia in the east. Population pressure, expressed through intensified grazing and collection of firewood, has caused extensive deforestation of the entire region. As humans and domesticated animals clear vegetation over such vast areas, climate is affected. Rainfall decreases because less water vapour is recirculated to the atmosphere and heating and cooling patterns of the land surface are changed. Even when plentiful rains result in good yields from crops, the abundance may be snatched away by plagues of grasshoppers and locusts, whose populations grow rapidly under the moister conditions. The Sahel is at present truly devastated.

In wetter parts of the world, tropical rainforests are being cleared at an alarming rate for forest products and largely unsuitable agriculture, bringing profit to a few in the short term and yet another disaster in the long term. The destruction of rainforests dismantles an intricate, self-sustaining system that cannot be rebuilt. Species, most of them unknown to science and many of them with potential commercial value, are being driven to extinction. Burning of cut vegetation adds greatly to the already dangerous level of carbon dioxide in the atmosphere. In this instance, the ecological principles that should govern human use of the rainforest are generally understood. 'As Michael Robinson, director of the National Zoo, Washington, D.C., has pointed out, "The problems are not (the result of) ignorance and stupidity. The problems derive from the poverty of the poor and the greed of the rich".

The destruction of rainforests is scarcely alone in its far-reaching effect. Automobile emissions reduce tree and crop production over large areas. Sulfurous gases from burning coal have acidified the rain and snow over large areas of eastern North America and Europe, killing forests and degrading freshwater ecosystems. Chlorofluorocarbons (CFCs) used to pressurize spray cans and as refrigerants have reduced the ozone concentration in the upper atmosphere, increasing the amount of dangerous ultraviolet radiation that reaches the earth's surface.

As a result of industrial growth and deforestation, levels of carbon dioxide, methane, and CFCs have increased in the atmosphere so much that they are changing the earth's climate. In the so-called greenhouse effect, long-wave radiation from the earth's surface is absorbed and retained by the atmosphere. As a result the temperature of the earth is increasing and is expected to rise by 2 to 6 degrees Celsius during the next century. Such change will cause a rise in sea level as the polar ice caps melt, it will disrupt agriculture and shift ecological habitats across the landscape as temperature and moisture patterns change, and it will perhaps cause widespread extinction of forms of life unable to adjust to the change and its effect.

Our ability to wreak havoc upon ourselves and the world in which we live is virtually unlimited. The most appalling evidence of this ability is our means of waging global nuclear war, wherein destruction of ecological systems and extinction of species, possibly including ourselves, would be virtually without parallel in the history of the earth. Nuclear war remains unthinkable even if we ignore the ecological consequencies. What we are gradually accomplishing in the degradation of a livable planet may end by being just as punishing. Merely to understand ecology will not solve our environmental problems in all their political, economic, and social dimensions. But as we contemplate the need for global management of natural systems, we are doomed to fail if we do not understand their structure and function, an understanding that depends upon the principles of ecology.

3

GRADING NATURAL WORLD

At first, we are bewildered by the diversity and complexity of nature. Every species plays a different role in the ecological play. Some connections make themselves apparent to us from the start. The caterpillar depends upon the tree whose leaves it eats. The tree sends its roots deeply in the soil, from which it obtains water and mineral nutrients needed for growth and maintenance. Feeding relationships unite all species into a system. But other connections also govern the structure and function of ecological systems. What, for example determines how many species live in a given patch of forest? Do their mutual interactions somehow limit the kinds of organisms that can coexist? What is responsible for the particular form and behaviour of each species? How do the forest, the grassland, and the oyster bed replenish the resources they take from the environment?

PATTERNS IN NATURAL WORLD

What better place to experience nature than in the tropical rainforest, where life is most luxurious and diverse? Nowhere else is one so acutely aware of nature. At night, especially, the multitude of active creatures make the life of the greatest cities seem paltry by comparison. In the Panamanian rainforest, the delicate first light of a November day is shattered by the explosive cries of a nearby troop of howler monkeys. The howling subsides, but it is presently answered by that of another troop farther away, and then another and another. We can easily locate the nearby troops of monkeys and so, we would guess, can the howlers themselves. Are we witnessing a morning proclamation of territory by each group? It is now light enough so that we can begin to wend our way along the narrow forest trails, being careful not to brush against the long sharp spines that viciously ornament the trunk of the black palm. We make our way around the giant buttresses that spread out from the bases of trees over the forest floor. There is a faint snapping sound, and our eyes quickly search for some small animal moving in the bushes. Another snap. A small object catapults in front of us and strikes the ground nearby. Finally we locate the source—a small bush whose fruits have opened and are shooting their seeds over the forest floor. As the fruit, which resembles a three-sided pea pod, dries out, the edges of the pod come together, squeezing the seeds with greater and greater force until the stalk of the seed finally breaks and the seed is ejected from the pod.

The drab browns and greens of the inner rainforest are occasionally broken by the flashing of iridescent butterfly wings. A blue morpho butterfly streaks by us, just out of reach. How striking it is that these butterflies should be so conspicuous. How different from the moths that are attracted at night to the lights around buildings; they possess every conceivable device for being unobtrusive. Their brown and gray colours look like dead leaves and bark. Many of them conceal their legs, normally a sure betrayal of any animal, beneath their wings; others benefit from legs modified to resemble bark or lichens. One species characteristically protrudes one leg or another from underneath its regular outline to break up the symmetry so characteristic of animal forms. Others have wings that, although intact, often have the broken and contorted appearance of dead leaves. The hind edge of the wing of another species bears the unmistakable picture of a rolled-up leaf complete with its shadow. How perfectly these moths blend with their backgrounds when they are resting on leaves or the bark of trees. How have these moths come to be that way, and why are other kinds of butterflies and moths so conspicuous?

We are constantly impressed by a sense of purpose in the forms of animals and plants we encounter. As we walk through the forest we wonder about the purpose of nature. Whose purpose? What purpose? Our thoughts are suddenly interrupted as we step, as if going from one room to another through a door, into a large clearing where we are blinded momentarily by the sun. A recent treefall caused this break in the forest canopy. The giant did not exit gracefully, but took with it at least a dozen other trees directly in its path or bound to it by vines. The clearing gives us our first glimpse of the sky in several hours. A long, graceful ribbon of hawks silently glides southward many hundreds of feet above. Their flight seems effortless—none of the birds moves its wings. The narrow ribbon is caught by a small updraft, sending it spiraling upward in wide, lazy circles until it breaks away again toward South America, seemingly impatient to be gone. Are these the same birds that we saw earlier in the fall, gliding along the ridges of the Appalachians in Pennsylvania? What has sent them south? Why are they going all the way to Argentina? What cue will they experience there that tells them spring is coming once again in the north and it is time to make the return journey?

Our own journey through the forest is coming to an end. By the time we return to our house, our eyes and ears have experienced more than our minds can sort out in a day, and we are tired after the long walk. But a shower and a good meal start us on an evening of conversation that takes us back over the day's journey to recount our observations and to ask questions—mostly to ask questions—far into the night.

Diversity

The naturalist stands in awe of the diversity of forms in nature and of the intricate interactions of these forms with one another and with their environments. In our walk through the tropical forest, we passed several hundred varieties of trees, and yet to untrained eyes they all seemed to be "doing" about the same thing. They were using the energy of sunlight assimilated by their green leaves to convert carbon dioxide from the air, and water and minerals from the soil, into the organic molecules that make up their structure. Why should there be so many kinds of plants when one seemingly could perform the same functions as all the others? Indeed, we know that vast forests consisting of one or, at most, a very few species of coniferous trees stretch across the middle latitudes of Canada.

There are hundreds of species of butterflies and moths in the forest, and thousands of species of other insects. Their appearance varies so much that one wonders if there are as many different environments within one forest as there are kinds of insects. In spite of how diverse the butterflies and moths look to us and, presumably, to their natural predators, they are remarkably similar in their feeding habits. As adults, they all have long tubular snouts, which normally lie curled beneath their heads, but which can extend to the depths of tubular flowers for feeding. And as larvae—caterpillars—they all eat green vegetation. Most of the species are picky eaters and can be found on only one or a few species of plants. A few feed very widely from the forest's smorgasbord. How do adults know which are suitable food plants to lay their eggs on? Why are some specialists and some generalists? Adult butterflies and moths, as well as many other insects, birds, and bats, perform a vital function in the forest by carrying pollen from flower to flower, thus ensuring that the plants will set seed. How do they decide which flowers to visit? Why is it that some species of moths—the silk moths—do not feed as adults and die within a few days of emerging from the pupa?

Many species are mutually bound to others for their survival. The delicately coloured harlequin beetle, for example, carries on its back a small community of mites and pseudoscorpions that feed upon the mites. This particular kind of pseudoscorpion is found nowhere else in nature. The mites perform a beneficial function for the beetle by scouring fungi from its delicate membranous hind wings (fungi seem to grow on almost everything in the tropics). The pseudoscorpions take advantage of this ready food supply, but the mites are at least partly protected when they hide in the numerous small pits that dot the forewings of the beetle. Examples of organisms living in intimate, mutually interdependent relationships abound. Lichens are a combination of alga and fungus—the first makes carbohydrates by photosynthesis, the second obtains water and minerals from the rock or tree trunk on which the lichen lives. Much of the digestion that occurs in the guts of animals is carried out by specialized bacteria and protozoa living there. Plants are aided in the uptake of minerals from the soil by specialized fungi intimately associated with their roots. Insects pollinate their flowers. Birds and mammals disperse their seeds.

Dynamic World

The large tree that falls in a forest may be several hundred years old when it finally topples; its life is exceptionally long compared to most living things. Death comes in many ways to organisms. Some fall victim to predators and parasites, while others die of exposure to the physical environment. A cold snap in spring or a pond's drying up late in the summer casts death's net very broadly. The natural community, whose presence seems so stable, actually undergoes constant turnover, with replacement by new individuals, just as the organic structure of our body is continually replaced during our lifetime. In spite of this dynamic aspect, the natural world is also measurably stable—an equilibrium is maintained. The forest we walk through today is very much as it was 5, 10, or even 100 years ago. The same kinds of plants and animals persist to the present, though most of these are not the same individuals that were here earlier.

The dead bodies of organisms and the wastes of biological processes do not pile up. They are broken down and their component parts are recycled by the community. The dead leaves rustling under our feet cover the decomposing remains of other leaves in the soil beneath them. Soil organisms transform their elegant shapes into an amorphous mass of decaying and decomposing plant tissues, finally reducing them to the mineral elements from which they were once, in part, synthesized.

Populations of organisms also are continually replaced. An insect may lay thousands of eggs each year, and some marine organisms shed millions of eggs into the water—more than necessary to compensate losses of individuals from populations. Yet in spite of tremendous potential for growth, various checks and balances keep most populations within rather narrow limits.

One must not assume, however, that the world is unvarying. Quite the opposite. Variation and equilibrium are in constant tension. All systems suffer disturbance—from weather, fire, tree falls, even a cow pat creates a major disturbance for some organisms—from which they continually recover. Patches of disturbance may range from a few millimeters, as an earthworm eats its way through the soil, to large portions of the earth, as global weather patterns shift. Much variation is imposed by the physical environment—climate, ocean currents, landscape evolution—but much is also generated by the biological world. Many organisms create disturbances for others as they forage and generally move about. The dynamics of population interactions can establish cycles of population change that send ripples throughout the biological community. Variation must be considered a part of the equilibrium of the natural world.

Living Processes

We constantly recognize patterns, organization, and interrelationships in the complexity of our surroundings. We are able to perceive these patterns because of the predictability of their elements. We can anticipate nature in various ways, but most efficiently by the generalization of past experience. For example, experience with the movement of objects thrown into the air enables us to predict their trajectories with accuracy. A good outfielder can predict where a baseball will land long before it begins to drop.

Our lives are organized around patterns of our environment. Only if nature is predictable can we respond properly to it. Birds live in the woods, fish inhabit the sea. Without ever having been in a particular forest, we could expect to find birds rather than fish simply on the basis of past experience with birds and fish and forests and seas. By experiencing the unnaturalness of surrealism and dreams, we realize how completely our minds are bound up in the various patterns we recognize in nature. As ecologist G. Evelyn Hutchinson once pointed out, "(if) we imagine ourselves encountering in the middle of a desert a rock crystal carving of a sewing machine associated with a dead fish to which postage stamps are stuck, we may suspect that we have entered a region of the imagination in which ordinary concepts have become completely disordered."

Patterns have two sources of predictability, one achieved through observation, and the second by understanding the mechanism that produces the pattern. In the first case, predictions are based on extrapolating observations to new but similar situations. We do this when we predict the flight path of a ball. But by applying the laws of motion, we could have predicted the trajectory of the ball without any previous experience with the phenomenon, knowing only its initial speed and direction. Similarly, by understanding the principles governing the form and functioning of animals, we would know that we would not be likely to find an organism resembling a fish living in a forest.

In the development of science, empirically observed patterns almost always preceded the discovery of the causative principles that produce the patterns. After detailed observations, the German astronomer Johannes Kepler discovered that the time required for a planet to revolve around the sun is inversely related to its distance from the sun. Only later did the English physicist Isaac Newton formulate laws of motion whose predictive powers are so great that they made possible the detection of unseen planets on the basis of their gravitational effects on the motion of some of the known planets. Alfred Wegener did not understand the underlying mechanism when in 1915 he proposed, from observations of geological and geographical relationships of the continents, that the major land masses slowly drift over the surface of the earth. In fact, for many years Wegener was ridiculed and his ideas rejected for lack of a plausible mechanism. (In the last two decades, however, irrefutable evidence for continental drift has been discovered, its rate has been measured directly by satellite observations, and plausible mechanisms have been proposed.) The same has been true in the biological sciences. Early naturalists classified organisms into a regular hierarchy of species and other taxonomic groupings based on similarities. But only after Charles Darwin proposed his theory of evolution was the basis for these patterns understood.

Perception

Our world view is very much a product of the information we receive. Overwhelmed by the sights and sounds in our surroundings, we often forget that other organisms perceive the world differently than we do. We mostly see and hear the world around us. Smells drift by largely unnoticed. Our sense of taste is dull. We also become accustomed to interpreting natural phenomena in terms of surroundings that we are familiar with. Islanders know water better than natives of Kansas, for example.

Because of our size, mobility, and the pace of our lives, we are sensitive to particular scales of variation in time and space. Our perception of distance differs from that of the collembolan, which finds a jungle in a pinch of soil, or the water flea, which perceives a drop of pond water as we would the entire pond. For us, a few tens of years are a lifetime. But our lives are instants in the clock of evolution and ecosystem development, and eternities to microorganisms.

Our limited perception of nature can encumber our ability to understand natural phenomena. For years, most ecologists were trained and worked in Europe and temperate North America, where seasonal fluctuations in temperature are a major aspect of the environment. When naturalists visited the tropics, they were impressed by the constant year-round temperature and they assumed that biological communities are less variable in tropical regions than in temperate regions. Only recently have ecologists bothered to count individuals of tropical species over long periods. Surprisingly, they have found that populations undergo marked seasonal fluctuations in the tropics, and additionally that they can vary considerably from year to year. Temperature defines the seasons in a temperate climate and forms a familiar pattern. But the patterns are different in the tropics, where the seasons are marked by wet and dry periods, and where rainfall is notoriously unpredictable at many times of the year.

Our interpretation of the natural world is also plagued by the way that size determines pertinent scales of time and space. A flea can jump a hundred times its length. We immediately react by translating that distance to a familiar scale—comparing flea lengths to human lengths. The comparable human jump would clear two football fields. How incredibly strong fleas are! But the flea does not actually perform athletic miracles. As size changes, the relationships between distance, power, and time change accordingly. Consider the wing beat of flying organisms. Try to move your arms up and down as rapidly as possible. How fast? Two or three times per second? Most large birds can flap their wings between 2 and 20 times per second, and some small insects can do so up to 500 times per second. This may seem amazing, but as the size scale changes, so does the relevant time scale. You can demonstrate this very simply for yourself. Tie a weight to a string, and start it swinging like a pendulum. As you shorten the length of the string, you will notice that the weight swings back and forth much faster. In fact, the rate of swinging varies inversely with the length of the string. Halve the length of the pendulum, and its swinging frequency doubles. The wings of most small insects are less than a centimeter long, so we should not be surprised that an insect can beat its wings more than a hundred times faster than we can flap our arms.

Scientific Basis

Scientists look at the natural world from many different viewpoints, depending on their training and temperament, the problems they study, and the systems with which they work. All perspectives and approaches are valid to the extent that they can help us to understand the natural world. But in spite of their differences, all scientists employ similar methods of study. A question is the starting point of any inquiry. Phenomena—patterns in nature—prompt us to inquire how or why a pattern came to be. We form several hypotheses to answer our question, and then test each hypothesis by suitable experiments or further observations. If the results are consistent with a hypothesis, we may begin to generalize our understanding of a phenomenon and make valid predictions based on this understanding. More importantly, if the results are not consistent with a hypothesis, we may reject that hypothesis as flawed and thus narrow our search for the "truth" of nature. Experimental results—tests of ideas—provide new observations, which in turn may prompt us to ask new questions or rephrase old ones. Scientific inquiry is thus self-perpetuating; once a new inquiry is begun, it snowballs, and questions lead to new questions.

The principal features of scientific inquiry—the detection of pattern in nature, the proposal of ideas to explain patterns (theories), and the testing of theories by evaluating predictions they permit us to make about previously unappreciated phenomena. Although there are accepted guidelines for developing and testing theory, there is no single best way to approach science. Scientific inquiry is as much an art form as painting or the composition of music. While our culture dictates certain aesthetic standards, individual compositions or paintings are unique products of the human spirit and imagination. So it is with science. Differences in intuition and approach among individual practitioners make science exciting and guarantee its progress. Much of this variation arises from different experiences with the diversity of nature.

4

LIFE HISTORY ADAPTATION

Adaptation is a form of response to the environment. It also resolves conflicting requirements of organisms. An individual cannot be everywhere at once; it cannot do everything at once. It must divide limited time, energy, and nutrients among competing demands. A mammal's fur conserves heat when at rest in the cold, but impedes the dissipation of heat when active, especially in a warm climate. As an adaptation, the thickness of fur optimally resolves this conflict between heat conservation and dissipation for a particular type of organism in a particular climate.

As already discussed decisions concerning prey, patch, and habitat choices. Each choice an organism makes influences its allocation of time between activities having different costs and benefits. Thus a decision to pursue an item of food yields a potential increment of energy and nutrients consumed but exacts a price in the currency of potential prey passed by. Each decision resolves a choice; the rules according to which each individual makes decisions express evolved properties of its nervous system.

Body insulation and prey choice seem remote to the adaptationist's measure of evolutionary fitness, partly because each constitutes a small part of the overall phenotype exposed to selection and only indirectly influences survival and number of offspring. Other characteristics, such as number of eggs, age at first reproduction, and parental care, seem more directly tied to fitness components and so they have received special attention from evolutionary ecologists. These so-called *life history adaptations* resolve conflicts over the allocation of limited time and resources between activities and processes that more or less directly enhance fecundity and those that enhance survival. In this chapter, we shall examine the Contributions of particular life history adaptations to fitness, the linking of such adaptations through competing demands on limited time and resources, and criteria by which such conflicts are resolved in the best evolutionary interests of the organism.

Study of Adaptation

R. J. Lincoln et al. (1982), in their *Dictionary of Ecology, Evolution and Systematics*, define a life history as "the significant features of the life cycle through which an organism passes, with particular reference to strategies influencing survival and reproduction." One of my skeptical colleagues unmasked the meaning of this definition when he quipped, "A life history is everything an organism is and does." While such a broad definition deprives the term of any meaning whatsoever, there has nonetheless developed a substantial and important literature dealing with the evolution of life histories. A perusal of this literature conveys the impression that by life history ecologists mean traits whose variation directly influences life table schedules of fecundity and survival, hence the population growth rate of individuals having particular values of those characters. Age at first reproduction is directly visualized in the life table itself; number of eggs directly influences fecundity, all other things being equal. The challenge to understanding such adaptations derives, of course, from the fact that all other things are not equal. In a world of limited resources, one must pay for more eggs by less growth, storage, and maintenance, leading to increased risk of death and fewer eggs produced in the future by survivors.

The study of life histories is largely a legacy of David Lack, whose influence is ubiquitous in population biology and evolutionary ecology. During the early 1940s, Reginald Moreau, a British colleague of Lack who had worked in Africa for many years, called attention to the fact that songbirds in the tropics laid fewer eggs—2 or 3, on average—than their counterparts at higher latitudes—generally 4 to 10, depending on the species. Lack, who had been thinking much about the role played by adaptations for feeding in the process of speciation and the coexistence of ecologically similar organisms in communities, turned his attention to this latitudinal gradient in clutch size and to other variations in clutch size both within and among species. In his first paper on the subject, published in 1947, Lack clearly recognized that any genetically determined increase in clutch size would be strongly selected owing to the greater fecundity of the bearer, unless counteracting forces reduced survival of the offspring. He presumed that the ability of adults to gather food for their young was limited, and so broods with more than a certain critical number of offspring, determined by availability of food, would be undernourished and the survival of the chicks thereby reduced. In Lack's (1947) words:

> The average clutch-size is ultimately determined by the average maximum number of young which the parents can successfully raise in the region and at the season in question, i.e. . . . natural selection eliminates a disproportionately large number of young in those clutches which are higher than the average, through the inability of the parents to get enough food for their young, so that some or all of the brood die before or soon after fledging, with the result that few or no descendants are left with their parent's propensity to lay a larger clutch.

Lack further suggested that because of the longer day length at high latitudes, temperate and arctic-zone birds could gather more food and therefore rear more offspring than birds breeding in the tropics, where day length remains close to 12 hours year round.

Further developments in the evolutionary study of life histories will be reviewed a bit later. But before going on, we must clearly distinguish between life history adaptations and life table variables. The population growth rate (fitness) of individuals bearing particular life history traits can be determined from their age-specific schedules of survival and fecundity. These life table entries express the interaction between the adaptations of the organism and its environment; they are not themselves adaptations, even though their values vary among genotypes. The distinction between adaptation and life table becomes apparent when one recognizes that identical individuals would have different life tables in different environments.

If life history adaptations are not equivalent to life table variables, then the question arises once more. What is a life history? Fitness is a rate—that of population growth. Rates are inversely related to time. In the life table, time's place is taken by the age of the individual. Thus adaptations that influence the age of births and deaths also influence the rate of increase, or fitness. They thus comprise one set of life history adaptations. Among these are rates of development, including the timing of such discrete events in the development program as metamorphosis (tadpole to frog, veliger to snail, caterpillar to moth, and so on) and the onset of sexual maturity. The life table entries themselves are age-specific expectations of fecundity and survival; these are influenced by sets of partially overlapping adaptations. For example, parental care increases realized fecundity but also increases the risk of parental death. Fecundity at each age reflects the allocation of resources between growth and reproduction at an earlier age. Adaptations that influence more than one element in the life table pose the greatest challenge because their modification alters fitness in a complicated fashion. Finally, there is the problem of termination of life. In some species, such as the salmon, chambered nautilus, and century plant (agave), individuals reproduce once and die. In others, reproductive vigor and survival probability fall off with age as the individual undergoes physiological senescence. But because fitness is enhanced by long life and procreation, genetically determined decline and death are enigmatic.

Variation between Species

Life history traits and life table parameters vary with respect to environmental conditions and to each other. Consistent patterns of variation, such as the general decrease in reproductive rate of animals from the poles to the tropics, have aroused curiosity and at the same time have suggested possible relationships between adaptation and environment that govern the pattern. Lack noted that clutch size and day length vary together, although plausible explanations for their correlation have been proposed based on many other covarying environmental conditions. The manner in which traits vary with respect to each other has also suggested mechanisms of functional integration of the phenotype. The observation that fecundity and adult mortality

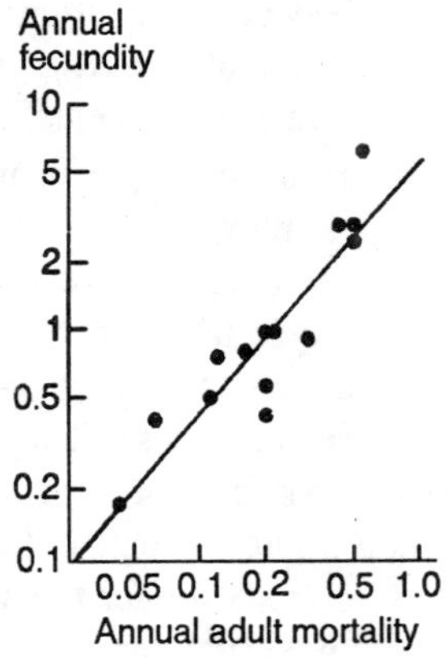

Fig. 4.1. Relationship between annual fecundity and adult mortality in several populations of birds ranging from albatross (low) to sparrow (high).

are strongly correlated has suggested to some (perhaps reinforced by their own experience) that parental care burdens the individual with risks.

Some sets of life history traits are generally associated. At one extreme, elephants, albatrosses, giant tortoises, and oak trees exhibit long life, slow development, delayed maturity, high parental investment, and low reproductive rate; at the other extreme, one finds mice, fruit flies, and weedy plants. In broad comparisons within the plant and animal kingdoms, such associations of traits vary in close relation to body size and undoubtedly reflect the relative slowness of all life processes in large organisms. But even among organisms of similar size and body plan, different environments produce widely divergent life histories. Storm petrels, which are seabirds the size of thrushes, rear at most a single chick each year, do not begin to reproduce until 4 or 5 years of age, and may live to 30 or 40 years. Thrushes, themselves, may produce several broods of 3 or 4 young each year beginning with their first birthday, but rarely live beyond 3 or 4 years. Similarly varied life histories may be found even among different populations of the same species.

Beyond such strong associations of life history traits with environmental conditions, many taxonomic groups also exhibit characteristic values of life history adaptations. Thus ducks (Anseriformes) usually lay 8 to 10 eggs per clutch, shorebirds (Charadriiformes) 4, hummingbirds (Apodiformes) 2, and petrels (Procellariiformes) 1. These differences probably reflect ways in which taxonomically conservative traits affect the selection of habitats by organisms and their particular interactions with the environment. Ducklings feed themselves and so their number is not limited by the ability of parents to gather food for them. Shorebirds typically lay very large eggs whose number may be limited by the ability of parents to incubate them successfully. Taxonomic affinity of a trait does not, however, reveal the significance of large egg size to shorebirds (why don't other birds lay large eggs?) or why shorebirds don't modify nest structure (usually a shallow depression in the ground) to accommodate more eggs (ducks incubate much larger egg masses than shorebirds).

No one knows why all species of hummingbirds, from the tropics to the arctic, lay 2 eggs per clutch. Apparently clutch size is constrained by some other adaptation or set of adaptations of these birds to their peculiar way of life. The single-egg clutch of all petrels, representatives of which are found in all oceans of the world and which vary in size from 30-gram storm petrels to 10-kilogram albatrosses, is thought to reflect a sparse and unpredictable food supply, which greatly limits the ability of adults to provide food to their chicks. But other seabirds that lay only 1 egg per year, such as gannets and swallow-tailed gulls, appear to have little difficulty rearing twins when provided an additional egg or chick.

Life history variation in every group of organisms provides a similarly rich phenomenology that both raises one's curiosity and suggests tentative explanations for the adaptive basis of life history traits. This natural history provides the setting within which theoretical and experimental studies have attempted to resolve how adaptation and environment interact to determine the life table of a population.

Theory of Origin of Life

The history of evolutionary ecology begins with Charles Darwin, who both identified and provided tentative explanations for some of the phenomena treated in this part of the book. But with regard to the types of life history traits discussed in this chapter, Darwin was uncharacteristically silent. He recognized the great reproductive capacity of all living beings but regarded the environment as an agent of selective mortality, undiscriminating with regard to birth rate.

Darwin did not have the benefit of life table analysis to guide his thinking. But even its development by A. J. Lotka and others, and its application to evolutionary phenomena by the population geneticist Ronald Fisher in his classic book *The Genetical Theory of Natural Selection* (1930), did not generate sufficient interest to inaugurate a new field of inquiry. In fact, the next major step toward the establishment of life history as an important focus for ecologists was the publication of Moreau's and Lack's papers on clutch size in birds during the mid-1940s. Although these attracted considerable attention, they still did not spark the flame that was to begin burning two decades later. In part. Lack's theory was too simplistic; it isolated clutch size (or reproductive rate more generally) as a single adaptation unrelated to other aspects of the phenotype. The ability of adults to deliver food to their offspring was accepted as determined by food availability in the environment. Neither the effort devoted to gathering food nor the time and effort devoted to caring for the young entered into the equation for fitness. Indeed, Lack's approach was decidedly nonquantitative. In addition, population biology during the late 1940s and most of the 1950s was embroiled in a controversy over the role of density dependence in the regulation of population size. Absorbed as ecologists were over the "balance of nature", their attention was diverted from inquiry into life histories.

The early 1960s marked a turning point in population studies and saw the birth of modern evolutionary ecology. It is sometimes difficult to know what forces urge a discipline in one direction or another. The year 1959 was the centennial of the publication of *On the Origin of Species*. Dover Publications reprinted Ronald A. Fisher's *Genetical Theory of Natural Selection* in 1958. At that time George Williams, at the State University of New York at Stonybrook, was pondering the adaptive bases for the evolution of senescence and insect societies. In 1960, papers were published by A. W. F. Edwards, W. A. Kolman, and H. Kalmus and C. A. B. Smith on the adaptive significance of the I : I ratio of males to females in most populations, a topic not touched since Fisher's treatment in 1930. This was a period of reunification of ecology and evolution.

Life history study burst upon this arena in 1966 with the publication of papers by Martin Cody and George Williams. Cody made two points of lasting significance. First, he applied Levin's ideas about fitness sets to life history evolution, calling attention to the fact that different components of fitness may be under conflicting selective pressures. Adaptation, he said, is largely the resolution of compromises in the allocation of time and energy to competing demands. Second, Cody introduced the concept that different life history adaptations are favoured under conditions of high and low population density, relative to the carrying capacity of the environment. At high density, selection favours adaptations that enable individuals to survive and reproduce with few resources; hence efficiency carries a premium. At low density, adaptations promoting rapid population increase are selected; hence high rates of productivity, regardless of efficiency, increase fitness. These contrasting strategies were referred to as *K*-selected and *r*-selected traits, respectively, after the variables of the logistic equation for population growth.

George C. Williams's (1966) paper explored the demographic coupling between life history traits. He pointed out that each increment of reproductive effort influenced both contemporary fecundity and survival to reproduce in the future. Moreau (1944) had recognized 20 years earlier that fecundity and adult survival could be linked:

> It is not unreasonable to suggest that (the number of broods the parents can produce during their reproductive lives) may be affected by both the number of broods in the season and the number of

young in the brood. It is possible, for example, that B/5 (a brood size of 5 offspring), at least in some circumstances, might put a significantly bigger strain on the parents than B/4, so that they were prevented from raising a larger total number as the product of the smaller broods in the same season; or that a succession of B/5 would so shorten the reproductive lives of the parents that their total of offspring, produced in smaller, less exacting broods, would be greater.

But Williams quantified present and future components of fitness in the uniform currency of reproductive value, based on life table calculations, and indicated how the conflict between the effect of a life history modification on present and future reproduction was resolved according to the relative values of present offspring and the expectation of future offspring. A simple illustration of this principle compares evolution in two populations, one in which individuals have a high probability of survival between breeding seasons and the other, a low probability. In the first population, the expectation of future reproduction has a high value and selection tends to diminish reproductive investment in favour of protecting an otherwise high survival rate. In the second, because most individuals die before having further opportunity to breed, selection favours a high investment in the current crop of offspring.

W. D. Hamilton completed the 1966 triptych of life history papers with a theoretical contribution showing explicitly how variation in each of the life table entries—that is, fecundity and survival at each age—results in variation in fitness. This result and other related derivations formed the mathematical basis for subsequent development of most life history theory. Papers by Madhav Gadgil and William H. Bossert (1970), showing that the outcome of selection depended on the form of the relationship between life table values determined by particular life history adaptations, and by Garth Murphy (1968) and William M. Schaffer (1974), raising the issue of life history evolution in a variable environment, pretty much completed the groundwork.

Natural Selection

Because life history evolution is now seen as an optimal resolution of conflicting demands on the organism, a critical part of the study of life histories has been to understand the allocation of limited time and resources to competing functions. Time spent searching for food cannot be used to care for offspring directly nor to watch for predators. Energy and nutrients allocated to reproduction cannot be used to support growth. Besides time, energy, and nutrients, organisms also must partition their body structure—even cells within tissues—between competing functions. In many species, eggs are produced in direct proportion to the size of gonads, seeds in direct proportion to flower number. Photosynthetic rate depends in part on how much of its production a plant has allocated to photosynthetic tissue at the expense of root and support tissue. Plants also are built upon a modular body plan. In some species, nodes at points of leaf attachment may produce either lateral branches or flowers, but not both, thus trading off growth form against reproduction.

Selection of an increase in one or another function in an organism places demands on the individual to increase delivery of energy and nutrients or to increase allocation of production toward that function. We think of energy, nutrients, and production as being in limited supply because selection drives these functions to the physiological limit. Hence any increase in one component of demand, as by increased reproductive effort, results in a decrease in delivery to another component of demand.

Demonstrating the validity of this assumption of trade-offs that is built into most life history theory has been difficult. Adding and subtracting eggs in nests of several species of birds has, in some cases, revealed an inverse relationship between the number of chicks in the nest and their survival. This often results in maximum productivity from intermediate brood sizes, as predicted by David Lack. For example, Goren Hogsted (1980) showed that the clutch laid by individual female magpies produced the maximum number of chicks that she could nourish. Either adding or subtracting eggs resulted in fewer offspring fledged.

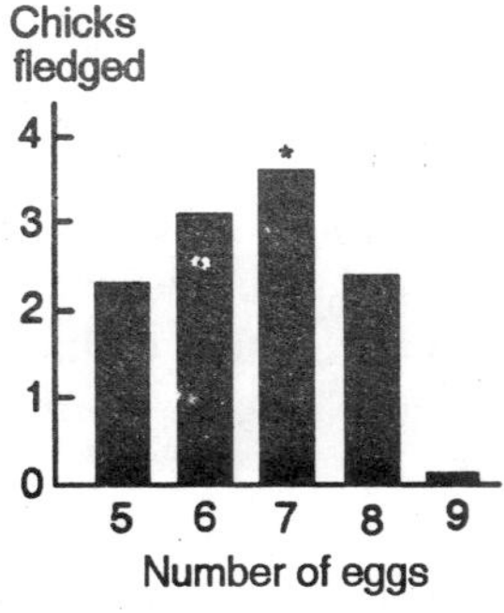

Fig. 4.2. Number of chicks fledged from nests of European magpies in which 7 eggs were laid but the experimenter added or removed eggs to make up manipulated clutches between 5 and 9.

When Diane DeSteven (1980) performed a similar clutch addition experiment with tree swallows, she found that adults were able to feed two extra chicks. The same proportion of young left the nest, and at about the same weight, in enlarged broods as in normal broods. Furthermore, DeSteven was unable to detect any difference in year-to-year survival of adults depending on the number of chicks they reared, although her samples were too small to detect subtle changes. However, Nadav Nur (1988), in similar experiments with the blue tit, found that among parents rearing enlarged broods, fewer survived to the following breeding season and these produced fewer offspring.

David Reznik (1983) altered allocation of resources to reproduction in guppies by preventing females from mating with males. If growth and reproduction competed for allocation of assimilated resources, the experimental fish should have attained larger size by the end of the study period. But, in fact, little of the difference in reproductive tissue accumulated between mated and unmated females was converted to growth.

Reznick (1985) summarized the literature purporting to address the question of trade-offs between life history traits and found few studies whose results could be interpreted unambiguously. The evidence presented in published studies falls into four categories:

1. Phenotypic correlations between traits within populations. These are difficult to interpret because each trait may respond independently to some third character not measured in the study. For example, low reproductive rate and long life span might be consequences of low metabolism, rather than having any direct interrelationship.
2. Experimental manipulations of the sort performed by Hogsted, DeSteven, Nur, and Reznick. These may indicate direct functional relationships between traits, because other variables can be controlled in an experiment. But because manipulated values are not genetically determined, the outcomes of the experiments do not directly address the evolution of life history patterns. They instead assess phenotypic plasticity, which may or may not reflect evolutionary tradeoffs.
3. Genetic correlations between traits. These indicate the degree to which two traits will respond in concert to selection on either or both. But, as we have seen, selection can produce an evolutionary trajectory perpendicular to genetic correlations, given enough time. Thus while genetic correlations can predict the short-term response,

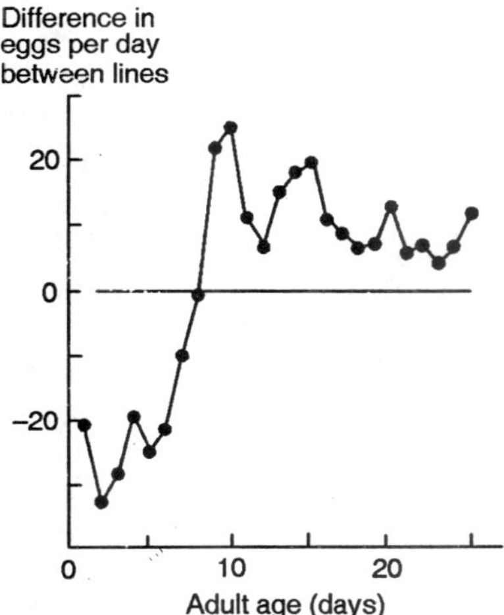

Fig. 4.3. Difference in daily egg production as a function of age between a control line of fruit flies and a line in which life span was terminated at an early age. Negative values indicate increased egg production rate among short-lived flies.

they do not necessarily predict the long-term course of life history evolution.

4. Correlated responses to selection. Selection experiments certainly are the most direct approach to studying life history evolution. The difficulty of performing such experiments has greatly limited their application, as one might expect. Revealing examples, however, are the experiments of Michael Rose and Brian Charlesworth (1981) and Rose (1984), who artificially terminated the life span of *Drosophila* flies at an early age and found a heritable increase in rate of egg production early in life and a heritable decrease in natural life span.

According to Reznick's tabulation, the evidence of available studies revealed costs of reproduction more often than not, but the case needs strengthening by more direct experimentation. Most issues concerning the evolution of life histories can be phrased in terms of three questions: When should I begin to produce offspring? How often should I breed? How many offspring should I attempt to produce in each breeding episode? Each of these questions expresses in a different way the fundamental trade-off between fecundity and adult survival.

Life Span

When should an animal or plant begin to breed? Long-lived organisms typically begin to reproduce at an older age than short-lived ones. What selective forces could produce this result? We shall assume that age at first reproduction has genetic variation and can be selected independently of other life history characteristics, although it also may reflect rate of development and physiological processes selected for other reasons. At every age, an individual must choose between attempting to reproduce and abstaining from breeding. When young individuals resolve this choice in favour of abstention, they may delay the onset of sexual maturity. Thus age at first reproduction can be understood in terms of the benefits and costs of breeding at a particular age. The benefit appears in the life table as an increase in fecundity at that age. The cost may appear as reduced survival to older ages, reduced fecundity at older ages, or both.

Consider the following example. A type of fish continues to grow only until sexual maturity. Its fecundity is directly proportional to body size. Suppose that the number of eggs laid per year increases by 10 for each year that reproduction is delayed, so that individuals breeding in their first year produce 10 eggs and the same number each year thereafter, individuals first breeding in their second year produce 20 eggs, and so on. Comparing the cumulative egg production of early- and late-maturing individuals, one can see that the optimal age at first reproduction varies in direct proportion to expected life span.

For organisms that do not grow after their first year (most birds, for example), the choice between breeding or not depends on balancing current reproduction against survival. Nonbreeding individuals avoid the risks of preparing for reproduction—courtship, nest building, migration to breeding areas. Presumably, life experience gained with age also reduces the risks of breeding, increases the realized fecundity of a certain level of parental investment, or both, favouring delayed reproduction. Balancing this are many factors that reduce expectation of future reproduction, including high predation rates, encroaching senescence at old age, and, for organisms that live a single year or less (annuals) in seasonal environments, the end of the productive season.

Perennial Life

Another issue in the evolution of life history adaptations has been the number of breeding episodes that the individual undertakes. Plants and animals either reproduce during a single season and die (annuals) or have the potential to reproduce over a span of many seasons (perennials). Population biologists have pondered the relative advantages of each habit in terms of the trade-off between survival probability and fecundity. To survive the nonreproductive "winter" period, a perennial plant must allocate resources to storage of materials in roots and formation of freeze-resistant or drought-resistant buds, presumably at the expense of production. But at some point the advantages of the perennial habit must outweigh the costs in reduced fecundity relative to annuals.

Following the earlier lead of Lamont Cole (1954), Charnov and Schaffer (1973) compared annuals and perennials in an algebraic model. Suppose a population of plants contains some individuals that produce a large number of seeds at the end of the first growing season and then die (annual), and others that produce fewer seeds but survive through the winter to reproduce in subsequent growing seasons (perennial). Which has the greater fitness? For the purposes of their model, Charnov and Schaffer assumed that annual and perennial plants have the same probability of survival during their first (in the annual's case, only) growing season (S_0) and that perennials have a constant probability of survival thereafter (S_p).

The factor by which a population of an annual plant grows (λ) equals the number of seeds each individual produces (B_a) times their survival to reproductive age (S_0), or $\lambda_a = B_a S_0$. The increase of a population of a perennial plant equals the number of seeds (B_p) times their survival (S_0), plus the probability of survival of the parent (S_p), hence $\lambda_p = B_p S_0 + S_p$. The population growth rate of the annual exceeds that of the perennial ($\lambda_a > \lambda_p$) when $B_a S_0 > B_p S_0 + S_p$. Dividing both sides of the inequality by S_0, we obtain $B_a > B_p + S_p / S_0$ [or $B_a - B_p > S_p/S_0$]. Accordingly, an annual life history is favoured when the number of seeds produced by the annual exceeds the fecundity of the perennial by the ratio S_p/S_0. When few perennials survive from one breeding season to the next or when perennials produce relatively few seeds, the annual habit is favoured. Where individuals survive well once established but seedlings survive poorly (high S_p/S_0), the annual habit must result in extremely high fecundity to be favoured. It is no wonder that annuals predominate among the floras of deserts, where few adult plants can survive drought periods, and perennials predominate among the floras of the tropics, where competition and predator pressure make difficult the establishment of seedlings. Adding complexity to the model, by incorporating growth from year-to-year and annual variation in survival probabilities, does not destroy the basic qualitative conclusion that the life history strategy is determined primarily by the ratio of adult survival to juvenile.

SURVIVAL OF ADULTS

For annual plants, expectation of life beyond the first breeding season is so small that all resources are devoted to current reproduction. Perennials, however, must allocate resources between current reproduction and adaptations that prolong life. When particular adaptations affect both fecundity and survival, the trade-off between the two must be optimized. Intuitively, when life span is short regardless of the consequences of reproduction, the balance of allocation should tip in favour of current fecundity. When potential life span is great, current fecundity should not unduly jeopardize future reproduction. This can be shown algebraically quite simply. First, we partition adult survival into two components, one directly related to reproduction (SR), and the other independent of reproduction (*S*). Now, fitness may be expressed as

$$\lambda = SS_R + S_0 B$$

Certain reproductive adaptations that cause small changes in the values of survival (ΔS_R) and fecundity (ΔB) will influence fitness according to

$$\Delta\lambda = S\Delta S_R + S_0 \Delta B$$

When changes that enhance fecundity (ΔB positive) also reduce survival (ΔS_R negative), their effects on $\Delta\lambda$ depend on the relative values of S and S_0. In general, when S is large compared to S_0, selection favours adaptations that increase adult survival at the expense of fecundity, and vice versa. Thus one expects parental investment in offspring to decrease with increasing adult life span.

SURVIVAL AND FECUNDITY

In the preceding models, we assumed that fecundity and adult survival were constant values, unvarying over age. In reality, however, among animals and plants that reproduce repeatedly, rates of survival and fecundity vary with age within the reproductive period. To the degree that differences in these variables represent the outcome of genetically determined modifications of the life history, it is important to understand the relative strengths of selection acting on changes in life table variables at different ages.

The characteristic equation of a population relates the rate of geometric or exponential increase in population size to the life table variables by

$$1 = \sum \lambda^{-x} l_x b_x \qquad \text{...(1)}$$

where survivorship to age x (l_x) is the product of the individual survival rates up to that age ($l_x = s_0 s_1 \ldots s_{x-1} = \Pi_{i=0}^{x-1} s_i$). This equation allows us to determine how small changes in s_x and b_x, change fitness, and therefore indicates the strength of selection on adaptations that affect these life table variables. As William D. Hamilton (1966) and John Merrit Emlen (1970) have shown, a small change in fecundity at age x influences λ according to

$$\Delta\lambda = \frac{\lambda^{-x} l_x \Delta b_x}{\lambda^{-1} \sum x\lambda^{-x} l_x b_x} \qquad \ldots(2)$$

This equation shows that the strength of selection on b_x diminishes with age in direct proportion to the decrease in survivorship. For example, if 50 per cent of individuals survived to age 1 and only 25 per cent at age 2, then selection on change in fecundity at age I would be twice as strong as that on the same change in fecundity at age 2. If, under these conditions, an increase in fecundity at the first age caused that at the second to decrease, so long as the gain in b, was more than twice the loss in b_2, the modification would be selected.

The relative strength of selection at different ages also is influenced by the rate of growth of the population (λ). When population size is approximately constant ($\lambda = 1$), the term λ^{-x} is 1 at all ages. But in an increasing population ($\lambda > 1$), the term λ^{-x} falls off with age and modifications affecting fecundity at young ages have relatively greater effect on fitness. Symmetrically, the amount by which adaptive changes in fecundity at younger ages are favoured is reduced in declining populations.

In any population, fewer individuals live to older ages. Hence a smaller proportion of genes that affect life table variables at older ages are expressed—exposed to selection—and therefore strength of selection declines with l_x. In a growing population, each individual born today is a larger fraction of the total population than is each individual born in the future; as the population expands the value of each individual diminishes. Therefore, from the standpoint of the life table, offspring born late in an individual's life have relatively less value than offspring born to the same individual early in its life, when they constitute a larger proportion of the total population.

The effect of a change in survival rate at age x (Δs_x) on fitness is given by

$$\Delta\lambda = \frac{\lambda^{-x} l_x s_x \sum_{i=x+1} \lambda^{x-i} b_i \prod_{j=x}^{i-1} s_j}{\lambda^{-1} \sum x \lambda^{-x} l_x b_x} \qquad \text{...(3)}$$

Not surprisingly, the strength of selection on a fractional change in survival rate ($\Delta s_x/s_x$) varies in direct proportion to the survivorship to age x, but also in proportion to the expectation of reproduction at older ages. As with changes in fecundity, the relative strength of selection is diminished with age in increasing populations and augmented in decreasing populations.

Optimization of the life history is a matter of resolving conflicts in the expression of adaptations in different components of fitness, the life table variables. Equations (2) and (3) give us a quantitative basis for evaluating conflicting selective forces and predicting general patterns of adaptation expressed in the life table. For example, in conflicts between early and late reproduction, which might come about through use of resources for current reproduction that might otherwise be stored and used for future reproduction, conflict will be resolved in favour of reproduction at younger ages and fecundity might be expected to decline with age. This pattern should be more pronounced in populations with low adult survival rates and following selection during phases of rapid population growth. When conflict arises between fecundity at age x and survival to the following breeding period, the relative strengths of selection on b_x and s_x balance the value of current offspring against the expected value of future offspring according to

$$\Delta\lambda = \frac{\lambda^{-x} l_x \left(\Delta b_x + \Delta s_x \sum_{i=x+1} \lambda^{x-i} b_i \prod_{j=x}^{i=1} s_j \right)}{\lambda^{-1} \sum x \lambda^{-x} l_x b_x} \qquad \text{...(4)}$$

(The term $\Sigma\lambda^{x-i} b_i \Pi s_j$ is simply the expectation of future reproduction of an individual of age $x + 1$ weighted by population growth.) Therefore, adaptations that increase current fecundity prevail over those that increase survival when adult survival is low and when the population is rapidly growing owing to high fecundity. By this reasoning, species with characteristically long life spans should put less effort into producing offspring, and more into avoiding predation and other sources of mortality, than species that, by their nature and habitat, are characteristically short-lived.

Many plants and invertebrates, and some fish, reptiles, and amphibians do not have a characteristic adult size. They grow, at a continually decreasing rate, throughout their adult lives (indeterminate growth). Fecundity is directly related to body size in most species with indeterminate growth. Because egg production and growth draw upon the same resources of assimilated energy and nutrients, increased fecundity during one year must be weighted against reduced expectation of fecundity in subsequent years. For organisms having longer life expectancies, growth should be favoured over fecundity during each year. For organisms with less chance of living to reproduce in future years, resources allocated to growth instead of to eggs are largely wasted.

Consider two hypothetical fish, each weighing 10 grams at sexual maturity, but which allocate resources to growth and reproduction differently. Both gather enough food each year to reproduce their weight in new tissue or eggs. Fish A allocates two-tenths of its production to growth and eight-tenths to eggs, whereas fish B allocates one-half each to growth and eggs. Calculated growth, fecundity, and accumulated fecundity show that for fish living 4 or fewer years, on average, high fecundity and slow growth give the greater overall productivity, whereas for fish living longer than 4 years, more rapid growth and lower fecundity are superior. Adult mortality, therefore, determines the optimum allocation of resources between growth and reproduction.

Reproductive Episode

Some species of salmon have adopted a course of rapid growth for several years, culminating in a single immense reproductive effort, in which a large portion of the body tissues is converted to eggs, followed shortly after spawning by death. Gadgil and Bossert (1970) reasoned that because salmon make so great an effort to migrate upriver just to reach their spawning grounds, it may be to their advantage to make the trip just once, at which time they should produce as many eggs as possible, even if this supreme reproductive effort requires the conversion of muscle and digestive tissue to eggs and ensures their death.

The salmon live history pattern is sometimes referred to as "big-bang" reproduction but more properly as semelparity. This term comes from the Latin *semel* (once) and *pario* (to beget); it is contrasted with iteroparous, from *itero* (to repeat). Semelparity is rarely encountered among animals and plants that live for more than 1 or 2 years. Usually, the effort and allocation of resources required to survive between growing seasons are so much greater than those used to prepare for breeding that, once a perennial life form has been adopted, reproduction every year seems the most productive pattern.

The best-known cases of semelparous reproduction in plants occur in the agaves and the bamboos, two distinctly different groups. Most bamboos are tropical or warm-temperate-zone plants that form dense stands in disturbed habitats. Reproduction does not appear to require substantial preparation or resources, as needed to grow a heavy flowering stalk. But opportunities for successful seed germination probably are rare. Once established, a bamboo plant increases by asexual reproduction, continually sending up new stalks, until the habitat in which it germinated is fairly packed with bamboo. Only at this point, when vegetative growth becomes severely limited, do plants benefit from producing seeds, which may colonize disturbed sites.

The environment and habits of agaves are at opposite ends of the spectrum from those of bamboos. Most species of agave inhabit arid climates with sparse and erratic rainfall. Plants grow vegetatively for several years, the number varying from species to species, then send up a gigantic flowering stalk. After producing its seeds, the agave dies. One curious fact about agaves is that they frequently live side by side with yuccas, a group of plants with a similar growth form, but which flower year after year. The root systems of agaves do, however, differ from those of yuccas; yucca roots descend deeply to tap persistent sources of groundwater. Agaves have shallow, fibrous roots that catch water percolating through the surface layers of desert soils after rain showers, but that are left high and dry during drought periods. The erratic water supply of the agave may prevent successful seed production or seedling establishment every year, and the period between suitable years may be very long. Under these conditions, it may be most advantageous for the agave to grow and store nutrients until an unusually wet year comes—perhaps 1 in 10 or even 1 in 100—and then to put all resources into reproduction.

Senescence

While few organisms exhibit programmed death associated with reproduction, most do experience a gradual increase in mortality and decline in fecundity resulting from deterioration of physiological function, known as senescence. For example, rates of most physiological functions in humans decrease in a roughly linear fashion between the ages of 30 and 85 years, to 80 to 85 per cent of the value in 30-year-old individuals in nerve conduction and basal metabolism, 40 to 45 per cent in the volume of blood circulated through the kidneys, and 37 per cent for maximum breathing capacity. Birth defects and infertility generally occur with increasing prevalence in women progressively older than 30 years. Senescence, reproductive decline, and death in old age do not result from abrupt physiological change. Rather, the demographic consequences of senescence result from a gradual decrease in physiological function with age. Such changes are found throughout the animal kingdom.

How can senescence evolve? Why is senescence not eliminated by selection when survival presumably is advantageous to an individual at any age? The answer to these questions is generally thought to originate in the declining strength of selection on genes expressed at progressively greater age, owing to the fact that fewer individuals bearing those genes survive to express them. Given this age-dependence of selection, senescence can be thought of arising in two ways. First, deleterious genes are constantly being added to the population by mutation, whose rate probably varies little with respect to age of expression of the gene. The ability of selection to remove these alleles declines with age, and so deleterious alleles with later age of expression will build up to higher levels in the population. Individuals that do survive to old age will be more likely to express bad genes as reduced physiological function.

Second, some alleles may act pleiotropically to enhance fitness at early ages but reduce fitness in later life. Such alleles will tend to be incorporated into the gene pool because effects expressed at young age contribute more to fitness than do those expressed at old age. Although there is relatively little evidence for such pleiotropic alleles (indeed, for some genes, alleles beneficial at one age ought to be so at all ages), fruit flies selected for increased early survival have reduced survival rates later in life.

When should we expect senescence to begin? How rapidly should it encroach upon old age? The answers to these questions depend upon differences in the strength of selection between ages. Changes in fitness caused by changes in survival rate at a given age (x) are equal to

$$\Delta\lambda = \frac{s_x^{-1}\Delta s_x \sum\limits_{i=x+1} \lambda^{-i} l_i b_i}{\lambda^{-1}\sum x\lambda^{-x} l_x b_x} \qquad \ldots(5)$$

which is another way of expressing equation (3). Until the age at first reproduction, the sum of the $\lambda^{-i}l_i b_i$ terms is constant because all

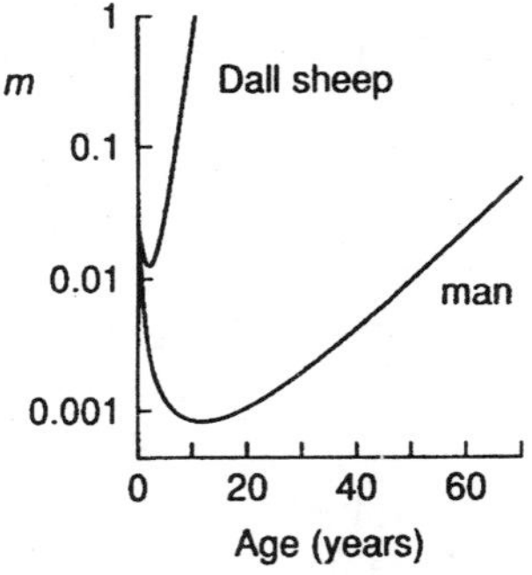

Fig. 4.4. Relationship between annual mortality rate and age in the Dall mountain sheep and a human population showing that senescence encroaches more rapidly in a population having higher minimum adult mortality.

reproduction (terms in b_i) lies in the future. Hence the strength of selection on changes in survival rate remains constant until the onset of reproduction, and senescence should not manifest itself until after that point, which is certainly the case in humans. Furthermore, senescence should increase faster in populations with higher characteristic mortality rates because the sum of the $\lambda^{-i}l_i b_i$ terms drops off faster with age after reproduction begins. Thus, in the Dall sheep, whose minimum annual adult mortality rate is about ten times that of humans, senescence encroaches much more rapidly as well.

Growth Rate

The relative strength of selection on life history traits expressed at different ages depends on the growth rate of the population, as we have seen. This has had important consequences for thinking about life history evolution. For example, reproductive rate has been linked to the growth rate of populations to explain latitudinal variation in fecundity. The argument runs as follows. In temperate and arctic regions, populations are periodically reduced by catastrophic weather and individuals die with little regard to their genotypes. Population crashes are followed by longer periods of population increase during which adaptations that increase intrinsic population growth rate (r)—including increased fecundity and earlier maturity—are selected. In "constant" tropical environments, where populations fluctuate little, populations remain near the limit imposed by resources (K), and adaptations that improve competitive ability and efficiency of resource utilization are selected.

The distinction between temperate and tropical patterns has been described as the r- and K-selection spectrum. The term r refers to the growth capacity (exponential growth rate) of the population, and K denotes the carrying capacity of the environment for population—the upper resource limit to population size. Although the naming of the concept set off a minor semantic battle among population ecologists, r- and K-selection occupy an important place in current thinking about life history patterns.

Eric Pianka (1970) listed a variety of traits that could be considered as either r-selected or K-selected. Selection favouring r-selected traits under conditions of population growth could arise in two ways. First, individuals in populations reduced below their carrying capacities, and therefore presented with abundant resources, should be able to grow more rapidly, reproduce at an earlier age, and produce more progeny than individuals in populations at the carrying capacity. This provides a resource-based explanation for r- and K-selected traits in different populations. In populations regulated by density-dependent processes, all modifications of the phenotype influence the relationship between population growth rate and density. Modifications that enhance growth rate at low population density but reduce growth rate at high density are favoured only when population density is low (and presumably growing); hence these are distinctively r-selected traits. Conversely, modifications that enhance population growth rate at high density, even at the expense of growth rate at low density, are K-selected traits.

A second mechanism for generating divergent r- and K-selected traits derives from the dependence of strength of selection at different ages on the rate of population growth (λ). As we have seen, in a growing population modifications of traits expressed at later ages are relatively more weakly selected than those expressed at earlier ages. As a result, in a growing population selection favours early reproduction at the expense of longevity and continued fecundity. Early reproduction and high reproductive rates are traits listed by Pianka (1970) as r-selected.

Although the theory is plausible, a direct relationship between population growth rate or population fluctuations and life history characteristics has not been established. Pianka (1970) placed insects at the r-selected end of the spectrum and mammals at the K-selected end, reasoning that insect populations fluctuate more than mammal populations. But the differences in life history traits between the two groups could be attributable to differences in body size, over which differences arise in the time and power scale of all physiological processes. Small organisms move more rapidly relative to body length, use more energy relative to body weight, and have more rapid development and shorter generations than large animals. These traits may be inherently correlated to size through physical and physiological relationships—just as a pendulum swings at a rate inversely related to its length—and thus largely insensitive to environmental influence. The importance of *r*- and *K*-selection theory, relative to other sources of variation in life histories, depends on demonstrating a direct link between differences in population fluctuations and life history traits in pairs of otherwise similar organisms. A further difficulty with the interpretation of adaptations according to *r*- and *K*-selection theory is that the traits attributed to different levels of population fluctuation are similar to those predicted for different levels of adult mortality and population turnover, even among populations with constant size.

Several investigators have attempted to contrast genetic responses to *r*-selected and *K*-selected regimes in laboratory populations. Francisco Ayala (1965) found that when populations of *Drosophila* were maintained for long periods under crowded conditions the numbers of adults per cage gradually increased, presumably owing to selection of traits that improved fecundity and survival at high density. Further experiments in which *Drosophila* populations were kept considerably below carrying capacity by removing adults confounded the selective effects of low density with those of high mortality. Similar experiments on laboratory populations of bacteria and protozoa have also produced ambiguous or negative results.

Bet Hedging

When the environment varies unpredictably over the life span of the individual, selection may favour the spreading of reproduction over many seasons or concentrating it early in life, depending on the circumstances. When recruitment of offspring is unpredictable from year to year, selection favours adult survival at the expense of present fecundity, a strategy referred to as "bet-hedging" by Stephen Steams (1976). The logic of the strategy is best appreciated by considering the extreme case, breeding only once. If conditions fluctuated such that in some years breeding success were zero, semelparous breeders would occasionally fail to reproduce and their lines would die out. Spreading reproduction over several years, even at the expense of annual fecundity, would be favoured under such conditions.

The strength of selection for bet-hedging strategies is difficult to calculate analytically; it depends on the amount of variation in life table variables and the distribution of that variation. Because populations increase geometrically, some authors have proposed that the geometric mean of population growth rate over years provides a better measure of fitness than the arithmetic mean. For example, a population that alternated between growth rates (λ) of 1 and 3 would grow more slowly than one that had a consistent growth rate of 2 (the same arithmetic average) every year. In 4 years, the first would grow to 9 times its present size ($1 \times 3 \times 1 \times 3$; a geometric mean of 1.73), the second, 16 ($2 \times 2 \times 2 \times 2$; geometric mean of 2). But the importance of bet-hedging in natural populations will not be fully appreciated until more data are collected on variation in life table parameters and the biological constraints involved in bet-hedging adaptations are understood. At present, it is not possible to draw any conclusions about the role of environmental variation in molding the life histories of plants and animals.

5

Competition and Coexistence

For many biologists, implicit in Darwin's theory of natural selection was a view of nature "red in tooth and claw" in which species scrambled to outcompete each other and leave the most offspring. How true a picture is this?

It is first worthwhile to be specific about what types of competitive event may occur in nature. *Competition* may be *intraspecific* (between individuals of the same species) or *interspecific* (between individuals of different species). Competition can also be characterized as scramble competition or contest competition. In scramble (*resource*) competition, organisms compete directly for the *limiting resource*, each obtaining as much as it can. Under severe stress, for example when fly maggots compete in a bottle of medium, few individuals can command enough of the resource to survive or reproduce.

Such competition is most evident between invertebrates. In *contest* (interference) competition, individuals harm one another directly by physical force. Often, this force is ritualized into threatening behaviour associated with territories. In these cases, strong individuals survive and take the best territories, and weaker ones perish or at best survive under suboptimal conditions. Such behaviour is most common in vertebrates.

Mathematical Models

Mathematically, competitive interactions can be described by equations derived independently by Lotka (1925) in the United States and by Volterra (1926) in Italy, commonly called the Lotka-Volterra equations. For two species growing independently,

$$\frac{dN_1}{dt} = r_1 N_1 \left(\frac{K_1 - N_1}{K_1} \right)$$

and
$$\frac{dN_2}{dt} = r_2 N_2 \left(\frac{K_2 - N_2}{K_2} \right)$$

where *r*, *N*, and *K* represent the variables namely per-capita rate of population growth, population size, and carrying capacity.

Another term must be introduced to allow for the effect of each population on the other. In most cases, individuals of one species are larger than those of the other, and a conversion factor is needed to convert species 2 into units of species 1, such that

$$N_1 = \alpha N_2$$

where α is the conversion factor. Thus, in the diagram below one individual of species 2 uses the same amount of resources as four individuals of species 1.

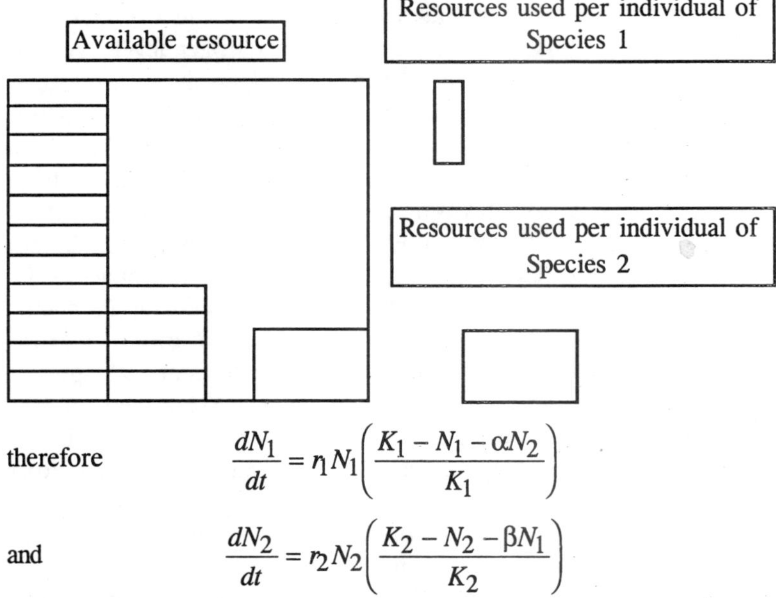

therefore
$$\frac{dN_1}{dt} = r_1 N_1 \left(\frac{K_1 - N_1 - \alpha N_2}{K_1} \right)$$

and
$$\frac{dN_2}{dt} = r_2 N_2 \left(\frac{K_2 - N_2 - \beta N_1}{K_2} \right)$$

where β is the conversion factor to convert N_1 into units of N_2.

Population growth of N_1 continues to the carrying capacity of the environment K_1 in the absence of N_2. If there are K_1/α individuals of N_2 present, no population growth of N_1 is possible. Between these two extremes are many combinations of N_2 and N_1 at which no further growth of N_1 is possible. These points fall on the diagonal $dN_1/dt = r_1 = 0$, which is often called the *zero isocline*. Population growth of N_2 can be represented by a similar diagram. Combining the two figures and adding the arrows by vector addition illustrates what happens when the species co-occur. Essentially there are four possible outcomes: species 1 goes extinct; species 2 goes extinct; either species 1 or species 2 goes extinct, depending on the initial densities; or the two species *coexist*.

A drawback to the Lotka-Volterra model of competition is that no mechanisms are specified that drive the competitive process. Tilman (1982, 1987) criticized this approach and emphasized that we need to know the mechanism by which competition (occurs. Knowing the mechanism will enable better predictions of the outcome. Tilman began by considering the responses of two competing plant species to environmental variables, say nitrogen and light again. As with the Lotka-Volterra models, we can draw zero-growth isoclines for both species, this time based on their responses to light and nitrogen. If light levels are too low, a species will not grow; above a certain light level, growth proceeds. The same happens for nitrogen levels. A sort of all-or-nothing response is envisaged. We can do the same for the second species. By superimposing the two zero-growth isoclines, we can determine the outcome if the two species coexist in the same habitat. Once again, four different outcomes are possible. In the first instance, species A will always need more of both resources than species B, and species B wins out while species A goes extinct. In the second instance, the roles are reversed. In the two remaining cases, the zero-growth isoclines cross and there may be an equilibrium point. To determine what happens in these other two scenarios, an extra piece of information is needed: the consumption curves of each species and the ratio of existing resources, that is, the position of the resource supply point. The resource supply point can be located in any position, I through 6 in Figure 5.1. In region 4 in Figure 5.1 (c) and (d). Species A is limited more by resource 2, and Species B is limited more by resource 1. If species A consumes relatively more of resource 1 than species B, the equilibrium point is unstable and one or other of the species will go extinct. On the other hand, if species B consumes relatively more of resource 1 than species A, there is a stable equilibrium point. In this situation, each species consumes more of the resource that limits its own growth and the system equilibrates at the intersection point of the two zero-growth isoclines. Thus, competitors

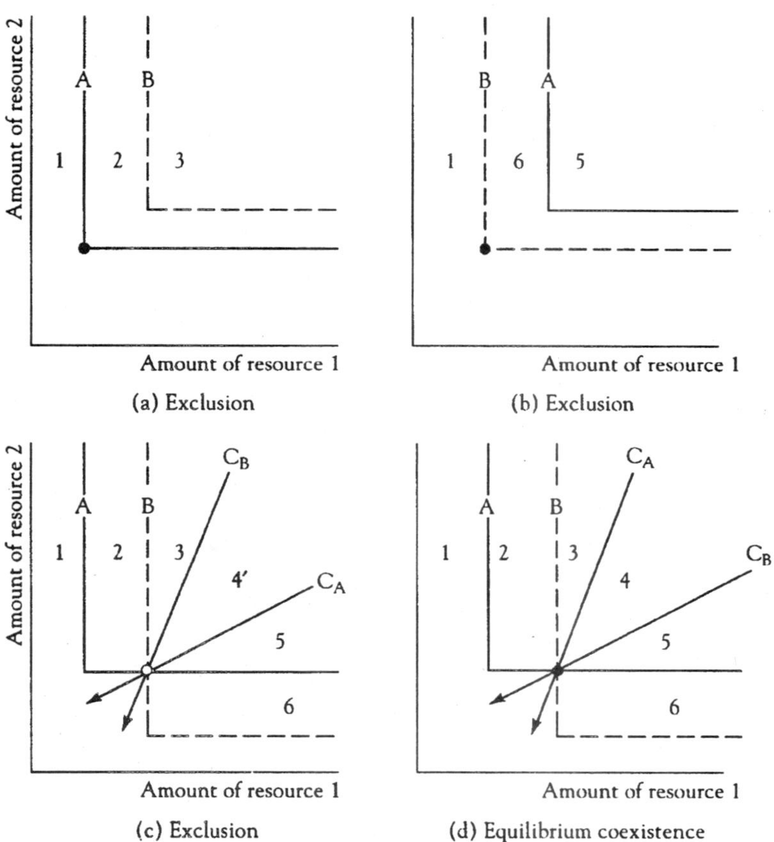

Fig. 5.1. Tilman's model of competition for two essential resources.

may coexist depending on their resource utilization. Tilman tested this model by growing two species of diatoms, *Asterionella formosa* and *Cyclotella meneghiniana*, in chemostats under controlled rates of nutrient supply. *Asterionella* required higher levels of silicon, and *Cyclotella* higher levels of phosphorus. In Tilman's experiments, there were four levels of supply: low phosphate, high silicate to high phosphate, low silicate. As predicted by his theory, Tilman could get either coexistence or one or the other species to go extinct by varying the nutrient supply.

Laboratory Studies of Competition

One must ask whether mathematical formulations represent real biological systems. One of the first and most important tests of these equations was performed in 1932 by a Russian microbiologist, who studied competition between two species of yeast, *Saccharomyces*

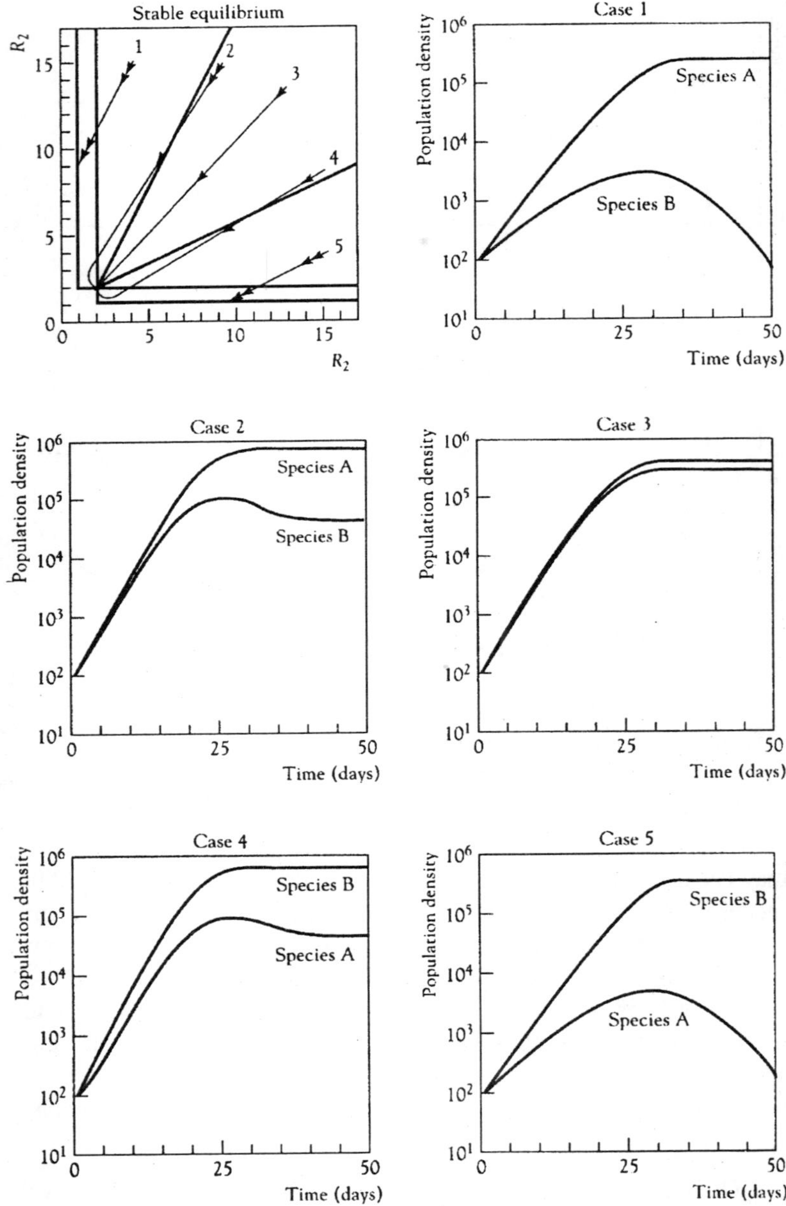

Fig. 5.2. The outcome of competition for Tilman's model for five different resource supply point for Figure 5.1, equilibrium coexistence. The resulting five population curves for the two species show the time course of competition.

cervisiae and *Schizosaccharomyces kephir* (species renamed since 1932). Alone, both species grew according to the logistic curve; the asymptote reached was a function of ethyl alcohol concentration. Ethyl alcohol is a by-product of sugar breakdown under anaerobic conditions and can kill new yeast buds just after they separate from the mother cell. In cultures where the two yeasts grew together, population densities were lower than they were under single-species conditions. From these data,

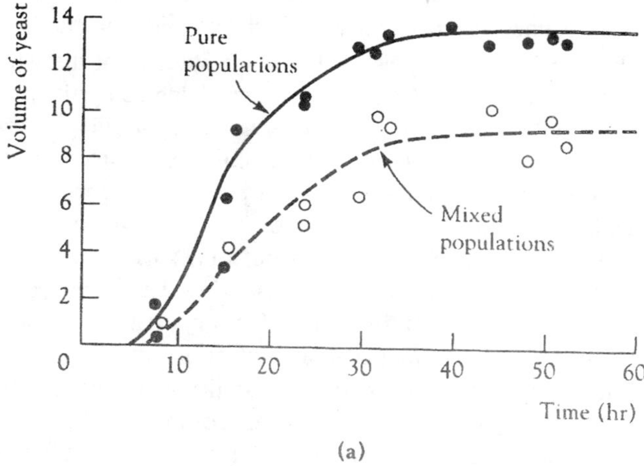

(a)

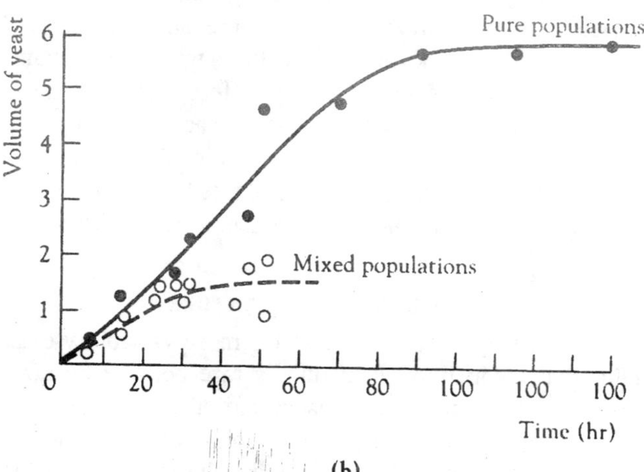

(b)

Fig. 5.3. ***(a) Growth of population of the yeast Saccharomyces in pure cultures and in mixed cultures with Schizosacchoromyces. (b) Growth of population of the yeast Schizosacchoromyces in pure culture and in mixed cultures with Sacchoromyces.***

Gause was able to calculate that $\alpha = 3.15$ and $\beta = 0.44$; that is, 1 volume of *Saccharomyces* = 3.15 volumes of *Schizosaccharomyces*. Because alcohol is the limiting factor, Gause argued that he could determine α and β by measuring alcohol production of the two yeasts, which turned out to be 0.113 percent EtOH/cc yeast for *Saccharomyces* and 0.247 percent for *Schizosaccharomyces*. Thus, $\alpha = 0.247/0.113 = 2.18$, and $\beta = 0.113/0.247 = 0.46$. The values of α and β obtained from the Lotka-Volterra equations were indeed in general agreement with those obtained independently by a physiological method.

In the late 1940s, Thomas Park and his students at the University at Chicago began a series of experiments examining competition between two flour beetles, *Tribolium confusum* and *Tribolium castaneum*. *Tribolium confusum* usually won, but in initial experiments the beetle cultures were infested with a sporozoan parasite, *Adelina*, that killed some beetles, particularly individuals of *T. castaneum*. In these early experiments, *T. confusum* won in 66 out of 74 trials because it was more resistant to the parasite. Later, *Adelina* was removed, and *T. castaneum* won in 12 out of 18 trials. Most important, with or without the parasite, there was no absolute victor; some stochasticity was evident. It was evident that competitive ability was greatly influenced by climate; each species was a better competitor in a different microclimate. However, single-species rearings in a given climate could not always be relied on to predict the outcome of mixed-species rearings (examine the entry for cold-moist climate). Later, it was found that the mechanism of competition was largely predation on eggs and pupae by larvae and adults. Park then varied the aggressive or cannibalistic tendencies of the beetles by selecting different strains; he obtained different results according to the strain of each beetle used although again the results could not always be predicted from the particular strains used. However, Park had demonstrated a complete reversal of competitive outcome as a function of temperature, moisture, parasites, and genetic strains.

Competition in Nature

What of systems in nature where far more variability exists? One view holds that competition in nature is rare because by now, of all potential competitors, one has displaced the other. An alternative view holds that competition is a common enough force in nature to be a major factor influencing evolution. A third alternative is that predation and other factors hold populations below competitive levels. The question is important in applied situations—for example, in biological-control campaigns—because it is vital to know whether releasing one natural enemy against a pest is likely to be more effective than the release of many, where competition between enemies might reduce their overall effectiveness. Circumstantial evidence suggests that fewer natural enemies become established where many are released; yet sometimes this phenomenon does little to reduce overall effectiveness of control. Competitive effects between plants are also often thought to be of paramount importance in influencing crop yields, and many applied ecologists immediately assume all plants compete it resources are limiting. Again, this is important in applied ecology because while agronomists may strive to reduce competition, entomologists argue that more than one crop is valuable to encourage a wide variety of natural enemies and thus so reduce insect pest densities. The most direct method of assessing the importance of competition is to remove individuals of one species A and to measure the responses on species B. Often, however, such manipulations are difficult to make outside the laboratory. If individuals of species A are removed, what's to stop them migrating back into the area of removal? If cages are used to stop immigration of species A, emigration of species B is prevented, and the numbers of species B may go unnaturally high (the so-called cage effect). Three of the most cited examples of competition in nature involve barnacles, parasitic wasps, and chaparral shrubs.

Two barnacles, *Chthamalus stellatus* and *Balanus balanoides*, dominate the British coasts. Their distribution on the intertidal rock faces are often well defined. Joseph Connell (1961) showed that *Chthamalus* could survive in the *Balanus* zone when *Balanus* was removed. In nature, *Balanus* grows faster on rocks of the middle intertidal zone, squeezing *Chthamalus* out. The limits of *Balanus* distribution are determined in the upper zone by desiccation and in the lower zone by predation and competition for space with algae. *Chthamalus* is more resistant to desiccation than *Balanus* and is normally found only high on the rock face. Thirty-four years later, in 1988, Connell repeated his competition experiments in the same area of the Scottish coast and once more observed strong evidence for competition.

Pest control is big business in the United States, and for some biological control projects, there is much labor available in the form of trained "scouts" to survey an area for evidence of successfully reproducing released enemies. Thus, when three parasitic wasps of the genus *Aphytis* were introduced into southern California to help control the red scale (*Aonidiella aurantii*)', an insect pest of orange trees,

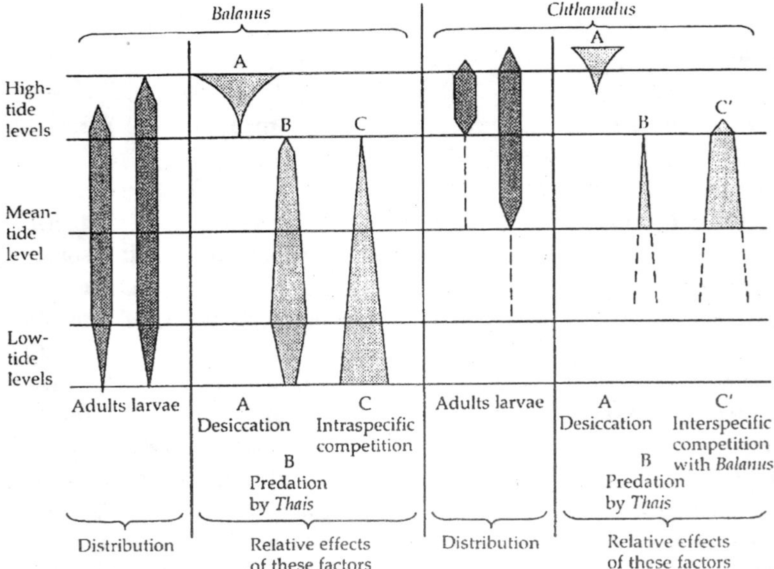

Fig. 5.4. Intertidal distribution of adults and newly settled larvae of Balanus balanoides and Chthamalus stellatus.

there were unprecedented amounts of data on the results. *Aphytis chrysomphali* was introduced accidentally from the Mediterranean in 1900 and became widely distributed. In 1948, *A. lignanensis* was introduced from south China and began to replace *A. chrysomphali* in many areas.

In 1956-1957 another species, *A. melinus*, was imported from India and immediately displaced *A. lignanensis* from the hotter interior areas.

The mechanism by which competitive displacement occurred was that female *A. melinus* could use smaller scale insects as hosts and could lay a higher proportion of "female" eggs in them than the other two species. Thus, *A. melinus* preempted most scales as hosts before the other parasites could use them. There are probably very few other examples of such well-documented competitive displacement in nature. However, the caveat here is that all species were exotic and may have been expected to compete more than native species that have evolved together over millions of years.

Plants are often thought to suffer more from competition than do animal populations because plants are rooted in the ground and cannot move to escape competitive effects. In southern California chaparral, grassland shrubs such as the aromatic *Salvia leucophylla* and *Artemisia californica* are often separated from adjacent grassland by bare sand 1 to 2 m wide. Volatile terpenes are released from the leaves of the aromatic shrub; these inhibit the growth of nearby grasses. Some plants, such as black walnut, *Juglans nigra*, produce similar chemicals (here, juglone) from their roots, which leach into the soil, killing neighbouring roots. This phenomenon is termed *allelopathy*; the action of penicillin among microorganisms is a classic case. In many cases, such allelopathic chemicals are toxic to some competitors but not to others. Competitive interactions between plants are of paramount importance in *agroforestry*, a relatively new concept in which crops are grown under forest cover so that the land will yield both food and timber. Many species of tree, especially *Eucalyptus* in tropical regimes, are not suited to the practice of agroforestry because of their adverse effects on plants growing beneath them.

Other good but more anecdotal instances of competitive displacement in nature include the deliberate introductions of animals into areas for economic gain and the effects of habitat alteration. Although such examples do not represent well-conceived scientific experiments because of the economic impacts involved, more observations over a larger scale were available than for most scientific experiments. The introduction to Gatun Lake, Panama, of the cichlid fish *Cichla ocellaris* (a native of the Amazon) is thought to have led to the elimination of six of the eight previously common fish species within five years. Also, the introduction of the game fish *Micropterus salmoides* (large-mouth bass) and *Pomoxis nigromaculatus* (black crappie) led to the diminution of local fish and crab populations. Similarly, after construction of the Welland Canal linking the Atlantic Ocean with the Great Lakes, much of the native fish fauna was displaced by the alewife (*Alosa pseudoharengus*) through competition for food. Introduced African dung beetles were successful in reducing the numbers of pest flies that competed for dung in Australia.

Despite the difficulties of demonstrating competition, it is valuable to know how frequent it might be in nature. Two reviews of the literature attempted to answer this question. Connell (1983) reviewed 72 studies on active competition as reported in the literature. Competition was found in 55 percent of the 215 species, and in 40 percent of 527 experiments involved. Connell suggested that his result appears logical if one takes the following view. Imagine a resource set with, say, four species distributed along it. Then if only adjacent species competed, competitive effects would be expected in only three

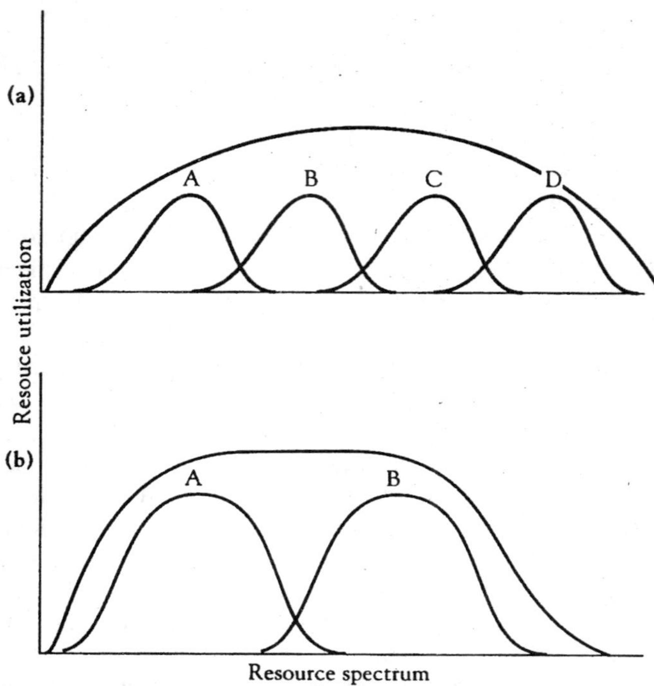

Fig. 5.5. (a) Resource supply and utilization curves of 4 species, A, B, C and D along a resource gradient. (b) When only two species utilize a resource set, competition would nearly always be expected between them.

out of the six species pairs (50 percent). Of course, the mathematics would be drastically different according to the number of species on the axis. For any given pair of adjacent species, however, competition would be expected, and indeed Connell found that, in studies of single pairs of species, competition was almost always reported (90 percent), whereas in studies involving more species, the frequency was only 50 percent.

In a parallel but independent review of 150 field experiments, Schoener (1983) reported competition in more than 90 percent of 164 studies and in 75 percent of the species studied. Why the difference between Connell's and Schooner's studies? It is because of slightly different samples of studies, methods of analysis, and perhaps predispositions of the authors. Connell, perhaps believing predation to be a more important force in nature, was more rigorous in what he accepted as a satisfactory experiment. For example, Hairston (1989) points out that at least one of the experiments accepted as evidence of competition by Schoener did not meet the necessary requirements of experimental design.

Both reviews may well overrepresent the actual frequency of competition in nature because of some common flaws:

1. "Positive" results demonstrating a phenomenon (here competition) may tend to be more readily accepted into the literature than "negative" results demonstrating patterns indistinguishable from randomness.
2. Scientists do not study systems at random; those interested in competition may well choose to work in a system where competition may be more likely to occur.

On the other hand, the reviews may fail to reveal the true importance of competition because

1. By now most organisms have evolved to escape competition and the lack of fitness it may confer.
2. Competition may only occur in certain "crunch" years where resources are scarce. Nevertheless, this competition is severe enough to structure the community. If a crunch year occurs only one in five years and a researcher does experiments in any of the other four, competition may go undetected.

Despite these drawbacks some general patterns are evident from Connell's and Schoener's work if one assumes there are no taxonomic biases in reporting the frequency of competition. Folivorous insects (leaf feeders) and fitter feeders (such as clams) showed less competition than plants, predators, scavengers, or grain feeders. Marine intertidal organisms tended to compete more than terrestrial ones, and large organisms more than small ones. Some patterns like this might be expected; for example, given limited intertidal space, it would not seem odd to detect competition for space between sessile organisms. Seeds and grains also provide a limiting but very important nutrient-rich resource for desert granivores. Brown and co-workers provide good evidence that all members of the grain-feeding guild from rodents and birds to ants compete for this resource.

Lawton (1984) argued that there is much evidence of vacant *niche* space on the plants of the world. As evidence, he showed that bracken fern (*Pteridium aquilinum*) in Europe has a large array of chewing, sucking, mining, and galling insects in a wide range of habitats, but in the United States and especially in Papua, New Guinea, whole guilds (for example, gall formers) are missing. With such vacant niches available, insects cannot be expected to compete so fiercely. The fact that so many introduced insects, other animals, and plants have become established and thrive when introduced into new countries and novel habitats suggests that few niches in natural ecosystems are filled. The implication is that many empty ecological niches exist, but it is hard for a human observer to tell what constitutes available niches independently of the species that occupy them. For example, could vampire bats later evolve in Africa to take advantage of the big game there? Could sea snakes, present now in the tropical parts of the Indian and Pacific oceans, evolve in the Atlantic region?

Lawton and McNeill (1979) also suggested that insects often "lie between the devil and the deep blue sea," that is, between a huge array of predators and parasites on the one hand and a deep blue sea of abundant but low-quality food on the other. As a result, they could scarcely become abundant enough to compete. Schoener himself had noted a dearth of competitive effects between *herbivores*, as compared to those between plants. Thus, plants and perhaps other groups with a lack of control from the next *trophic level* above could be reasoned to compete more. The foundations of this argument had been laid years earlier by Hairston, Smith, and Slobodkin (1960), who had essentially argued that the "earth is green" and that the phytophagous insects that could eat this greenery must therefore be held in check by animals from the next trophic level up. Herbivores, of course, constitute a huge fraction of the Earth's biota—more than 25 percent of the Earths species are phytophagous insects alone—so such patterns must be taken seriously. Interestingly, the latest review suggests that, in the mid-1980s to early 1990s, competition in insects has actually been found more frequently than in previous decades. An examination of 193 pair-wise species interactions showed interspecific competition in 76 percent of the cases. Perhaps more modern methods have been able to detect competition more easily. Clearly, we may have to reevaluate our position on competition in general, and insect competition in particular, as more modern data become available.

Finally, it is valuable to examine the precise mechanisms of the competitive process. Schoener (1985) divided these mechanisms into six categories:

1. Consumptive or exploitative—using resources
2. Preemptive—using space
3. Overgrowth—one species growing over another and blocking light or depriving the other of a resource.
4. Chemical—by production of toxins

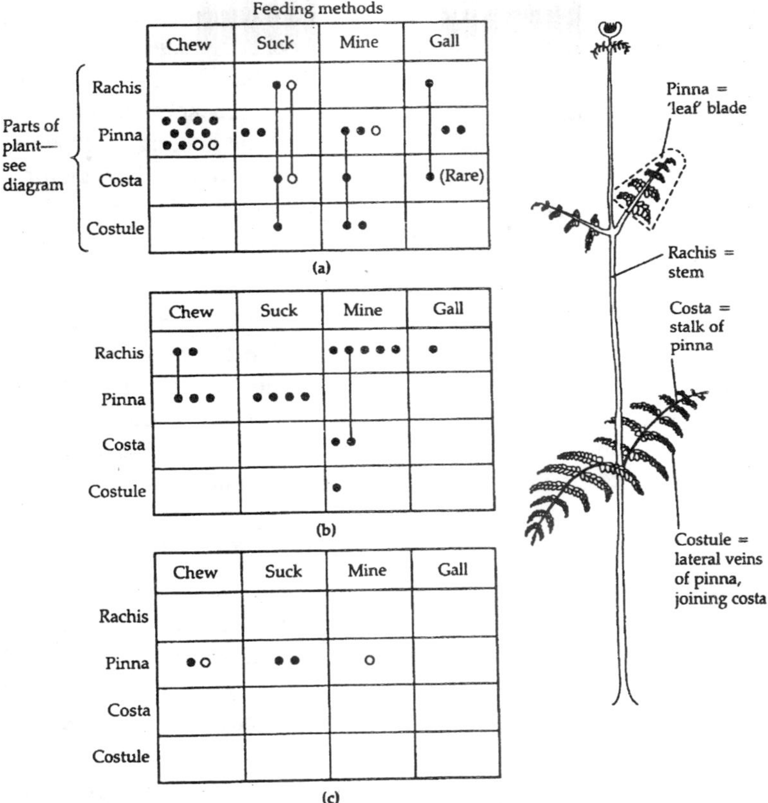

Fig. 5.6. Feeding sites and feeding method of herbivorous insects attacking bracken on three continents.

5. Territorial—behaviour or fighting in defense of space
6. Encounter—transient interactions directly over specific resources.

Exploitative competition is by far the most common, occurring in 71/188 = 37.8 percent of cases. This has led some observers to underscore the frequency of indirect effects in nature because, in exploitative competition, species only interact via a third species, which is the shared-food source. Again, we will return to indirect effects later. Some of Schoener's findings about mechanisms are easy to interpret—preemptive and overgrowth competition appear among sessile space users, primarily terrestrial plants and marine macrophytes and animals living on hard substrates.

Territorial and encounter competition occur among actively moving animals, especially birds and mammals. Chemical competition occurs among terrestrial plants; toxins become too dilute in aquatic systems. A final tidbit from both Connell's and Schoener's studies is that most often only one member of a species pair responded to the addition or removal of individuals of the other. The logic here is that such asymmetric competition should be expected; the superior competitor will probably be more strongly limited by some other factor—environmental tolerances or predators. Usually, the larger organism has the competitive advantage.

COEXISTANCE

Active competition may not always lead to competitive displacement. It is conceivable, for example, that in areas of overlap, species change their lifestyles or feeding habits so that competition is minimized. If species do compete in nature, the important question is perhaps not how much competition goes on but how similar can competing species be and still live together. This question has received more attention in ecology than any other single topic, but Lewin (1983) suggests it has led ecologists into futile works and blind alleys.

Theories Based on Morphology

In a seminal paper entitled "Homage to Santa Rosalia, or why are there so many kinds of animals?" G. Evelyn Hutchinson (1959) looked at size differences, particularly in feeding apparatus, between congeneric species when they were *sympatric* (occurring together) and *allopatric* (occurring alone).

Ratios between characters studied when species were sympatric ranged between 1.1 and 1.43, and Hutchinson tentatively argued that the mean value of 1.28 could be used as an indication of the amount of difference necessary to permit coexistence at the same trophic level but in different niches. Some authors extrapolated that ratios of between 1.3 and 2.0 indicated sufficient differences to permit coexistence because weight varies to the third power of length and $1.3^3 = 2.2$. Hutchinson's idea came under heavy fire because

1. In a large series of examples purporting to support this hypothesis, statistical analysis showed no more differences between species than would occur by chance alone.
2. Size-ratio differences have too loosely been asserted to represent the ghost of competition past when, in fact, they could have evolved for other reasons.
3. Biological significance cannot always be attached to ratios, particularly those of structures not used to gather food: ratios of

1:3 have been found to occur between members of sets of kitchen skillets, musical recorders, and children's bicycles.

4. Maiorana (1978) argued that, in such cases, these values may simply reflect something about our perceptual abilities.
5. Ratios of between 1:1 and 2:2 often result when things are lognormally distributed in nature and have small variances.

Needless to say, followers of Hutchinson's ideas have been swift to rebut some of these ideas. Losos, Naeem, and Colwell (1989) argued that the Simberloff-Boecklen statistics had deficiency of statistical power. In other words, they erred too strongly on the side of accepting a null hypothesis of no effect of competition when such an hypothesis was false. The paper also refutes the claim of Eadie, Broekhoven, and Colgan that the effects of competition cannot be distinguished from the effects of lognomal distributions. Doubtless, variations of these arguments will continue to sway back and forth in the future. It is noteworthy, however, that even one of the chief protagonists in the debate, Daniel Simberloff, has recently published articles that support the idea of character displacement in mammals.

There are some authors who have tried to use natural experiments to support the validity of Hutchinson's ratios. In the primeval forests of Canada, four indigenous parasitoids attacked the wood-boring siricid larva *Tremex columba*. Each had a different ovipositor length and laid an egg only when the ovipositor was fully extended. The ovipositor can be regarded as a food-provisioning apparatus for the larva; each species layed eggs in *Tremex* cocoons at a different depth in logs.

In the 1950s, a fifth species, *Pleolophus basizonus* was introduced into the area in an attempt to control another pest, the European sawfly. Its ovipositor length was intermediate between those of two of the existing species:

Even before the introduction of *P. basizonus*, the first three species were tightly packed but still maintained the minimum separation of about 1:1 noted by Hutchinson. However, when *P. basizonus* was introduced, strong competition ensued. *Pleolophus basizonus* or either of the two other species could have been displaced. In fact, *M. aciculatus* and *M. indistinctus* were forced out of the more favourable high-host-density sites.

Besides separation in size, species may also differ in their use of a particular set of resources, such as food or space. Consider three species normally distributed on a resource set, where $K(x)$ is the resource availability of x or carrying capacity, d is the distance between

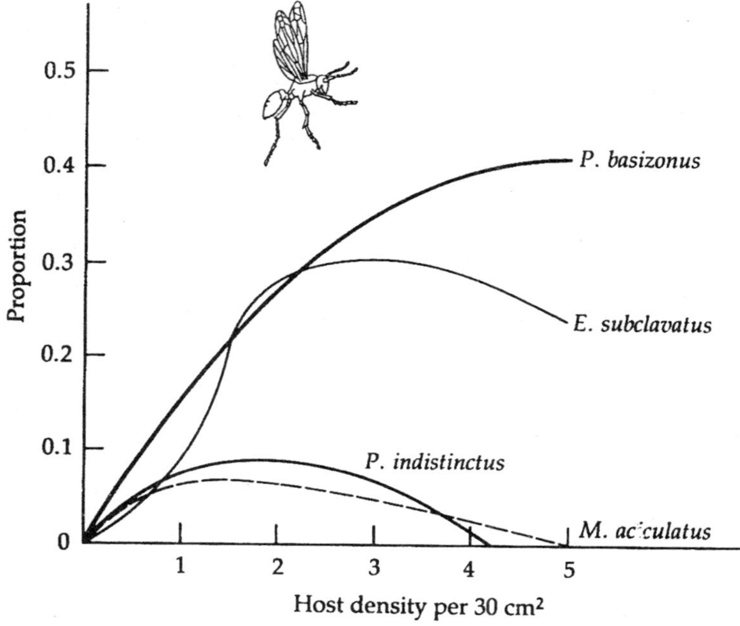

Fig. 5.7. The response of parasitoids increasing host density illustrated by the change in the proportion of each species in the total parasitoid complex.

abundance maxima, and w represents one standard deviation, approximately 68 percent of the area on x one side of the curve. It has been argued mathematically that, if $d/w < 1$, species cannot coexist; if d/w is < 3, there will be some interaction between species; and if $d/w > 3$, species coexist harmoniously. The problem is that species abundancies are often not normally distributed.

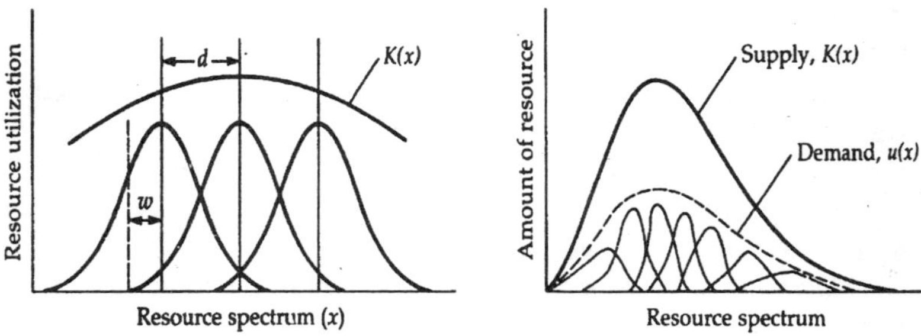

Fig. 5.8. Theoretical resource-utilization relationships. (a) The "simplest case" of three species with similar (and normal) resource-utilization curves. (b) The more typical case with varying resource-utilization curves broadest in the region of fewer resources and less interspecific competition.

Where the resource has a discontinuous distribution or occurs in distinct units, like leaves on a shrub. The niche breadth of a species can then be quantified by Levins's (1968) formula:

$$\text{niche breadth} = \frac{1}{\sum_{i=1}^{S} p_i^2 (S)}$$

where p_i = proportion of species found in the *i*th unit of a resource set of S units, such that B_{max} = 1.0 and B_{min} = 1/S. Proportional similarity between species is then given by

$$PS = \sum_{i=1}^{n} p_{mi}$$

where p_{mi}, is the proportion of the less-abundant species of the pair in the *i*th unit of a resource set with n units. Finally, the niche overlap of one species with another is represented by α_{ij} where

$$\alpha_{ij} = \sum_{h=1}^{n} p_{ih}\, p_{ij}\, (B_i)$$

where p_{ih} and p_{ij} are the proportion of each species in the *i*th unit of the resource set. Niche over- laps calculated in this way have been used as competition coefficients in classical Lotka-Volterra equations,

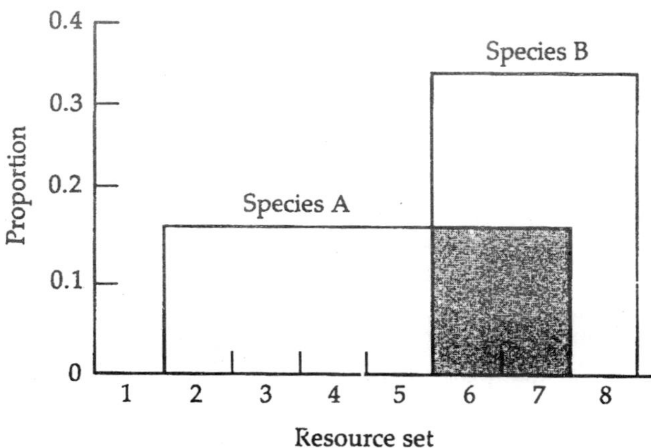

Fig. 5.9. Hypothetical distribution of a species, A, with a broad niche and a species, B, with a narrower niche, on a resource set subdivided into eight resource units. The species have the same proportional similarity, but species A overlaps B more than B overlaps A.

although a safer method is to measure the effect of one species on another experimentally.

In accordance with Hutchinson's ideas, *PS* values of less than 0.70 have been taken to indicate possible coexistence, and those greater than 0.70, competitive exclusion. However, species may differ not only along one resource axis, but along many, such as food, temperature, and moisture. For two resource axes, proportional similarity indices can be combined—proportional similarity values of 0.8 and 0.6 on two axes combine to give an overall *PS* of 0.48. Theoretically, coexistence would be permitted in cases where combined *PS* values $0.7 \times 0.7 = 0.49$ or less. Such analyses become more and more complex and less biologically meaningful as new axes are included. Perhaps the most serious criticism of both *d/w* and *PS* treatment is that resource axes identified by the researcher as important may not accurately reflect limiting resources for organisms. Can sweep-net samples of insects on foliage reliably indicate food availability for birds? Furthermore, faced with the apparent contradiction that many ecologically similar species coexist with no apparent differences in biology, many researchers would argue that the correct niche dimensions had not yet been examined.

It is worth noting that, in many situations, competing species have been found to differ hardly at all in morphology and yet still are found together. Two reasons have been proposed. First, in the presence of high levels of predation, competitively *dominant* species are likely to be selected by predators over less-abundant prey. Thus, good competitors will probably never be able to eliminate poor competitors totally if predation occurs; this is the idea of predator- mediated coexistence. Second, it is important to realize that many real populations in nature exist not in closed systems but in open areas where migration is possible and where there is good connectance between populations. Caswell (1978) has theorized that, given good connectance between areas, immigration into an area of competitively inferior species, from areas where they do well, will be sufficient to maintain reasonable population sizes of both competitors for an indefinite time. Thus, if predators open up resources in an environment by killing members of the competitively dominant species, high connectance between populations means that competitively inferior species may first appear there by immigrating from other areas.

r and K Selection

One of the most popular concepts to come out of competition theory is the idea of the *r-K continuum*. Not all organisms are well suited to compete with others; some are better able to live in more hostile environments, often in a competitive vacuum. There is a continuum of reproductive strategies that encompass so-called *r*-selected species at one end and *K*-selected species at the other, *r*-selected species are fugitives with a high rate of per-capita population growth, *r*, but poor competitive ability. An example is a weed that quickly colonizes vacant habitats (such as barren land), passes through several generations, and then disappears or is competitively excluded by individuals of more *K*-selected species. *K*-selected species compete well but tend to increase more slowly to the carrying capacity, *K*, of the environment. It is interesting to note that biological control—the control of pests by natural enemies (usually insect parasitoids or predators)—has proved more successful on pests that are closer to the *K* type than to the *r* type. Conway (1976) has proposed different control techniques for pests that lie toward either the *r* or the *K* end of the spectrum.

More recently, alternatives to the *r* and *K* continuum have been proposed. Gill (1974) suggested a three-way classification scheme with *r*, *K*, and α strategies, the last being characterized by high competitive ability. For plants, Grime (1977, 1979) proposed the *R*, *C*, and *S* strategies, where *R* strategists (ruderals) are adapted to cope with habitat disturbance (especially man-made); *C* strategists (competitors) arc adapted to live in supposed highly competitive environments such as the tropics; and *S* strategists (tolerators) arc adapted to cope with severe abiotic environmental parameters. Finally, Greenslade (1983) proposed the *r*, *K*, and *A* strategies, where A species are adapted to tolerate adverse environmental conditions. Useful though each of these schemes is, MacArthur and Wilson's original concept remains the rock on which each is based. Sometimes, however, the MacArthur-Wilson foundation seems a little shaky. Despite the apparently broad array of support for the *r* and *K* concept from a wide variety of taxa, on closer examination the actual empirical evidence for this idea is difficult to assess. Different authors have used the terms *r* and *K* selection too loosely and in different senses. It has thus become an "omnibus" term, a term with such different intuitive definitions as to be ambiguous. For many people, *r* and *K* selection is what Hardin (1957) termed a *panchreston*—something that can explain almost anything.

6

COEVOLUTION

The organism evolves as a set of integrated, cooperative organs and activities because interactions between components of the organism-system determine the individual's fitness. Interactions between individuals within populations also influence the evolution of social behaviour and other adaptations, and may lead to highly structured societies based on mutual cooperation. Extrapolating these phenomena to higher levels of ecological organization, one is led to ask whether biological communities have evolved properties of structure and function arising from interactions among their component species.

We have seen that one species may apply selection upon another, as a predator does upon its prey, and that this interaction may result in an adaptive response—perhaps protective colouration or the production of a defensive chemical. Such responses in the second species alter the environment of the first; adaptations selected among prey make the prey population more difficult to exploit on the whole, thereby applying pressure on the predator to modify its tactics or alter its diet. Evolutionary responses would also seem to link species involved in mutualistic interactions, which depend on specialized adaptations of each participant to respond appropriately to the other.

Reciprocal evolutionary responses between populations are referred to as coevolution, although, as we shall see below, there is considerable disagreement over what coevolution is. The importance of the topic in the minds of ecologists is emphasized by recent reviews and edited volumes. Several issues are involved. First, do pairs of populations undergo reciprocal evolution, or do "coevolved" traits arise from the response of populations to general selective pressures exerted by a variety of other species, followed by ecological sorting out of species with compatible features? Second, are species organized into interacting sets based on their evolved adaptations, whether "coevolved" or not? And third, do such adaptations enhance such system properties as productivity of the biological community and its resistance to perturbation?

Ecological Interaction

This definition refers to a process of mutual selection and adaptive response within a restricted set of species leading to properties that are unique to the system. Coevolution is more frequently ascribed to apparent matching of adaptations between pairs or small groups of species, without evidence bearing upon the evolutionary history of the relationship itself. The logical difficulty of inferring coevolution from matching adaptations is revealed by considering adaptations of organisms to physical characteristics of the environment. Adaptation and environment are clearly matched, yet the physical environment does not respond adaptatively and the match between organism and environment cannot be called coevolution, Much of the controversy over what coevolution is has resulted from inferring a process from patterns that such a process, as well as others, might have produced.

Modern discussion of coevolution stems from a paper by Paul Ehrlich and Peter Raven (1964), entitled "Butterflies and Plants: A Study in Coevolution." The paper opens with an admonition that reciprocal evolution, generally ignored before then, holds a key to interpreting patterns of organic diversity:

> One of the least understood aspects of population biology is community evolution—the evolutionary interactions found among different kinds of organisms where exchange of genetic information among the kinds is assumed to be minimal or absent. Studies of community evolution have, in general, tended to be narrow in scope and to ignore the reciprocal aspects of these interactions. Indeed, one group of organisms is all too often viewed as a kind of physical constant. In an extreme example a parasitologist might not consider the evolutionary history and responses of hosts, while a specialist in vertebrates might assume species of vertebrate parasites to be invariate entities. This viewpoint is one factor in the general lack of progress toward the understanding of organic diversification.

Ehrlich and Raven devote most of the paper to the relationships of groups of butterflies specialized to feed on particular groups of plants, and speculate on how feeding preferences are based upon chemical characteristics, particularly defensive compounds, of the leaves. The patterns suggest coevolution; plants evolving new chemicals to improve their defenses, insects evolving detoxification mechanisms. This interpretation was accepted by most ecologists. Coevolutionary interpretations of predator-prey, host-pathogen, and competitive relationships, and, of course, mutualisms, became commonplace.

It is interesting to note that data similar to those set forth by Ehrlich and Raven had suggested a similar interpretation at an earlier time, but the idea did not take hold. C. T. Brues (1920), in an important paper on "The Selection of Food Plants by Insects, With Special Reference to Lepidopterous Larvae," recognized the same patterns of food plant specialization detailed by Ehrlich and Raven:

> If we examine the food-plants of the genera or higher groups of butterflies, we find that most of them exhibit well-marked preference for certain, usually related plants. . . . This must not be understood to mean that the individual species of insects affect indiscriminantly many or all members of the plant group, but that their normal food-plant or plants do not fall outside the group. . . [T]he fixity of the instinct to feed on only certain kinds of plants is all the more extraordinary, for we cannot readily dismiss it as a physiological or nutritional necessity.

Brues then specifically addressed the possibility of coevolution:

> On account of the very close biological association between insects and plants in many ways it is true that the two have been mutually specialized until they have become highly modified in reference to one another, but this is not the case with food-plants, as no benefit ordinarily accrues to the plants and any idea of parallel evolution must be restricted to a development of undesirable attributes on the part of the plants and adaptations on the part of the insects to overcome such barriers to feeding.

The idea of undesirable attributes evolved as anti-herbivore defenses had been recognized many decades earlier by the German naturalist E. Stahl:

> We have long been accustomed to comprehend many manifestations of the morphology (of plants), of vegetative as well as reproductive organs, as being due to the relations between plants and animals, and nobody, in our special case here, will doubt that the external mechanical means of protection of plants were acquired in their struggle (for existence) with the animal world. The great diversity

> of mechanical protection does not appear to us incomprehensible, but is fully as understandable as the diversity in the formation of flowers. In the same sense, the great differences in the nature of chemical products, and consequently of metabolic processes, are brought nearer to our understanding, if we regard these compounds as means of protection, acquired in the struggle with the animal world. Thus, the animal world which surrounds the plants deeply influenced not only their morphology, but also their chemistry.

Brues (1922) added the possibility that animals might change in response to plant defensive adaptations, Fraenkel (1959) drove home the defensive nature of exotic plant chemicals, which were being discovered and characterized in great number at the time, and Ehrlich and Raven (1964) placed the system in the then new context provided by the merging of ecology and evolutionary biology during the late 1950s and early 1960s.

Mutualisms

The best illustrations of what appear to be coevolved traits come from obligate mutualisms in which two species are inextricably bound through mutual dependence. Daniel Janzen's (1966, 1967) study of interdependence between certain kinds of ants and swollen-thorn acacias in Central America is exemplary. The acacia plant provides food and nesting sites for ants in return for protection that the ants provide from insect pests. The bull's-horn acacia (*Acacia cornigera*) has large horn-like thorns with a tough woody covering and a soft pithy interior. To start a colony in the acacia, a queen ant of the species *Pseudomyrmex ferruginea* bores a hole in the base of one of the enlarged thorns and clears out some of the soft material inside to make room for her brood. In addition to housing the ants, the acacias provide food for the ants in nectaries at the bases of their leaves, and in the form of nodules, called Beltian bodies, at the tips of some leaves. As the colony grows, more and more of the thorns on the plant are filled; in return, the ants protect the plant from insect pests. A colony may grow to more than a thousand workers within a year, and eventually may have tens of thousands of workers. At any one time, about a quarter of the ants are outside the nest actively gathering food and defending the plant against herbivorous insects. The relationship between *Pseudomyrmex* and *Acacia* is obligatory: neither the ant nor the acacia can survive without the other. Other ant-acacia associations are facultative. That is, the ant and the acacia can co-occur to mutual benefit, but they can both exist independently as well. Species of acacia that lack the protection of ants altogether frequently produce toxic compounds that defend their leaves against herbivores.

The mutualism between ants and acacias has been accompanied by adaptations of both parties to increase the effectiveness of the association. For example, *Pseudomyrmex* is active both night and day, an unusual trait for ants, and thereby provides continuous protection for the acacia. In a similar adaptive gesture, the acacia retains its leaves throughout the year, and thereby provides a continuous source of food for the ants. Most related species lose their leaves during the dry season.

To test the influence of ants on the growth and survival of acacia plants, Janzen kept ants off new acacia shoots and compared their growth to shoots that had ants. After 10 months, the shoots lacking ants weighed less than one-tenth those with intact ant colonies, and produced fewer than half the number of leaves and a third the number of swollen thorns. The mutual benefits of the ant-acacia relationship and the highly specialized adaptations of both the plants and the ants provide a strong case for coevolution based upon a long evolutionary association between the two species.

A similar situation in which ants protect aphids and leafhoppers from predators and harvest the nutritious "honeydew" that they excrete is more difficult to interpret. One such system involving aphids and leafhoppers on ironweed (*Veronia noveboracensis*) in New York State has been studied experimentally by Catherine M. Bristow (1984). Aphids (*Aphis*) are small, sedentary, and form dense colonies on the inflorescences. Also occurring on ironweed is the larger membracid (leafhopper) *Publilia*, which sucks plant juices from the leaves. These insects are tended by three species of ants. One, in the genus *Tapinoma*, is tiny (2-3 mm) but abundant. The other two (*Myrmica*) are larger (4-6 mm) and more aggressive, but less common. The two genera of ants rarely co-occur on the same plant.

The presence of *Tapinoma* greatly enhances survival of aphid colonies but has less effect on the survival of leafhoppers. The larger *Myrmica* offers substantial protection to leafhoppers but is less effective in warding off predators of aphids. Where Bristow excluded both species of ants, predators were more numerous; where she added the predatory larvae of ladybird beetles, *Myrmica* and, to a lesser extent, *Tapinoma* effectively reduced their predation on leafhoppers.

The system has all the elements expected of coevolution, but it is not clear that the adaptations of the ant and homopteran participants evolved in response to each other. Most insects that suck plant juices produce large volumes of excreta from which they either do not or cannot extract all the nutrients. Thus honeydew production may reflect diet rather than being an adaptation to encourage protection by ants. Ants are voracious generalists, likely to attack any insect they encounter. Hence no special adaptation may be required to confer benefit on the aphids and leafhoppers upon whose excreta they also feed. The fact that the different genera of ants more effectively protect different honeydew sources may simply reflect their different sizes and levels of aggression, likely evolved in response to unrelated environmental factors.

Why don't ants eat the aphids and leafhoppers they tend? Perhaps this restraint is an evolved character of ants which facilitates the ant-homopteran mutualism. It may even have arisen as an extension of the common ant behaviour of defending plant structures that produce nectar—flowers or specialized nectaries. The point has not been addressed experimentally in this system, but a similar situation in the Cape region of South Africa highlights the importance of particular adaptations of ants, whether they are coevolved or not, to maintaining an ant-plant mutualism. There, many species of plants in the family Proteaceae have seeds with fleshy, edible, attached structures, called *elaiosomes*. Foraging ants pick up seeds and transport them to their underground nests, where the elaiosomes are eaten. The seeds themselves, which the ants cannot eat, are then discarded either in underground chambers or in refuse heaps on the surface. In many regions, these disposal sites are suitable for seed germination and seedling establishment. But in the fynbos (brushy, chaparral-like habitat), germination of many species of plants occurs only after fires have swept the habitat, and the only seeds that germinate are those stored by ants in underground nest chambers.

Recently, the Argentine ant *Iridomyrmex humilis* has invaded areas of fynbos shrublands and displaced many of the less aggressive native ants. *Iridomyrmex* differs from the native ants in not storing seeds within its nests; the elaiosomes are removed on the surface where the seeds are dropped. W. Bond and P. Slingsby (1984) found that germination of one species of *Mimetes* following fire was drastically reduced in areas invaded by *Iridomyrmex*. With the continued persistence of the Argentine ant, it is likely that much of the native Cape flora will disappear as underground seed reserves are depleted. In a similar case, Stanley Temple (1977) has suggested that the virtual

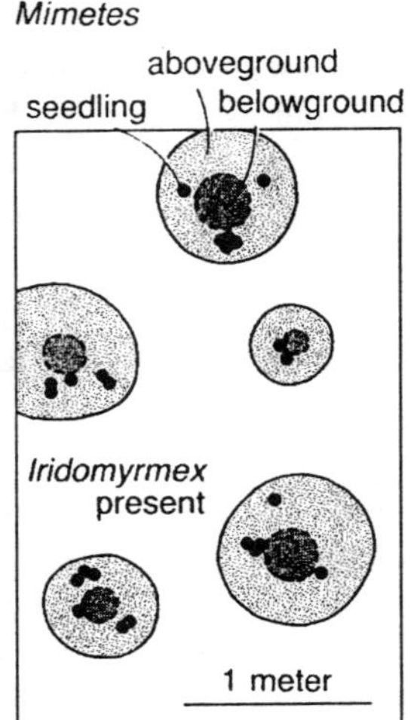

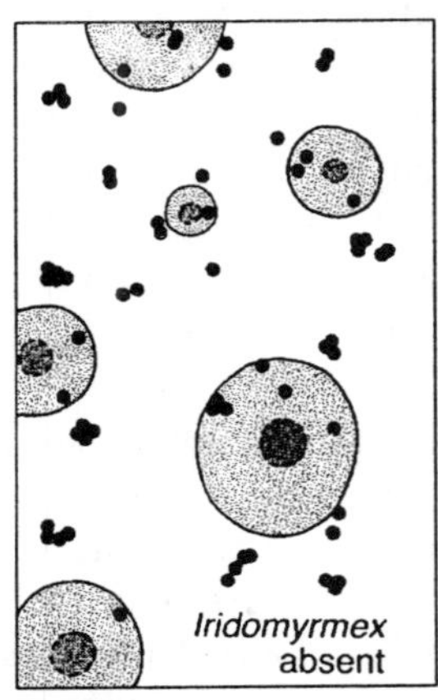

Fig. 6.1. Seedling dispersion in Mimetes cucullatus populations after a burn in the presence of Iridomyrmex (left) and in its absence (right).

extinction of the tree *Calvaria major* on the island of Mauritius followed upon the extinction more than 300 years ago of the dodo bird, the only native species capable of effectively dispersing *Calvaria*.

Clearly, particular adaptations of ants influence their effectiveness as seed dispersers. But these adaptations may be evolutionarily independent of the plants themselves. The ant-dispersers are diet generalists and their food-caching behaviour likely evolved for reasons unrelated to their role as seed-dispersers of *Mimetes* and its relatives. The plants may have merely evolved to take advantage of this fortuitous element in their environment without any reciprocal evolution on the part of the ant. Whereas some mutualisms, such as that between ant and acacia described above, clearly involve specialized adaptations of both parties and seem to represent cases of coevolution, many mutualisms may involve more serendipitous arrangements.

Returning to the problem of feeding relationships of herbivores on plants, and of parasites on hosts, Janzen (1980) questioned the usual coevolutionary interpretation of defensive behaviours, structures, and chemical, and their circumvention by consumers. He suggested that the defenses of plants and hosts were not usually selected by the herbivores and parasites that can circumvent them. Suppose, for example, that plant P is eaten by herbivore H. Mutation P* arises in the plant population, which results in the production of a substance toxic to H. P* quickly increases relative to P, and the population of H declines. Coevolutionists would suppose that H might respond by mutation H*, whose bearers rendered the toxin ineffective by metabolizing or sequestering it. The result would be a system of P* and H* whose populations were reciprocally adapted to one another.

Janzen suggested alternatively that mutation P* might be circumvented by some other species, say F, already capable of handling the new defense, perhaps even recognizing a potential food plant by its presence. Therefore, rather than leading to highly coevolved groups of small numbers of species, evolutionary responses to herbivores and plant defenses might instead lead to a reshuffling of the feeding relationships within a community or even the invasion of the community by new species.

Coevolution in Plants

Plant geneticists have developed strains of domestic crops, such as flax and wheat, that are resistant to particular genetic strains of various pathogens, such as rusts (teliomycetid fungi). The crop strains differ from one another by a few, perhaps single, genetic changes that make them either susceptible or resistant to infection by particular strains of rust. Over the course of crop improvement programs, when new strains of rust have appeared, either by mutation or by immigration from other areas, crop geneticists select new resistant strains of the crop by exposing experimental populations to the pathogen.

Mode (1958) used the crop-rust system as the basis for the first explicit genetic model of coevolution. He prefaced his work with the basic premise of coevolution:

> it seems reasonable to assume . . . that such obligate parasites as the rust, smut, and mildew fungi have evolved in association with their hosts. The genetic system of host and parasite, therefore, [has] very likely been established in response to two types of opposing selection pressures, namely, the selection pressure exerted on the host by the parasite, and the selection pressure exerted on the parasite by the host.

Mode's model assumed that virulence factors in the rust and resistance factors in flax were alleles of single genetic loci whose fitness relationships exhibited strong interactions between the genotypes of the rust and flax. Mode found that, providing certain conditions of fitness relationships were met,

> a host-pathogen system operating under complementary genetic systems of the host and parasite will eventually reach a state of

> stable equilibrium (which is) advantageous to both the host and parasite. In the first place, a stable pathogen population solves the host's problem of maintenance of resistance to disease, and secondly the pathogen is able to survive without eliminating its host.

The stable equilibrium achieved in Mode's model could easily be upset by the appearance of a new virulence gene in the pathogen or resistance gene in the host. Hence stability in the coevolved system depends upon constancy of genotypes within each population.

Perhaps a more usual situation is the appearance of new genotypes either by migration or mutation, creating continual evolutionary flux in such a system. This seems to have been the case in the rust-wheat system described by G. J. Green (1975) in which new virulence genes of the rust *Puccinia graminis* appeared from time to time and swept through the population. Genetic races of wheat rust are characterized both by physiological characteristics and by virulence when tested on lines of wheat containing different resistance alleles. Most of the virulence strains within a single physiological race of the rust differ by only one gene and it is possible to outline a plausible evolutionary relationship between the races. For example, strains 1 and 24 differ only by virulence on wheat with resistance gene 9b; strains 43 and 20, although in different physiological races, differ only in their virulence with respect to resistance gene 6.

The rust-wheat system contains the essential element of Mode's concept of coevolution, namely a strong interaction between the fitnesses of genotypes of the host and those of the pathogen. Rather than going to equilibrium, the system is kept in flux by the introduction of new virulence genes in the rust and, perhaps, by new resistance genes in

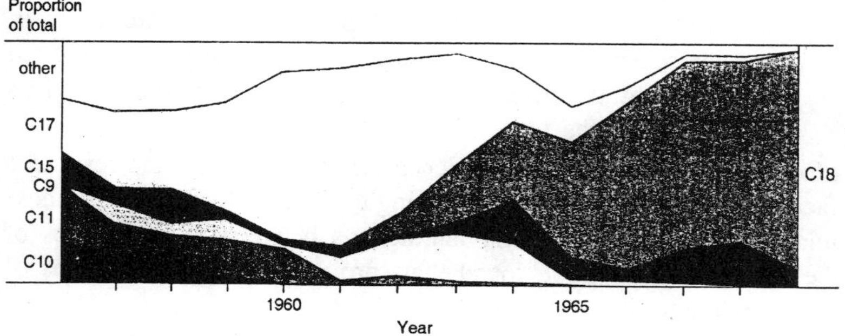

Fig. 6.2. Relative proportions of virulence genes in the rust Puccinia graminis infecting Canadian wheat.

the wheat, although the latter are pretty much controlled by plant geneticists.

Genotype-genotype interactions have been found in several natural systems and may turn out to be the rule in populations of plants and herbivores, or hosts and pathogens. George F. Edmunds and Donald N. Alstad (1978, 1981, 1983) demonstrated that variation between trees in the defenses of ponderosa pines were matched by variation in genotypes of scale insects that infest them. The scales are extremely sedentary, with so little migration from tree to tree that local populations (demes) on individual trees have evolved independently of those on other trees. This local adaptation is revealed when scales are transferred both between trees and between branches within the same tree. The survival of scales after inter-tree transplants is greatly reduced compared to control transfers within the same tree. It is reasonable to assume that the differences between trees and demes of scales are genetic, hence this represents a case of genotype-genotype interaction. C. Wiklund, Jr. (1981) and Michael Singer (1983) demonstrated intraspecific variation in preferences of lepidoptera among different species of host plants and Sara Via (1984) has shown genotype-host plant interactions in growth performance of larvae of the agromyzid fly *Liriomyza sativae* on cow pea and tomato. Hence the genetic background for coevolution seems to be in place.

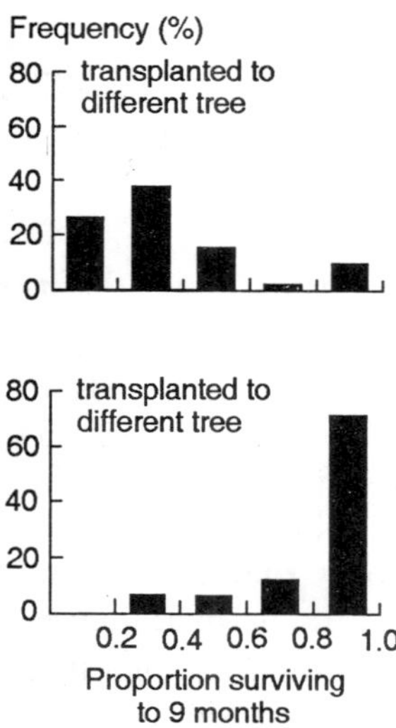

Fig. 6.3. Black pineleaf scale on needles of ponderosa pine illustrating the damage caused by feeding.

Plant Defense

Differences in the defensive chemicals of plants can be related to genetic changes, particularly when the pathways of biochemical synthesis and responsible enzymes are known. May Berenbaum (1978, 1981, 1983) has placed elements of the relationship between certain butterflies and their umbelliferous host plants in the context of coevolution. Umbellifers produce many noxious chemicals, among the most prominent of which are the furanocoumarins. The biosynthetic pathway leads from paracoumaric acid, which, being a precursor of lignin, is found in virtually all plants, to hydroxycoumarins, such as umbelliferone, and then to furanocoumarins. The last include linear and angular forms, which are produced directly from hydroxycoumarins by different enzyme reactions. As one proceeds down the biosynthetic pathway from *p*-coumaric acid to hydroxycoumarins, and linear or angular furanocoumarins, toxicity increases and occurrence among plant families decreases. Hydroxycoumarins possess some biocidal properties; linear furanocoumarins bind with pyrimidine bases and interfere with DNA replication in the presence of ultraviolet light; angular furanocoumarins interfere with growth and reproduction quite generally, although the mechanisms of action have not been detailed.

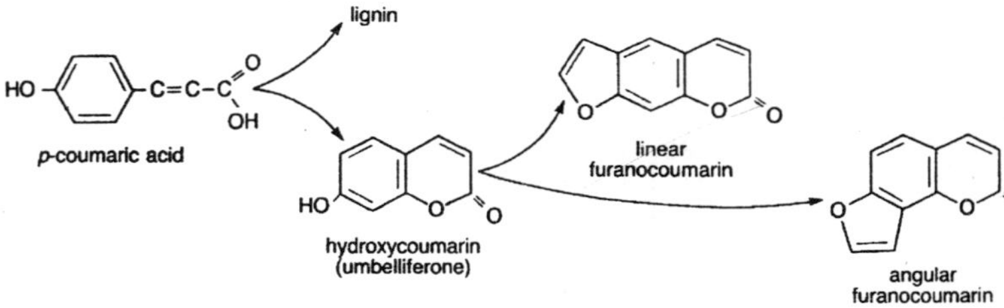

Fig. 6.4. Biosynthetic pathway of furanocoumarins.

Para-coumaric acid is widespread among plants, occurring in at least 100 families; Berenbaum (1983) lists only 31 families in which hydroxycoumarins have been found. Linear furanocoumarins (LFCs) are restricted to 8 plant families and are widely distributed only in 2—Umbelliferae (parsley family) and Rutaceae (citrus family). Angular furanocoumarins (AFCs) are known only from 2 genera of Leguminosae (pea family), and 10 genera of Umbelliferae.

Among species of herbaceous umbellifers in New York, some (especially those growing in woodland sites with low levels of UV light) lack furanocoumarins, others have linear furanocoumarins only, and some have both linear and angular furanocoumarins. From a survey of the herbivorous insects collected from these species, Berenbaum (1981) concluded that host plants containing angular and linear furanocoumarins were attacked by more species of insects than found on plants with only linear furanocoumarins, or none; that the herbivores on AFC/LFC plants tended to be extreme diet specialists, most having been found on no more than 3 genera of plants; and that these specialists tended to be abundant compared to the numbers of the few generalists found on AFC/LFC plants and compared to levels of any herbivores on either LFC plants or on umbellifers lacking furanocoumarins.

Although linear and, especially, angular furanocoumarins are extremely effective deterrents to most species of herbivorous insects, some genera that have evolved to tolerate these chemicals have become successful specialists. Berenbaum (1983) makes a strong case for coevolution here in Janzen's restricted sense. The taxonomic distribution of hydrocoumarins, linear furanocoumarins, and angular furanocoumarins across host plants suggests that plants containing LFCs are a subset of those containing hydroxycoumarins, and those containing AFCs are an even smaller subset of those containing LFCs. This is consistent with an evolutionary sequence of plant defenses progressing from hydroxycoumarins to LFCs and AFCs. Furthermore, insects specialized on plants containing LFCs belong to groups that characteristically feed on plants containing hydroxycoumarins, and those specialized on AFCs have close relatives that feed on plants containing LFCs. Although phylogenetic relationships and the history of host-plant utilization have not been matched, the taxonomic distributions of insects across host plants produce patterns that would be expected of a coevolved system.

Diffuse Coevolution

The controversy over whether coevolution takes place within small groups of organisms locked into either evolutionary struggles or cooperative ventures is giving way to a more generalized view of coevolution that recognizes far-reaching and overlapping evolutionary relationships between the species within a biological community. Although both members of mutualist and antagonist species pairs inevitably exert selection on each other, the general overlapping of species relationships into webs of interaction undoubtedly results in corresponding webs of adaptation. Speaking about the evolution of mutualisms, including the interactions of plants with their pollinators and seed dispersers, Henry F. Howe (1984) states that "For coevolution to occur, organisms must have distinctive selective effects on each other. For obligate mutualism to evolve, effects must be both unique and of long duration." Howe suggests that most mutualisms are "diffuse," by which he means that adaptation occurs in the context of a matrix of species interactions.

This expanded view of coevolution leads naturally into the final part of this book, which deals with the ecology of entire communities (assemblages of species). Before moving on, however, I would like to recount briefly one case of an obligate mutualism, clearly involving coevolution, which emphasizes the strong interdependencies that can arise in nature and the problems of arriving at a satisfactory mutual "agreement" between coevolved parties.

The Yucca Moth

The curious pollination relationship that occurs between species of yucca plants (*Yucca*, in the lily family) and moths of the genus *Tegeticula* was first described by C. V. Riley nearly a century ago (1892) and has been considerably elaborated since. The moth enters the yucca flower and deposits 1 to 5 eggs on the ovary. Later, when the eggs hatch, the larvae burrow into the ovary, where they feed on the developing seeds. But after the moth has laid her eggs, she scrapes pollen off the anthers in the flower and rolls it into a small ball, which she grasps with specially modified mouthparts. She then flies to another plant, enters a flower, and proceeds to place the pollen ball onto the stigma of the flower before laying another batch of eggs.

The relationship between the moth and the yucca is obligatory, *Tegeticula* can grow nowhere else, *Yucca* has no other pollinator. In return for pollinating its flowers, the yucca seemingly tolerates the moth larvae feeding on its seeds, but the extent of this loss of potential reproduction is small, rarely exceeding 30 per cent, and more nearly half that value on average, in *Yucca whipplei*. *Yucca* and *Tegeticula* are specialized with respect to each other. The moth has a highly idiosyncratic pollination behaviour. The yucca is specially adapted to make the moth's behaviour effective: the pollen is sticky and can easily be formed into a ball and the stigma is specially modified as a receptable.

A puzzling aspect of the relationship is the restraint the moth exercises in laying a small number of eggs in each flower. Over the short term, it would seem that moths laying larger numbers of eggs per flower might have higher individual fitnesses, even though such behaviour over the long term might lead to extinction of the yucca. Interpretation of this system poses the same difficulty as altruistic behaviours within populations, without the option of falling back upon kin selection. One possibility is that yuccas can effectively regulate the relative fitness of moths laying different numbers of eggs per flower by selective abortion of developing fruits that are too highly infested by moth larvae to produce seed. Selective abortion of insect-damaged fruits is well known, and yuccas are known to possess mechanisms for fruit abortion. The obvious experiment of transferring large numbers of moth eggs to single flowers of *Yucca*, followed by measuring both fruit maturation and emergence of fully developed larvae, has not been tried. The particular evolutionary steps of moth behaviour leading to the present system are lost in time, but the advantage to the moth of ensuring that the flower within which it lays its eggs is properly pollinated should be obvious. Most pollination relationships are neither so intricate nor so specialized as that of the yucca and moth. What "preadaptations" of *Yucca* and *Tegeticula* ancestors, or what accidents of history, started them along their coevolutionary pathway are considerations for the next generation of ecologists.

7

MUTUALISM AND COMMENSALISM

Organisms do not exist alone in nature but instead co-occur in a matrix of many species where the interactions in the table below are possible.

Nature of interaction	*Species 1*	*Species 2*
Mutualism	+	+
Commensalism	+	0
Herbivory	+	–
Predation	+	–
Parasitism	–	–
Allelopathy	–	0
Competition	–	–

+ =positive effect; 0=no effect; – =deleterious effect

Herbivory, predation, and parasitism all have the same general effects, a positive effect on one population and a negative effect on the other. Competition affects both species negatively. *Mutualism* and *commensalism* are less commonly discussed in ecology but are tied together with *parasitism*, under the banner of *symbiotic* relationships. In symbiotic relationships, the partners in the association live in intimate association with one another; they are always found in close proximity. The effects of mutualism and commensalism are different from those of parasitism, however, and are better discussed separately here.

MUTUALISM

In mutualistic arrangements, both species benefit. The colour insert shows members of some mutualistic relationships. In mutualistic pollination systems, both plant and pollinator (insect, bird, or bat) benefit, one usually by a nectar meal and the other by the transfer of pollen. In one extraordinary case, male euglossine bees visiting orchids in the tropics do not collect nectar or pollen but are rewarded instead with a variety of floral fragrances, which they modify to attract females. The tightness of the relationship in some pollination systems is underlined by the phenomenon of *buzz pollination*. Certain flowers whose anthers open through pores at the top shed their pollen when subjected to vibrations emanating from the buzzing of bee wings. To ensure that some pollen falls on its target bee below, the pollen is negatively charged. As the pollen rains down, it is attracted electrostatically to the bees, which tend to have positive charges.

Both parties also benefit in mutualistic fashion from seed dispersal when fruits are eaten by frugivorous birds, bats, or other mammals; the consumer receives a meal, and the plant receives an effective means of progeny *dispersal*. The lack of plant mobility has made many of them dependent on animals for pollination and seed dispersal. Temple (1977) has argued that the tree *Calvaria major* on the island of Mauritius has produced no seedlings for the past 300 years because its seeds do not germinate unless they have first passed through the digestive system of the now-extinct dodo. However, Witmer (1991) exploded this myth by noting that (a) seeds can germinate without abrasion in bird guts and (b) some living trees that are less than 300 years old exist, whereas the dodo went extinct in the 1660s, more than 300 years ago. Corals are mutualists, too. The animal polyps contain unicellular algae. Some people even view the association of humans with domestic animals or crops as a mutualism.

Several recent articles have commented that mutualistic interactions are not covered in sufficient detail in modern ecology texts. This is a contrast to ecological textbooks of the 1920s—1940s where positive interactions were hypothesized to be important driving forces in communities. However, even in the ecological literature, the frequency of research articles on mutualism (14 percent) is much less than that for such other interaction types as competition (31 percent), predation (24 percent), or herbivory (30 percent). So textbooks may merely reflect the state of the ecological literature. This low level of representation could be because many mutualism studies have been descriptive and have focused on particular adaptations or life-history characteristics of organisms and not on theory, whereas studies on other interactions focus on the interaction itself and the mechanisms involved. Bertness and Hacker (1994) also suggest that studies of mutualism are not often experimental and are paid little attention by theorists. There are few appropriate models for mutualistic interactions. We could incorporate the positive effect of one species on the other by modifying the Lotka-Volterra equations such that

$$\frac{dN_1}{dt} = r_1 N_1 \left(\frac{K_1 - N_1 + \alpha N_2}{K_1} \right)$$

and

$$\frac{dN_2}{dt} = r_2 N_2 \left(\frac{K_2 - N_2 + \beta N_1}{K_2} \right)$$

where αN_2, and βN_1, are the positive effects of species 2 on species 1 and species 1 on species 2, respectively. Such modifications often lead to unrealistic solutions in which both populations increase to unlimited size. We could allow each species to increase the carrying capacity of the other but place a limit on the interaction such that $\alpha\beta < 1$. Again, the models are unstable and often lead to extinction of one species. However, some authors have suggested that is an accurate finding: few obligate mutualisms in nature are very stable in the face of environmental change.

Pollination

Pollination studies are booming in modern ecology. Nearly 45 percent of all studies of mutualism involve pollination systems. This may be partly because of the great diversity of apparently tight, *coevolved*, and interesting systems to study and partly because money is available to study them. More than 90 crops in the United States alone are pollinated by insects. One of the finest studies in obligate mutualism involves figs and fig wasps as studied by Janzen (1979a). More than 900 species of *Ficus* exist, and virtually every one must be pollinated by its own species of agaonid wasp. The fig that we actually eat has an enclosed inflorescence containing many flowers. A female wasp enters through a small opening, pollinates the flowers, lays eggs in their ovaries, and then dies. The progeny develop in tiny galls and hatch inside the fig. Males hatch first, locate female wasps within their galls, and thrust their abdomens inside the galls to mate with them. The males then die without ever having left the fig. Females collect pollen from the fig and then leave to search out new figs in which to lay their progeny. A similar mutualistic relationship occurs between yucca plants and yucca moths. Both mutualisms are highly coevolved. The distribution of each species of yucca or fig is controlled by the availability of its pollinator and vice versa. In the late nineteenth century, Smyrna figs were introduced into California, but they failed to produce fruit until the proper wasps were introduced to pollinate them.

It is interesting that such highly coevolved systems arose because on superficial examination, the needs of the plants and those of the pollinators seem to conflict sharply. From the plant's perspective, an ideal pollinator would move quickly among individuals but retain a high fidelity to a plant species, thus ensuring that little pollen is wasted as it is inadvertently brushed onto the pistils of other plants. The plant should provide just enough nectar to attract a pollinator's visit. From the pollinator's perspective, it would probably be best to be a generalist and to obtain nectar and pollen from flowers in a small area, thus minimizing energy spent on flight between patches. This casts doubt on whether such relationships are truly mutualistic in nature or whether both species are actually "trying to win" in an evolutionary arms race. One way in which the plants encourage the pollinator's species fidelity is by sequential flowering through the year of different plant species and by synchronous flowering within a species. There are cases where both flower and pollinator try to cheat. In the bogs of Maine, the grass pink orchid (*Calopogon pulchellus*) produces no nectar, but it mimics the nectar-producing rose pogonia (*Pogonia ophioglossoides*) and is therefore still visited by bees. Bee orchids have even gone so far as to mimic female bees; males pick up and transfer pollen while trying to copulate with the flowers. So effective are the stimuli of flowers of the orchid genus *Ophrys* that male bees prefer to mate with them even in the presence of real female bees! Conversely, some *Bombus* species cheat by biting through the petals at the base of flowers and robbing the plants of their nectar without entering through the tunnel of the corolla and picking up pollen.

Finally, it is interesting to speculate about why ants, usually the most abundant insects in a given area, are so rarely involved in pollination. One reason might be that the subterranean nesting behaviour of many ants exposes them to a wide range of pathogenic fungi and other dangerous microorganisms, to which they respond by producing large amounts of antibiotics. These antibiotics inhibit pollen function. Peakall, Beattie, and James (1987) demonstrated that ants without metapleural glands and, therefore, without these secretions do successfully pollinate orchids in Australia.

Dispersal Systems

Mutualistic relations are highly prevalent in seed-dispersal systems of plants. Studies on these dispersal systems are also highly prevalent in the mutualism literature with 38 percent of studies focusing on dispersal. In the tropics, some fruits are dispersed by birds that are strictly frugivorous. These fruits provide a balanced diet of proteins, fats, and vitamins. In return for this juicy meal, birds unwittingly disperse the enclosed seeds, which pass unharmed through the digestive tract. Some plants, instead of producing highly nutritious fruits to attract an efficient disperser, simply produce abundant mediocre fruit in the hope that some of it will be eaten by generalists. Fruits taken by birds and mammals often have attractive colours—red, yellow, black, or blue; those dispersed by nocturnal bats are not brightly coloured but instead give off a pungent odor to attract the bats. (In contrast, because birds do not have a keen sense of smell, fruits eaten by them are generally odorless.)

In general, the relationships are not as obligately mutual as are plant-pollinator systems because seed dispersal is performed by more generalist agents. Nevertheless, a wide array of adaptations exists; one has only to look at the impressive specialization of parrot beaks, strong and sharp to crack and peel fruits, to see that the mutualistic relationship between plant and seed disperser is strong in this case. Some bizarre strategies also exist; for example, in the floodplains of the Amazon, fruit- and seed-eating fish have evolved that disperse seeds. Microbes appear to be good at cheating the plant in this system, for they will readily attack the fruit without dispersing it. Janzen (1979c) has suggested that microbes deliberately cause fruit and other resources such as carcasses to "rot," reserving it for themselves, by manufacturing ethanol and rendering the medium distasteful to vertebrate consumers. Animals eating such food are selected against because they become drunk and are easy victims for predators. As a countermeasure, some vertebrates have "learned" evolutionarily to tolerate these microorganisms and even to use them in their own guts to digest food. Alternatively, they often possess the enzyme alcohol dehydrogenase to break down alcohol.

Other Examples of Mutualism

Numerous other types of mutualism exist, although most are not commonly studied. On coral reefs, "cleaner" fish nibble parasites and dead skin (which might otherwise cause disease) from their customer fish at specific cleaning stations. Such systems often leave their participants open to cheaters. Saber-toothed blennies of the genus *Aspidontus* bear a striking resemblance to the common cleaner wrasse, *Labroides dimidiatus*. Instead of performing a cleaning function, however, the blenny bites chunks out of the customers. Saber-tooths are protected from attack by their resemblance to the cleaners, though customers, in time, learn to avoid the cleaning stations that blennies frequent.

One of the oldest ideas about mutualisms is that they are more common under harsh physical conditions when neighbours buffer one another from limiting physical stresses.

Bertness and Hacker (1994) have described an interesting mutualism in the salt marshes of New England where only a few plant species exist, each generally in distinct zonation patterns according to tolerances of soil salinities and waterlogging. In southern New England, terrestrial marsh borders are dominated by the perennial shrub *Iva frutescens* (marsh elder), whereas at the seaward side blackgrass, *Juncus gerardi*, dominates. In between the two is a region of blackgrass mixed with stunted *Iva*. Bertness and Hacker experimentally removed *Juncus* from around some *Iva* plants and found that soil salinities doubled and soil oxygen levels decreased. The photosynthetic rate of *Iva* went down and, fourteen months later, the *Iva* were dead. *Juncus* neighbours clearly have strong positive effects on adult marsh elders by reducing the soil salinity and enabling *Iva* to survive at places in the salt marsh where, alone, they would die. Next, Bertness and Hacker moved *Iva* and *Juncus* together into three different habitat types: low marsh (high salt stress), transition zone, and high marsh (low salt stress). In the low marsh, the biomass of both species was high, suggesting a mutualistic association in regions of high stress. However, in the high marsh, where conditions were relatively benign, plants did not do better; in fact, they did worse as competition ensued. This type of experiment,

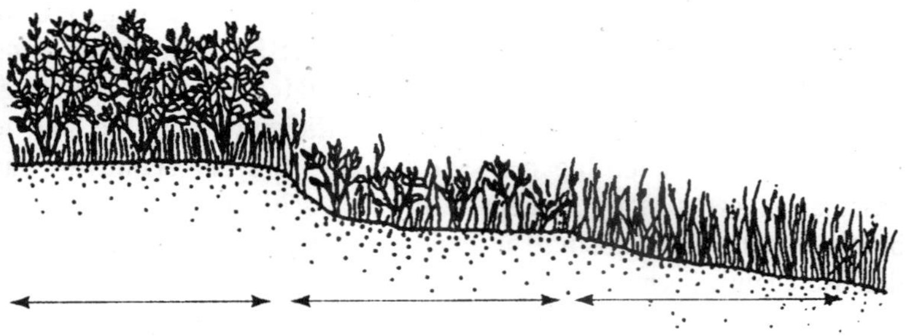

Fig. 7.1. Schematic diagram of the terrestrial border of a typical New England salt marsh.

predicting when and where mutualisms will occur, will be of much value to ecologists in the future.

Protection

In terrestrial systems one of the most commonly observed mutualisms exists between ants and aphids. Aphids are fairly helpless creatures, easy prey to marauding ants. Yet in general, ants tend to farm aphids like so many cattle. The aphids secrete honeydew, a sticky exudate that is rich in sugars and that the ants enjoy. In return, ants protect aphids from an array of predators, such as syrphid larvae, such parasites as braconid wasps, and other competing insects, by vigorously attacking them.

Obligatory Mutualism

In some cases, the mutualistic relationship is so tight that neither participant could exist without the other and is called *obligatory mutualism*. Such is the case for many lichens, which are combinations of algae (which provide the photosynthate) and fungi. The "lichenized" fungi include within their bodies and near the surface a thin layer of algal cells, forming only 3 to 10 percent of the weight of the thallus body. Of the 70,000 or so species of fungi, 25 percent are lichenized. Lichenized forms occur in deserts, in alpine regions, and across a wide range of habitats. Nonlichenized fungi are usually restricted to being parasites of plants or animals or to being involved in decomposition. Many ruminants shelter symbiotic bacteria in their guts, which break down plant tissue to provide energy for their hosts; cellulose is otherwise indigestible for mammals. Likewise, the roots of most higher plants (except the Cruciferae) are actually a mutualistic association of fungus and root tissue—the mycorrhizae. The fungi require soluble carbohydrates from their host as a carbon source (up to 40 percent of the photosynthate produced), and they supply mineral resources, which they are able to extract efficiently from the soil, to the host. The relationship between systemic fungal endophytes and vascular plants, usually thought to be a parasitic infection, has also been viewed as mutualism. The fungi are thought to aid their hosts in defense against herbivory.

COMMENSALISM

In commensal relationships, one member derives benefit while the other is unaffected. Such is the case when sea anemones grow on hermit-crab shells. The crab is already well protected in its shell and gains nothing from the relationship, but the anemone gains continued access to new food sources. The same benefits accrue to members of a phoretic relationship (*phoresy*), in which the association involves the passive and more-temporary transport of one organism by another, as in the transfer of flower-inhabiting mites from bloom to bloom in the nares of hummingbirds. Some of the most numerous examples of commensalism are provided by plant mechanisms of seed dispersal. Many plants have essentially cheated their potential mutualistic seed-dispersal agents out of a meal by developing seeds with barbs or hooks to lodge in the animals' fur rather than their stomachs. In these cases, the plants receive free seed dispersal, and the animals receive nothing. This type of relationship is fairly common; most hikers have been plagued by "burrs" and "sticktights." Sometimes, these barbed seeds can cause great discomfort to the animal in whose fur they become entangled. Fruits of the genus *Pisonia* (cabbage tree), which grows in the Pacific region, are so sticky that they cling to bird feathers. On some islands, birds and reptiles can become so entangled with *Pisonia* fruits that they die.

8

Parasitism

When one organism feeds off another hut does not normally kill it outright, the predatory organism is termed a *parasite* and the prey a *host*. Some parasites remain attached to their hosts for most of their lives, like tapeworms, which remain inside the host's alimentary canal. Others, such as ticks and leeches, drop off after prolonged periods of feeding. Mosquitoes remain attached for relatively short periods. By this definition, many species of phytophagous insects are parasitic upon their "host" plants. Still, there remain many problems of definition. Should organisms that feed off more than one individual without killing them he known as parasites or predators? For example, saber-tooth blennies (*Plagiotremus*) on the Great Barrier Reef dash out and bite chunks out offish hosts/prey that swim by. Should the large ungulates of the Serengeti plains—wildebeest, zebra, and the like—be known as parasites? Although they feed off more than one individual host grass, the grass is not killed and will grow back later. Should we retain the term *parasite* for organisms that remain in intimate contact with their hosts? Mosquitoes develop as larvae in a nonparasitic manner in pools of water, and the adults only come into contact with hosts for short periods. Rhinoceroses live on top of their food supply for their entire lives. What about parasites of insects? Many of these develop as internal parasites of caterpillars or other immature stages. In these cases, the host almost never survives, and the term *parasitoid* is used to refer to these parasites, each of which uses only one host hut invariably kills it. Even in this case, further gradation between parasitoid/parasite and predator is evident, as when an egg parasitoid hatches from a host egg and has to devour several more in the clutch before it is mature. May and Anderson (1979) have tried to distinguish two types of parasites—microparasites, which multiply within their hosts, usually within the cells (bacteria and viruses), and macroparasites, which live in the host but release infective juvenile stages outside the host's body. For most microparasite infections, the host has a strong immunological response. For macroparasitic infections, the response is short lived, the infections tend to be persistent, and hosts are subject to continual reinfection.

Despite these problems of definition, the biology of host-parasite relationships has a rich history of interesting, *coevolved*, and complex life-history patterns. Parasites on animals include those of interest to the conventional parasitologist—viruses, bacteria, protozoa, flatworms (flukes and tapeworms), thorny-headed worms (Acanthocephala), nematodes, and various arthropods (ticks, mites, and so on). Parasitoids from the parasitic Hymenoptera and Diptera are of more interest to the entomologist and biological-control specialist. Such parasitoids are often arraigned against other insects and may contribute more than 70 percent of the insect fauna. Because about 75 percent of the known global fauna consists of insects, then at least 50 percent of the animals on Earth might be considered parasitic. When the other large groups of parasites are considered— nematodes, fungi, viruses, and bacteria— it is clear that parasitism is a very common way of life. A free-living organism that does not harbor several parasitic individuals of a number of species is a rarity. The frequency of human infection by parasites is staggering. There are 250 million cases of elephantiasis in the world and more than 200 million of bilharzia, and the list goes on and on.

Defenses Against Parasites

The defensive reactions developed by hosts to resist parasites are nearly as impressive as those to combat predation:

Cellular Defense Reactions

These reactions particularly are found in insect larvae as a defense against parasitoids, where eggs of the parasitoid are "encapsulated" or enclosed in a tough case rendering them inviable.

Immune Responses in Vertebrates

These responses are the vertebrate body's defense against the parasitic microbes that cause disease in humans and animals. Phagocytes may engulf and digest small alien bodies and encapsulate and isolate larger ones. For microparasites, the host may develop a "memory" that may make it immune to reinfection.

Defensive Displays or Maneuvers

These actions are intended to deter parasites or to carry organisms away from them. For example, gypsy moth pupae spin violently within their cocoons to deter pupal parasites, and syrphid larvae often drop to the ground from the foliage they forage on to escape parasites.

Grooming and Preening Behaviour

This behaviour is found in mammals and birds, respectively, to remove ectoparasites.

Spread of Disease

The dynamics of microparasites, the spread of disease, have been modeled by May (1981) and Anderson (1982). Instead of examining basic reproductive output, R_0 of a parasite, scientists more usually observe R_p, the average number of new cases of a disease that arise from each infected host. The reason is that in *epidemiology*, the study of the spread of disease, the number of infected hosts is the most important factor, not the number of parasites. The transmission threshold, which must be crossed if a disease is to spread, is therefore given by the condition $R_p = 1$. For a disease to spread, R_p must be greater than 1, and for a disease to die out, it must be less than 1. The term R_p is influenced by

N, the density of susceptibles in the population

B, the transmission rate of the disease (a quantity correlated with frequency of host contact and infectiousness of the disease)

f, the fraction of hosts that survive long enough to become infectious themselves

L, the average period of time over which the infected host remains infectious

The value R_p is related to these factors by the equation $R_p = BNfL$.

Two generalizations can thus be made.

1. As L, the period of the host's life when it is infectious, increases, R_p increases. Some hosts remain infectious long after they are dead. This is especially true in plant parasites, which leave a residue of resting spores.
2. If diseases are highly infectious (have large *Bs*) or are unlikely to kill their hosts (have large *fs*), R_p increases. An efficient parasite therefore keeps its host alive.

By rearranging the above equation, we can obtain the critical threshold density N_T (where $R_p = 1$), the number of infected hosts needed to maintain the parasite population:

$$N_T = \frac{1}{BfL}$$

Now, if B, f, or L is large, N_T is small. Conversely, if B, f, or L is small, the disease can only persist in a large population of infected hosts. Cockburn (1971) has provided some interesting medical and anthropological evidence to back up these ideas, at least for humans. Measles, rubella, smallpox, mumps, cholera, and chicken pox, for example, probably did not exist in ancient times (Black 1975) because the hunter-gatherer populations were small—bands of 200-300 persons at most. These hands were too small to constitute reservoirs for the maintenance of infectious diseases of the types described. In a small population, there should be no infections such as measles, which spreads rapidly and immunizes a majority of the population in one epidemic. Measles only occurs endemically in human populations larger than 500,000. Instead, typhoid, amoebic dysentery, pinta, trachoma, or leprosy were probably the common afflictions, diseases for which the host remained infective for long periods of time. Malaria and schistosomiasis would still have been very prevalent because of the presence of outside vectors to serve as additional reservoirs. Paradoxically, civilization has increased the kinds and frequencies of diseases suffered by humans by enlarging the source pools and by domestication of certain animals. Most modern diseases have arisen because of intimate association with animals and their viruses. Smallpox, for example, is very similar to the cowpox virus, measles belongs to the group containing dog distemper and cattle rinderpest, and human influenza viruses are closely related to those found in hogs. AIDS is similar to a virus found in monkeys in Africa.

For parasites that are spread from one host to another by a vector (for example, an insect), the life-cycle characteristics of both host and vectors become important in the calculation of R_p. The appropriate equation is:

$$R_p = \beta^2 \frac{N_v}{N_w} f_v f_n L_v L_n$$

N_v and N_h represent the density of vector and host respectively, for example, mosquito and man or aphid and tomato plant. f_v and f_h represent the fraction of infected vectors and hosts that survive to become infectious themselves. L_v and L_h represent the periods of time over which vectors and hosts remain infectious. β is the effective transmission rate, the rate of mosquito biting or aphid feeding that leads to infection. It appears as a squared term because infection occurs both to and from the host.

Another important feature of the dynamics of vector-borne parasites is that the transmission threshold ($R_p = 1$) is dependent on a ratio,

$$\frac{N_v}{N_L} = \frac{1}{\beta^2 \; f_v \; f_L \; L_v \; L_h}$$

For a disease to establish itself, and spread, the ratio of vectors to hosts must exceed a critical level—hence, disease control measures usually aim directly at reducing the numbers of vectors and are aimed indirectly at the parasite. Insecticides are used to kill aphids and mosquitos, which transmit virus diseases of crops and malaria, rather than directing chemicals at the parasite. Of course, this is not always true; for example, yellow fever was eradicated in the United States by inoculation rather than by extinction of all mosquitos.

One of the diseases causing most concern in the United States recently is Lyme disease, a disease caused by a spirochete that is transferred from animals to humans by the bite of hard- bodied ticks. The disease came to public attention in the 1980s and has been increasing ever since. The disease is not new—it has been known in Europe for more than a hundred years—but it is only in the 1980s and 1990s that it has become prevalent in the United States. More than 40,000 cases were reported from 1982 to 1991.

There are two species of disease carrying tick, one in the eastern United States, and one in the western United States. Both have many possible hosts, consisting of more than a hundred species of mammals, birds, and lizards. Adults of both species of tick feed preferentially on large and, to a lesser extent, medium-sized mammals. One of the problems in the increase of the disease appears to be related to land-use patterns and limited hunting programs that have allowed one of the main hosts, white-tailed deer, to increase. Most evidence suggests that disease will become more prevalent and the ticks will continue to expand their range and become more numerous.

Models of Host-parasitoid Interactions

Because the vast majority of parasites are actually parasitoids attacking insects, and because insect pests are such economically important organisms in agriculture, other population models of parasite-host relations have centered on parasitoid-host interactions. In many ways, these are similar to predator-prey relationships because only a single host is killed.

Early models of parasitoid-host interactions centered on the *numerical response* of a population of parasitoids to host density. This contrasts with a functional response, which concerns the behaviour of individual parasitoids to prey density. The first model was developed by the Australian entomologist Nicholson in the 1930s, who proposed that the success of randomly searching parasitoids would be limited not by their own egg supply but by their ability to find hosts. The average area that one parasitoid searched in its lifetime was deemed to be constant and was termed the area of discovery, a, as distinct from the area actually traversed. Thus, if 0.3 of the total habitat area is traversed by a parasitoid, the area of discovery may be only 0.254 because some ground may be covered twice. In other words, it may be expected that only 25.4 percent of the hosts will be parasitized instead of 30 percent. If this model is to be tested, a must he calculated because measuring it is usually quite impractical. If N is the total number of hosts, P the number of searching parasitoids, and S the proportion of hosts not parasitized, then

$$a = \frac{1}{P} \log_e \frac{N}{S}$$

Once a is known, the proportion of hosts parasitized can be predicted from the number of searching parasites:

$$\log N_{(n+1)} = \log N_{(n)} \frac{aP_n}{2.3} + \log F$$

where $N_{(n)}$ and $N_{(n+1)}$ represent successive host populations, and F the host reproductive rate. The outcome of this model is usually to produce increasing oscillations in the populations of both species, and laboratory tests have often been in good agreement with the theory. Nicholson was aware that increasing oscillations did not often occur under natural conditions, except for some species of alpine insects—pests of coniferous forests that show regular oscillations of outbreak densities over a number of years. Nicholson envisaged a fragmentation of large host populations into small ones at high densities and, thus, a proliferation of subpopulations. Although this sort of interaction does occur in Australia between the moth *Cactoblastis cactorum* and its food supply, the prickly pear, once again such events have not been commonly documented in the field.

The relationships between area of discovery and parasite density were investigated by M.P. Hassell and G.C. Varley (1969), who proposed

$$\log a = \log Q - m \log P$$

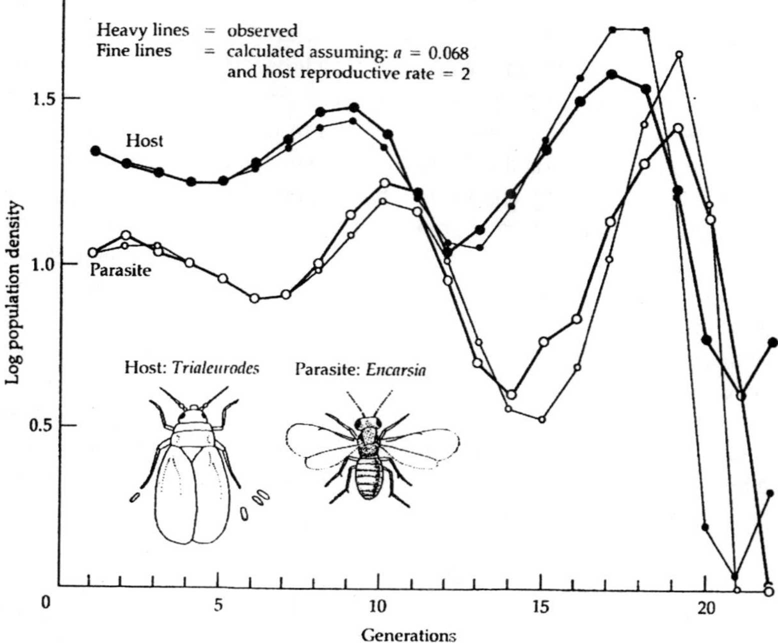

Fig. 8.1. Observed and calculated results of an interaction between Encarsia, a parasitoid, and its host, the greenhouse white-fly, Trialeurodes.

where Q is the "quest constant" or area of discovery when the parasite density is one, and m is the mutual interference constant. The outcome of some models based on quest theory are completely different from those of a Nicholsonian model because they include *interference*.

The stability of the quest models increases with greater values of m. More important, there is a wide range of values for Q and m that allow the coexistence of two or more parasite species on a single species of host, a very common phenomenon in nature, yet one not accounted for in Nicholson's models.

Despite the added realism of newer models, there remain some serious flaws, the most important of which is the assumption of random search by predators and parasites. There is little evidence that random search is generally prevalent. Most parasitoids are attracted by the scent of their prey, or they remain in areas where they have previously been successful. In either case, the searching population will tend to aggregate in areas of high host density. Unfortunately, nonrandom search is difficult to incorporate into models, though held studies give the impression that it tends to increase the stability of the system.

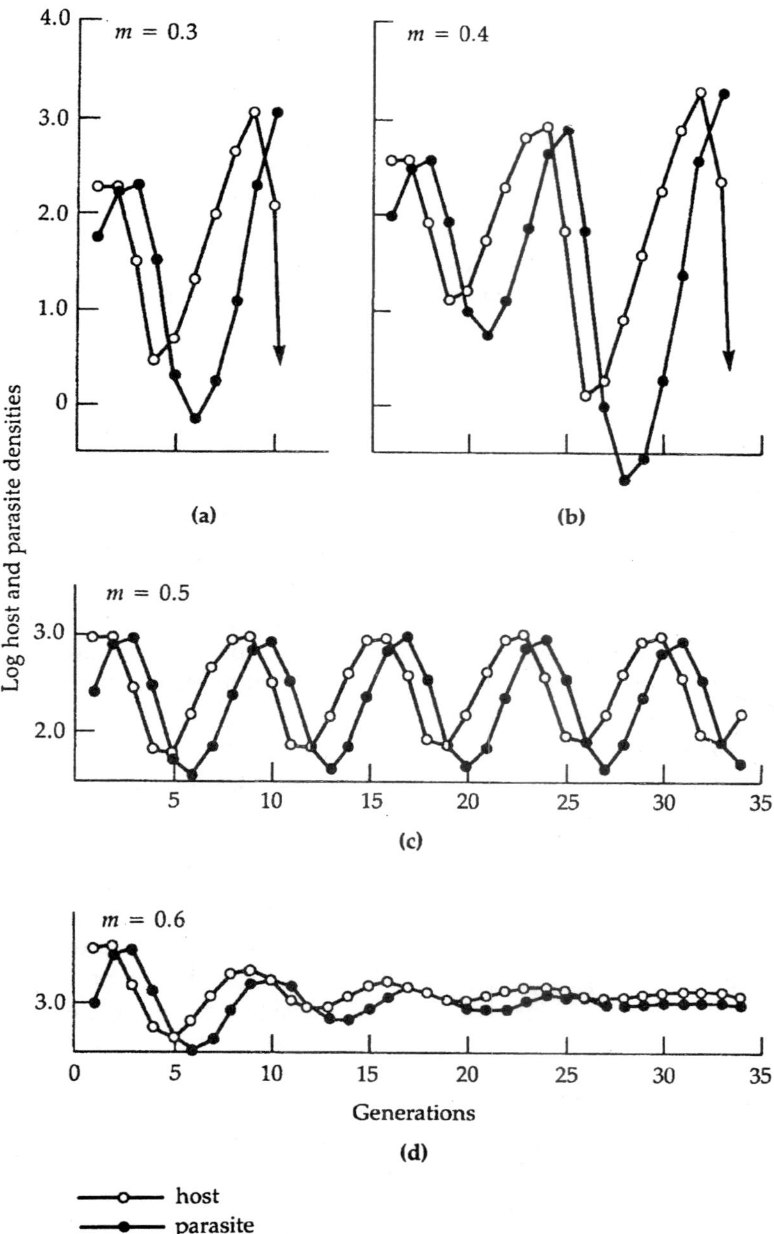

Fig. 8.2. Population models showing the increasing stability as the mutual interference constant m is increased from 0.3 in (a) to 0.6 in (d).

Effects of Parasites on Host Populations

Once again, the best way to find out the effect of parasites on the population abundances of their hosts is to remove the parasites and to reexamine the system. This has rarely been done, probably because of the small size and unusual life histories of many parasites, which makes them difficult to exclude. Also, as pointed out earlier, parasites may not always kill; they may merely impair the health of their hosts. This makes their effects even more difficult to gauge. However, inoculations which combat disease are known to save many lifes in humans and indeed in domestic animals. Agricultural sprays also reduce crop losses to disease.

Evidence from natural populations also suggests that parasites do have a substantial impact on their hosts. In North America, chestnut blight has virtually eliminated chestnut trees. In Europe and North America, Dutch elm disease has devastated elms. In Italy, canker has had severe effects on cypress. The population dynamics of bighorn sheep in North America are dominated by a massive mortality resulting from infection by the lungworms *Protostrongylus stilesi* and *P. rushi*. This parasite predisposes the animals to pathogens causing pneumonia. A fetus can become infected through the mother's placenta, and mortality in lambs can be enormous. The lungworm-pneumonia complex is regarded as one of the most influential mortality factors in many sheep populations with reported mortalities of up to 50-75 percent. Sometimes, epidemics can be even more severe when the parasite is less specific.

Rinderpest, caused by a virus, has at least 47 natural artiodactyl hosts, most of which occur in Africa. The disease is usually fatal in buffalo, eland, kudu, and warthog and less fatal in hushpig, giraffe, and wildebeest. Other species, such as impala, gazelle, and hippopotamus, appear to suffer little. A major epidemic swept through Africa in 1896, leaving vast areas uninhabited by certain species, and even in the 1970s distribution patterns reflected the impact. For example, zebra were exterminated in an area of the Elizabeth National Park in Uganda, and they still had not recolonized the area by 1954. Because wildlife was eliminated, other parasites were affected, too. Tsetse flies became absent from large areas of Africa south of the Zambesi River. Therefore, large areas became free from trypanosomiasis—sleeping sickness—a disease borne only by the tsetse fly. One parasite, rinderpest, thus had a severe impact on the pattern of life in an area. Furthermore, in the absence of tsetse flies, humans and cattle could move in, supplanting wildlife even further. Prins and Weyerhaeuser (1987) described two recent and major epidemics in wild mammals in a national park in Tanzania, east Africa. An anthrax outbreak lasted almost a year and killed more than 90 percent of the impala population, and a rinderpest outbreak of just a few weeks killed some 20 percent of the buffalo. The conclusion is that epidemics have a more severe impact on these populations than does predation. This is a disconcerting finding for conservation biology, for it means that certain populations on small reserves could be wiped out by disease unless recolonization is encouraged.

Another interesting situation exists in North America. The usual host of the meningeal worm *Parelaphostrongylus tenuis* is the white-tailed deer, *Odocoileus virginianus*, which is tolerant to the infection. All other cervids and the pronghorn antelope are, however, potential hosts, and in these species the worm causes severe neurological damage, even when very small numbers of the nematode are present in the brain. This differential pathogenicity of *P. tenuis* makes the white-tailed deer a potential competitor with other cervids because they cannot survive in the same area with white-tails. The deleterious effects of the parasite probably include direct mortality, increased predation, and reduced resistance to other disease. The activities of humans have altered the normal distribution pattern of the white-tailed deer. As northern forests were felled, the deer expanded their range from a stronghold in the eastern United States, eventually coming into contact with moose. In Maine and Nova Scotia, white-tailed deer have replaced moose as the major cervid, and they have also replaced mule deer and woodland caribou in some parts of their ranges. Whether or not reintroduction of caribou into regions now occupied by white-tailed deer is possible because of the action of parasites is now debatable. So, apart from direct mortality from parasites, competitive interactions between populations can be mediated by the action of parasites. This phenomenon is similar to Park (1948) compared competing populations of flour beetles with and without the parasite *Adelina triboli*. It is clear that parasites can have direct affects on the species they parasitize and indirect effects on other species. The subject of indirect affects is addressed later, but another interesting example concerns the parasites of the red grouse in England. More birds are killed by foxes as their burdens of the caecal nematode *Trichostrongylus tenuis* increase, suggesting that parasites increase susceptibility of red grouse to predation.

Cornell (1974) has argued that distributional gaps between bird species, where apparently favourable habitat exists, are maintained by the capacity of vectors to travel between populations. The rationale is that each population has a pathogen to which it is adapted but the other species is not. This concept has led to the idea that populations might compete for parasite-free space. Because the same type of phenomenon might operate for predators, the idea of enemy-free space is more widely circulated. Crosby (1986) argues that the Old World diseases smallpox, measles, typhus, and chicken pox were so devastating to peoples of the New World that the Old World invaders, who had a limited measure of immunity, found subjugation of the people much easier.

Parasites and Biological Control

Not all parasites are seen as detrimental by humans. Many are used as an effective line of defense against insect pests of crops, although only about 16 percent of classical biological-control attempts qualify as economic successes, so we cannot abandon chemical control as yet. The theory of pest control by means of the release of parasitoids is so far not very advanced. Huffaker and Kennett (1969) have suggested five necessary attributes of a good agent of *biological control*:

(i) General adaptation to the environment and host

(ii) High searching capacity

(iii) High rate of increase relative to the host's

(iv) General mobility adequate for dispersal

(v) Minimal lag effects in responding to changes in host numbers

Although these attributes seem necessary fur a good control agent, they are clearly not sufficient. So far, the application of biocontrol agents has been carried out by a hit-or-miss technique, rather than by a sound biological method. Some authors consider that this trial-and-error method probably makes best economic sense, given the high cost of research into the biology of natural enemies. Others have recommended new techniques, for example presenting novel parasite-host associations, as the most likely avenue for control where hosts have not had the opportunity to evolve complex defenses against these parasites from foreign lands. Arguments still rage as to whether it is better to introduce one parasite at a time or many. The problem is that, if more than one enemy is introduced, competition between parasites could ensue, lessening the overall level of control. Ehler and Hall (1982) provided evidence from a world review of 548 control projects to show that this phenomenon could be a problem; the more parasites released, the lower the rate of establishment, although this analysis was disputed by Keller (1984). It is probably fairest to say that in this, as in so many ecological situations, the jury is still out. Stiling (1990) reviewed the factors affecting success in biological control. The factor of greatest importance was the climatic match between the control agent's locality of origin and the region where it was to be released. This result stresses the value of studies in physiological ecology and that climatic variation is of vital importance in affecting biotic relations. This was underscored by a separate analysis of reasons for biological failures where reasons related to climate (34.5 percent) were more common than any other type of reason, including competition or parasitism by native insects.

It is noteworthy that the introduction of parasites for biological control can be a risky business for other, nontarget species and should be avoided unless strictly necessary. Howarth (1983), for example, lamented the reduction of native Hawaiian lepidopterans, partially due to wasp species introduced for biological control. He called for a more narrowly focused release effort rather than a hit-or-miss campaign. Such a concern is even more important when release of insect enemies to control weeds is considered. In this case, stringent host-specificity tests are performed to ensure introduced insects will not turn to feed on valuable crops, even in times of starvation. Nevertheless, *Cactoblastis* moths, the agents which so successfully controlled the pest cactus *Opuntia* in Australia, have recently arrived in Florida and are decimating native cacti there.

The release of genetically engineered parasites is a subject of even more concern. Australians are trying to use generically engineered viruses to control exotic pest mammals. When the British colonized Australia, they brought with them the English rabbit and the European red fox, both in the mid-1800s. The rabbits were for food, and the fox was to be hunted. Not long after these introductions, people realized they had made a mistake: the rabbits began to breed like rabbits, and the predatory foxes turned their attention to the native animals. So far, they have been implicated in the extinction of 20 species of local marsupials. The Cooperative Center for Biological Control of Vertebrate (VBC) pest populations (a government-and-university consortium) plans to release genetically designed viruses that will sterilize most foxes and rabbits by tricking the female's immune system into attacking male sperm. Some of the problems are that a live vector might get into other species and that it has the potential to travel abroad. Any travel of rabbits from Australia hack to Europe could facilitate movement of the disease. This could have serious implications back in the native habitat of foxes and rabbits. On the other hand, if the project is successful and safe, it could provide a model for wiping out pests in other fragile, threatened habitats. For example, the U.S. Department of Agriculture is considering using the technique to control feral pigs which are wreaking havoc in Hawaii. In New Zealand, introduced possums are causing trouble.

One of the key aspects of these programs is the simultaneous sterilization of both species. For example, because the rabbits are the main item on the foxes menu, any sudden population crash in the rabbits might force the foxes to seek alternative food, the likely choices being small, endangered marsupials. Clearly, getting the rate of spread of the diseases to be similar in foxes and rabbits would be the optimal solution.

9

PREDATION

Several types of predation can be recognized:

Herbivory : Animals feeding on green plants

Carnivory : Animals feeding on herbivores or other carnivores

Parasitism : (a) Animals or plants feeding on other organisms without killing them and (b) parasitoids, usually insects, laying eggs on other insects, which are subsequently totally devoured by the developing parasitoid larvae

Cannibalism : A special form of predation, predator and prey being of the same species

Although herbivory and parasitism could be viewed widely in the context of predation, each has special characteristics that tend to separate it out into a special subdivision of predation, and each subject will be discussed separately. Herbivory, for example, is often nonlethal predation on plants. Many leaves or parts of leaves of a plant can be eaten without serious damage to the "prey." In the animal world, on the other hand, predation generally means death for the prey.

Parasitism, too, is a unique type of event, in which one individual prey is commonly utilized for the development of one predator. In the special case of insect herbivores on plants, there is great debate about whether such insects can be called plant *parasites* because they seem to fit such a definition fairly well. Carnivory embodies the traditionally held view of predation, and cannibalism can be seen as a special case of it.

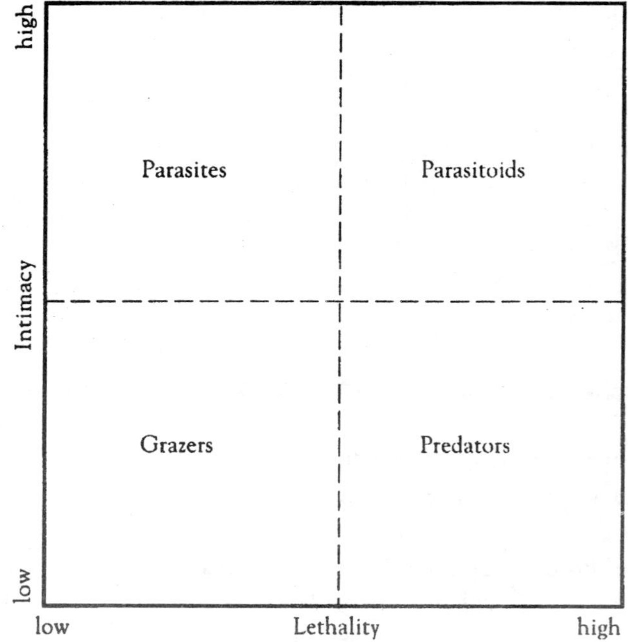

Fig. 9.1. Possible interactions between populations.

Strategies to Avoid Being Eaten: Evidence for the Strength of Predation as a Selective Force

In response to predation, many prey species have developed strategies to avoid being eaten. Edmunds (1974), Owen (1980), and Waldbauer (1988) have reviewed many case studies, which include the following.

Aposematic, or Warning, Colouration

Aposematic colouration advertises a distasteful nature. Lincoln Brower (1970) and coworkers (Brewer et al. 1968) showed how inexperienced blue jays took monarch butterflies, suffered a violent vomiting reaction, and learned to associate forever afterward the striking orange-and-black barred appearance of a monarch with a noxious reaction. The caterpillar of this butterfly gleans poison from its poisonous host plant, a milkweed. Many other species of animals, especially invertebrates, are also warningly coloured, for example, ladybird beetles. Caterpillars of many Lepidoptera are bright and conspicuous, too, because being noxious is their main line of defense—being soft-bodied, they would otherwise be very vulnerable to predators.

Crypsis and Catalepsis

Crypsis and *catalepsis* are camouflage and the development of a frozen posture with appendages retracted. This is another method of avoidance of detection by invertebrates. For example, many grasshoppers are green and blend in perfectly with the foliage on which they feed, exhibiting *apatetic colouration*. Often, even leaf veins are mimicked on grasshopper wings. Stick insects mimic branches and twigs with their long slender bodies. In most cases, these animals stay perfectly still when threatened because movement alerts a predator. Crypsis is prevalent in the vertebrate world, too. A zebra's stripes supposedly make it blend in with its grassy background, and the sargassum fish even adopts a grotesque body shape to mimic the sargassum weed in which it is found. Perhaps first prize should go to the chameleon, whose skin tones can be adjusted to match the background on which it is resting.

Mimicry

Though many organisms may try to blend into their background, mimicking the foliage or background around them, some animals mimic other animals instead. For example, some hoverflies mimic wasps. Several types of mimicry can be defined.

Mullerian

The convergence of many unpalatable species to look the same, thus reinforcing the basic distasteful design, as for example with wasps and some butterflies.

Batesian

Mimicry of an unpalatable species by a palatable one or of dangerous species (coral snakes) by innocuous ones (pseudocorals). Wickler (1968) has documented many cases in which flies, especially hoverflies of the family Syrphidae, are striped black-and-yellow to resemble stinging bees and wasps. Also, some butterflies painted to mimic distasteful models were recaptured more frequently than those painted to mimic palatable forms, suggesting the great survival value of mimicry. Until very recently, monarch and queen butterflies, themselves aposematically coloured, were considered models for mimic viceroy butterflies. However, David Ritland showed that all three species were unpalatable and thus coexisted as a Mullerian mimicry complex.

Aggressive

In this case, the body colouration permits individuals not so much to escape predation as to be better predators themselves. Many praying mantises mimic flowers so as to entrap insect prey. This strategy is shared by the crab spiders. Certain bottom-dwelling ocean fish also mimic the substrate so as to get in closer proximity to their prey. In extreme cases, such as with anglerfish, parts of the body are modified to act as lures for prey. Some species also use light-emitting organs to attract prey. On land, aggressive mimicry is practiced by female fireflies of the genus *Photuris*. Normally, males respond to the species-specific light flashes of their females and move toward them. *Photuris* females mimic the flashing patterns of females of other species to lure the males of those species and eat them.

Intimidation Displays

An example of intimidation display is a toad that swallows air to make itself appear larger. Frilled lizards extend their collars when intimidated to have the same effect.

Polymorphisms

Polymorphism is the occurrence together in the same population of two or more discrete forms of a species in proportions greater than can be maintained by recurrent *mutation* alone. Often, this phenomenon takes the form of a colour polymorphism; if a predator has a preference or *search image* for one colour form, usually the commoner, then the prey can proliferate in the rarer form until this form itself becomes the more common (so-called *apostatic selection*). Stiling (1980) advocated just this type of choice of prey, by a visually searching *parasitoid*, to maintain the difference between two distinct colour morphs, orange and black, in some leafhopper nymphs of the genus *Eupteryx*, though Stewart (1986a,b) also implicated thermal *melanism* as an important agent of selection in some species; that is, black morphs occur in cooler climates because they heat up faster in the sun. Owen and Whiteley (1986) have pointed out that in many species the form of the polymorphism is such that every individual is slightly different from all others. This is true in brittlestars, butterflies, moths, echinoderms, and gastropods. They suggest that such a staggering variety of form thwarts predators' learning processes, and they suggest the term *reflexive selection* for this type of phenomenon.

Phenological Separation of Prey from Predator

Fruit bats, normally nocturnal foragers, are active by day and at night on some small, species-poor Pacific islands such as Fiji. Wiens et al. (1986) suggest that the fruit bats are constrained elsewhere to fly only at night by the presence of predatory diurnal eagles.

Chemical Defenses

One of the classic defenses involves the bombardier beetle (*Bradinus crepitans*) as studied by Tom Eisner and coworkers. These beetles possess a reservoir of hydroquinone and hydrogen peroxide in their abdomens. When threatened, they eject these chemicals into an explosion chamber, where they mix with a peroxidase enzyme. The resultant release of oxygen causes the whole mixture to be violently ejected as a spray that can be directed at the beetle's attackers. Many other arthropods, such as millipedes, have chemical defenses too. This phenomenon is also found in vertebrates, as people who have had a close encounter with a skunk can testify.

Masting

Masting is the synchronous production of many progeny by all individuals in a population to satiate predators and to allow some progeny to survive. It is commonly documented in trees, which tend to have years of unusually high seed production. A similar phenomenon is exhibited by the emergence of 17-year and 13-year cicadas. It is worth noting in this context that both 13 and 17 are prime numbers, so no predator on a shorter multiannual cycle could repeatedly use this resource.

How common is each of these defense types? Brian Witz (1989) surveyed 354 papers that documented antipredator mechanisms in arthropods, mainly insects of 555 predator/prey interactions. By far, the most common antipredator mechanism was chemical defense, noted in at least 46 percent of the examples—at least because many categories were not mutually exclusive; for example, aposematic colouration is also usually coupled with noxious chemicals.

It is obvious that predation constitutes a great selective pressure on plant and animal populations. Despite the impressive array of defenses, predators still manage to survive by eating individuals of their chosen prey, often by circumventing the defenses in some way. The coevolution of defense and attack can be seen as an ongoing evolutionary arms race. According to Dawkins and Krebs (1979), the prey are always likely to be one step ahead. The reason is what they termed the *life-dinner principle*. In a race between a fox and a rabbit, the rabbit is usually faster because it is running for its life, whereas the fox is running "merely" for its dinner. A fox may still reproduce, even if it does not catch the rabbit. The rabbit never reproduces again if it loses. This arms race has been argued to be run not only between predators and prey but also between parasite and host and between plant and herbivore. In the latter case, the race often proceeds by the production of toxins by the host and detoxifying mechanisms by the predator. Usually, however, at least for plants and herbivorous insects, phylogenetic relationships among herbivores do not correspond well to those of their host plants. The result is that at each defensive turn by a plant (or other host), it is as likely to run into a new set of enemies as it is to escape the old ones.

Mathematical Models

What effect has the predator on its prey population? The answer depends on many things, including prey and predator density and predator efficiency. Rosenzweig and MacArthur (1963) modeled predator-prey dynamics using a graphical method. First, they assumed that, in the absence of predators, the prey population increases exponentially according to the Lotka-Volterra formula

$$\frac{dN}{dt} = rN$$

The rate of removal of prey increases with an increase in encounter rate by predators dependent upon the number of predators (*C*) and the number of prey (*N*). The rate of removal is also dependent upon a "searching efficiency" or attack rate of predators, termed *a*. The consumption of prey by predators is then $a'CN$, so

$$\frac{dn}{dt} = rN - a'CN$$

The rate of growth of the predators is given by

$$\frac{dC}{dt} = fa'CN - qC$$

where *q* is a mortality rate based on starvation in the absence of prey and $fa'CN$ is predator birth rate based on *f*, the predator's efficiency at turning food into offspring.

The properties of this model can be investigated by location of zero isoclines (regions of zero growth, neither positive nor negative). In the case of the prey, when

$$\frac{dN}{dt} = 0, rN = a'CN$$

or

$$C = \frac{r}{a'}$$

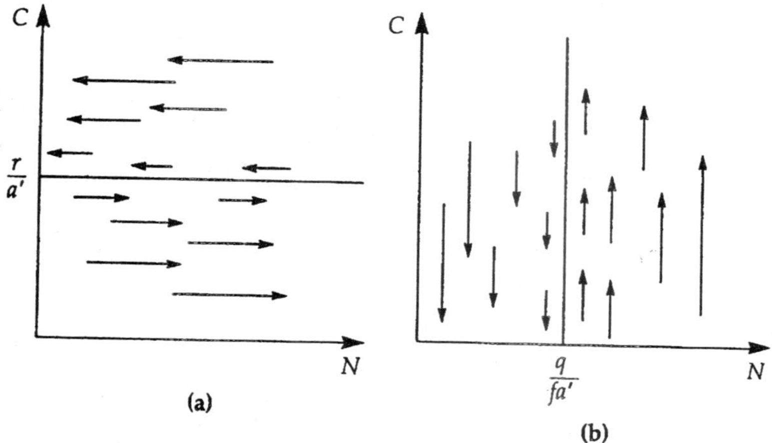

Fig. 9.2. A Lotka-Volterra type predator-model. (a) The prey zero isocline; (b) The predator zero isocline.

Because r and a' are constants, the prey zero isocline is a line for which C, the number of predators, is itself a constant. Along this line, prey neither increase nor decrease in abundance.

For the predators, when

$$\frac{dC}{dt} = 0, \; fa' CN = qC$$

or

$$N = \frac{q}{fa'}$$

The predator-zero isocline is also a straight line, one along which N, the number of prey, is constant. An assumption is that if there are enough prey, a population of predators will increase and that if there are not enough prey, they will starve. This is obviously a gross oversimplification. First, larger populations of predators require larger populations of prey to maintain them, so the zero isocline for predator growth should slant to the right. As predator density increases, so will mutual interference between predators, who will spend more time fighting one another and less time tackling prey. For zero growth, then, more realistically, there must be even more prey. Finally, at the highest densities of prey, it seems that the rate of growth of predators will be limited by something other than prey availability—say, social constraints on territory size or the availability of burrows or nest sites—so that the zero isocline will appear as in line D.

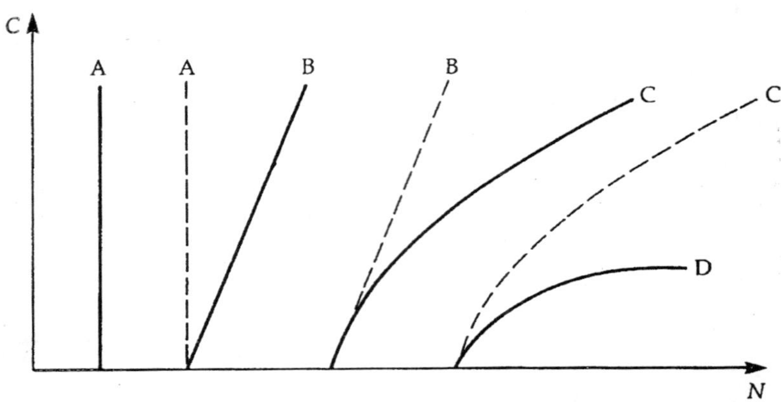

Fig. 9.3. Predator zero isoclines of increasing complexity, A to D.

Remember that at predator-prey combinations to the left of this line, predator numbers decrease, whereas to the right of it they increase.

The refinement of the prey-zero isocline is dependent on two concepts. The first is that the recruitment rate of prey into the population is high when N is low but low when N is high and near the carrying capacity of the environment. The logic is that strong competition for resources at levels of N near K severely reduces recruitment rate; hence, the recruitment curve is semicircular. The rate of growth of the prey population is also dependent on the level of predation, and different predation rates are represented by the steepness of the dashed lines. At each of the points where the consumption curve crosses the recruitment curve, the rate of population growth of the prey is zero, and these pairs of densities can be directly transferred to give the zero isocline for growth of the prey.

The effect of different densities of predators and prey can now be assessed by combination of the two isoclines. The highest levels of predation translate into large oscillations of predator and prey (i), sometimes so violent that one or both species goes extinct. Predators in this case are relatively efficient and abundant. Less-efficient predators (ii) give rise to small and often damped oscillations in the system. At the lowest levels of predation, the system is highly stable, but only low levels of predators are supported (iii). The problem with this analysis is that it is almost impossible to distinguish this type of predator-prey cycling from population fluctuations resulting from environmental factors such as bad weather. Furthermore, predator-prey isoclines can change dramatically with the incorporation of such factors as refuges from predators and the ability of predators to switch

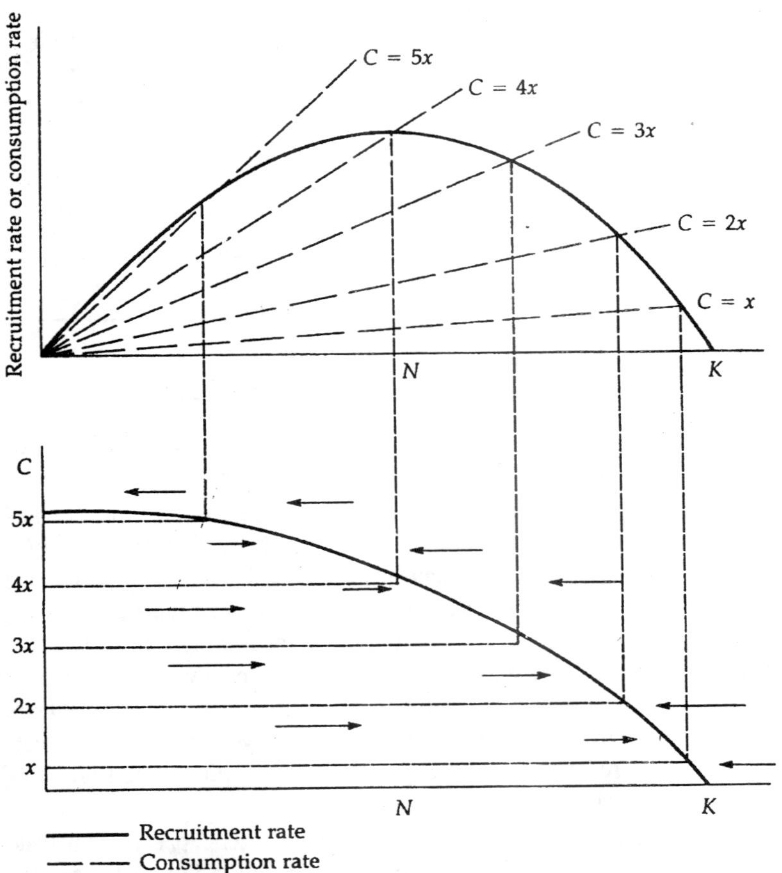

Fig. 9.4. Refinement of prey zero isocline.

back and forth between multiple species of prey. The predictive value of these models seems to decrease as their complexity increases.

How do the data on predator and prey abundances in the field compare with these types of graphical models? Predators of the Serengeti plains of eastern Africa (lions, cheetahs, leopards, wild dogs, and spotted hyenas) seem to have little impact on their large-mammal prey. Most of the prey taken are injured or senile and are likely to contribute little to future generations. In addition, most of the prey sources are migratory, and the predators—residents—are more likely to be limited in numbers by prey who also are resident in the dry season when migratory ungulates are elsewhere. Pimm (1979, 1980) has argued that the importance of predation is dependent on whether

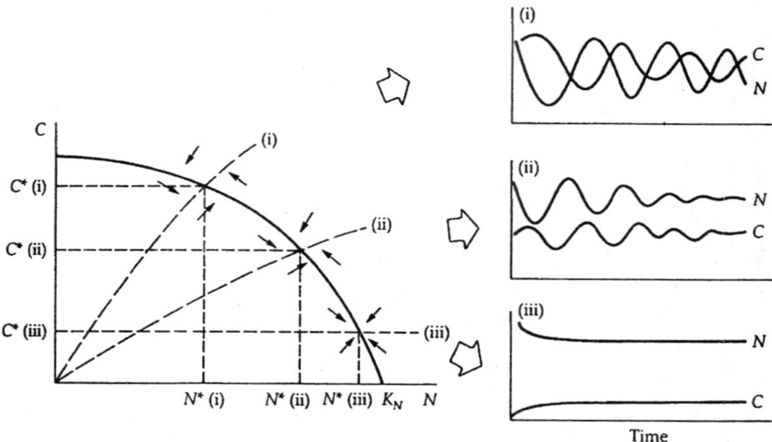

Fig. 9.5. A prey zero isocline with self-limitation, combined with predator zero isoclines with increasing levels of self-limitation: (i), (ii), and (iii).

the system is "donor controlled" or "predator controlled." In a donor-controlled system, prey supply is determined by factors other than predation, so removal of predators has no effect. Examples include consumers of fruit and seeds, consumers of dead animals and plants, and intertidal communities in which space is limiting. In a predator-controlled system, the action of predator feeding eventually reduces the supply of prey and their reproductive ability. Removal of predators in a donor-controlled system is obviously likely to have little effect, whereas in a predator-controlled system such an action would probably result in large changes in abundance.

Many predator-prey systems appear stable; others appear to fluctuate dramatically. Determining which mechanism works in which situation is not easy. For example, in the Arctic, there are two groups of primarily herbivorous rodents—the microtine varieties (lemmings and voles) and the ground squirrels. Ground squirrels exhibit the strongly self-limiting behaviour of aggressive territorial defense of burrows. Their populations are remarkably consistent from year to year. On the other hand, the microtines are renowned for their dramatic population fluctuations. Thus, even in the same habitat, results for different species vary dramatically. One series of population fluctuations analyzed in great detail is that of the Canada lynx and snowshoe hare.

The Canada lynx (*Lynx canadensis*) eats snowshoe hares (*Lepus americanus*) and shows dramatic cyclic oscillation every nine to eleven years. Charles Elton analyzed the records of furs traded by trappers

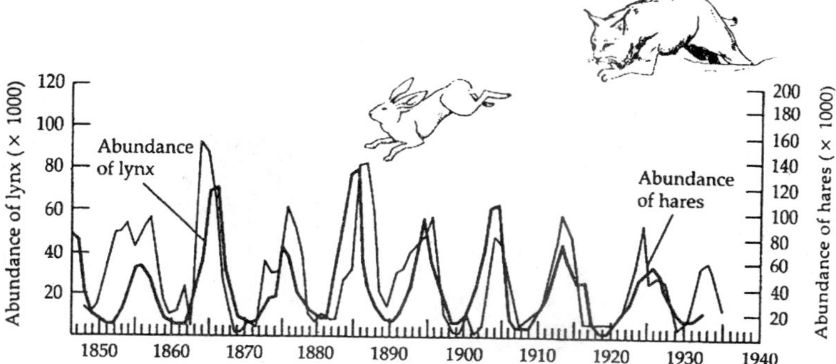

Fig. 9.6. The apparently coupled oscillations in abundance of the snowshoe hare (Lepus americanus) and Canada lynx (Lynx canadensis).

to the Hudson's Bay Company in Canada over a 200-year period and showed that a cycle has existed for as long as records have been kept. This cycle has been interpreted as an example of an intrinsically stable predator-prey relationship, but Keith (1983) has argued that it is winter food shortage and not predation that precipitates hare decline. He showed that heavily grazed grasses produce shoots with high levels of toxins, making them unpalatable to hares. Such chemical protection remains in effect for two to three years, precipitating further hare decline. Predators, he argued, simply exacerbate population reduction. Thus, although lynx cycles depend on snowshoe-hare numbers, hares fluctuate in response to their host plants. Subsequently, Smith et al. (1988) showed that, although food quality greatly affects hare *biomass*, most hares die of predation, not starvation. However, death due to predation is greatly exacerbated by poor quality of hares, which is of course greatly affected by food quality; thus, there seems to be a good deal of common ground between the predation and starvation camps. Most recently Sinclair et al. (1993) suggested a correlation between the frequency of sunspots, the level of herbivory on white spruce by snowshoe hares, and hare fur records stretching back 200 years.

Evidence from Experiments and Introductions

Perhaps the best way to find out whether predators determine the abundance of their prey is to remove predators from the system and to examine the response. One of the best examples involves dingo predation on kangaroos in Australia. The dingo, *Canis familiaris dingo*, is the largest naturally occurring carnivore in Australia and an important predator of imported sheep. Dingoes have been intensively hunted and poisoned in sheep country—southern and eastern Australia—and long

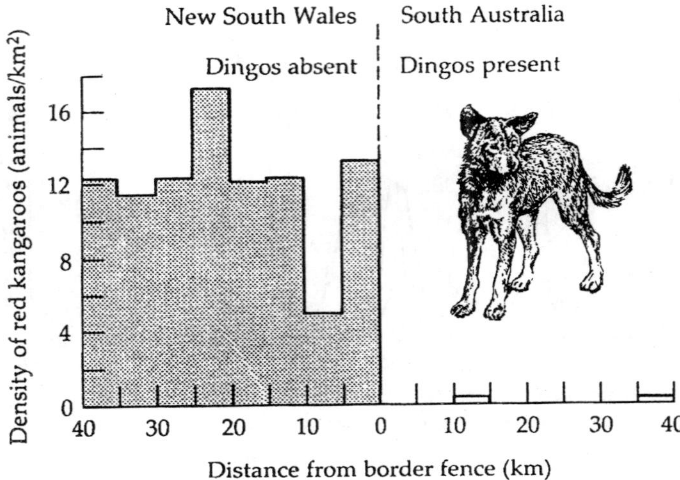

Fig. 9.7. Density of red kangaroos on a transect across the New South Wales-South Australia border.

fences (some up to 9,600 km) extend to prevent them from recolonizing areas, providing a classic experiment in predator control. The result has been a spectacular increase, 166-fold, of red kangaroos where the dingoes have been eliminated in New South Wales, over their density in south Australia, where dingoes have not been molested.

Emus (*Dromaius novaehollandiae*) are also more than 20 times more abundant in dingo-free areas. Dingoes are also frequent predators of feral pigs in tropical Australia. In Cape York, northern Queensland, there is a gross shortage of young pigs less than two years old on the mainland where there are dingoes. On neighbouring Prince of Wales Island where dingoes are absent, recruitment is considerable.

Other important exotic animals in Australia are European foxes and feral cats. Both can do damage to domestic livestock and are subject to eradication by shooting. In areas where these predators were shot, numbers of rabbits, also exotic in Australia, increased. Where rabbits increase, valuable rangeland may become overgrazed. The effects of predators on their prey is clearly a subject of interest to farmers as well as biologists.

Another striking example of predation pressure has been provided by an inadvertent introduction by humans. Marine sea lampreys (*Petromyzon marinus*) live on the Atlantic coast of North America and migrate into freshwater to spawn. Adult lampreys feed by attaching themselves to other fish, then rasping a hole, and finally sucking out

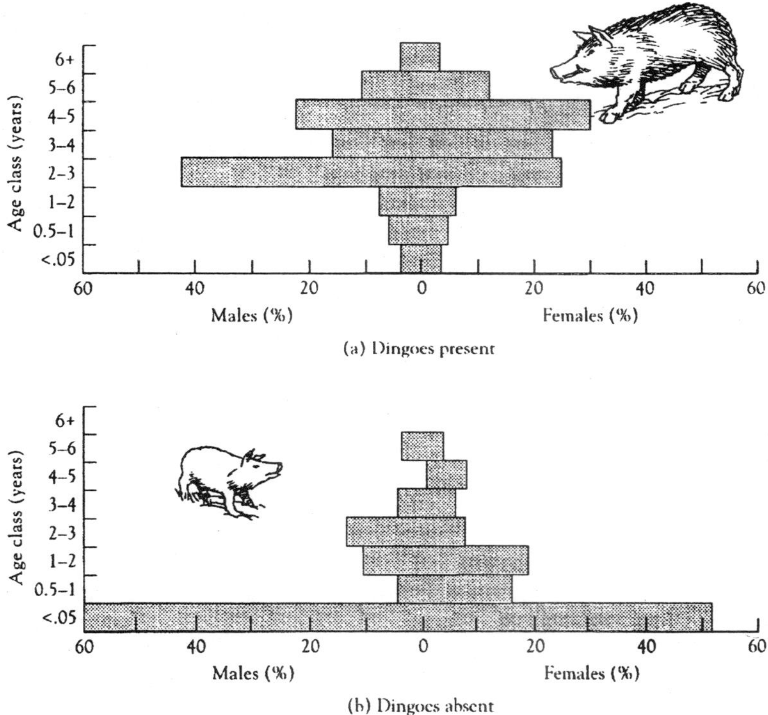

Fig. 9.8. Contrasting population structures of feral pigs where dingoes are (a) present and (b) absent in tropical northern Australia.

the body fluids. The passage of the lamprey to the upper Great Lakes was presumably blocked by Niagara Falls before the Welland Canal was built in 1829. The first sea lamprey was found in Lake Erie in 1921, in Lake Michigan in 1936, in Lake Huron in 1937, and in Lake Superior in 1945. Lake trout catches decreased to virtually zero within about 20 years of lamprey invasion. Control efforts have been applied since 1956 to reduce the lamprey population, and attempts are being made to rebuild the Great Lakes fishery.

Reviews and Natural Systems

Most of the examples discussed so far have focused on introduced species. Effects of such introductions are often very strong, hut what of totally natural predator-prey systems where predator and prey have evolved together for long periods? In 1903, lions were shot in Kruger National Park, South Africa, to allow numbers of large prey to increase. Shooting ceased in 1960, by which time wildebeest (*Connochaeter*

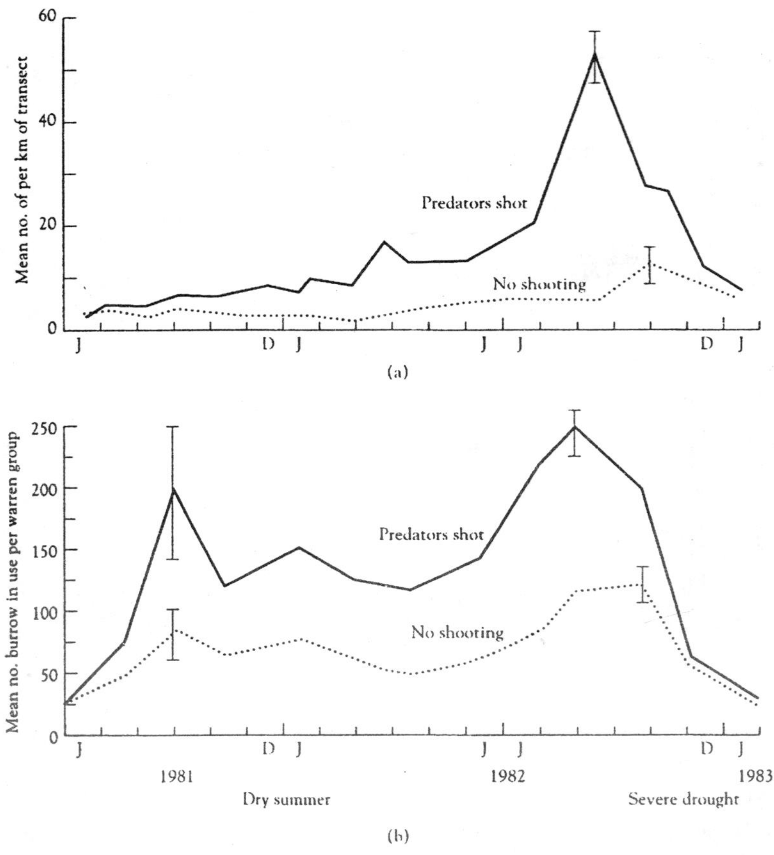

Fig. 9.9. Accelerated increase in rabbits with removal of European foxes and feral cats in an Australian field experiment. (a) Comparison of counts of rabbits per kilometer along transects where predator populations were continually shot (solid line) or left intact (dotted line). (b) Comparison of burrow use in warren groups where predator populations were continually shot (solid line) or left intact (dotted line).

taurinus) had increased so much that human culling was instigated from 1965 to 1972.

For many years, the moose population on Michigan's Isle Royale, a 45-mile long island enjoyed a wolf-free existence. Then, in 1949, during a particularly hard winter, a pair of Canadian wolves were able to colonize the island, walking across frozen Lake Superior. In 1958, wildlife biologist Durwood Allen of Purdue University began to track wolf and moose numbers. The wolf population peaked at 50

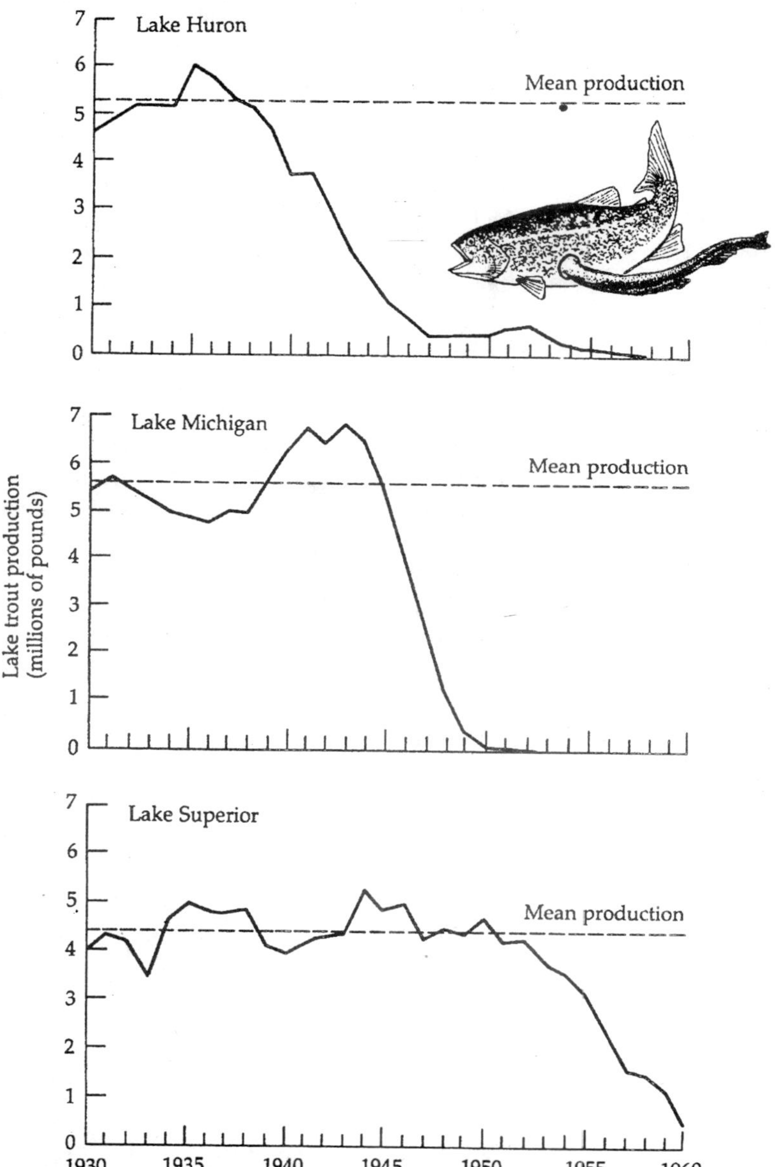

Fig. 9.10. Effect of sea lamprey introduction on the lake trout fishery of the upper Great Lakes.

individuals in 1980, and then in 1981 it took a severe nosedive. Wildlife ecologist Rolf Peterson of Michigan Technological University has followed this population in the most recent years. The wolf population has continued to decline. Only four pups were born in 1992 and 1993, all to the same female in one wolfpack. The other two packs on the island are down to just a pair of wolves each. Many observers feel the total wolf population is on its way to extinction. As for the moose population, it has reached a record level of about 1,900 moose in 1993. This seems good evidence that predation does have a strong effect on natural populations.

Why did the wolves decrease in numbers? First of all, there was a narrow genetic base to begin with. Restriction enzyme analysis of the wolves mitochondrial DNA turned up just a single pattern, indicating the wolves were all descended from a single female. They had only about half the genetic variability of the mainland wolves. Second, there was evidence of a deadly canine virus in 1981. This was probably a result of a parvovirus outbreak in nearby Houghton, Michigan, which was carried to the island on the hiking hoots of visitors.

In 1985 Andrew Sih and colleagues surveyed 20 years (1965-1984) of seven ecological journals for field experiments concerned with predation. Their survey yielded 139 papers involving 1,412 comparisons. Virtually every report, 132/139, or 95 percent, showed some significant effects. At least one comparison existed in 85.5 percent, where prey showed a large response to predator manipulations. Can we conclude that in most cases predators influence the abundance of their prey in the field? Probably, although the same caveats remain as in the reviews

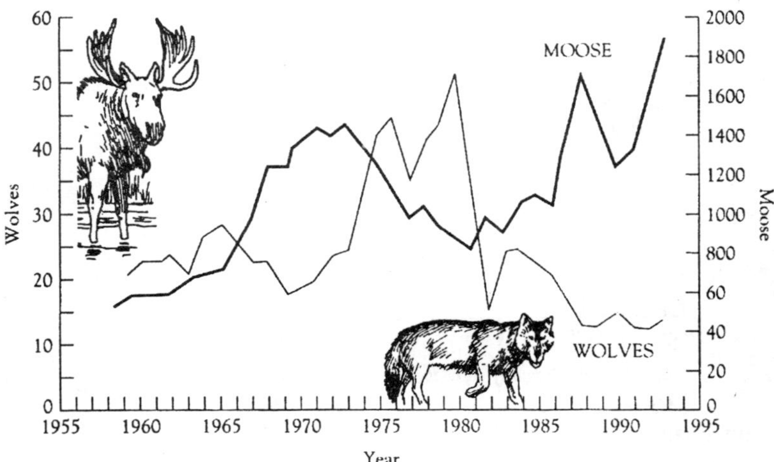

Fig. 9.11. The effects of wolf-predation on moose numbers in Isle Royale.

of field experiments investigating competition, that is, under reporting of negative results, biased views of investigating scientists, and so on.

Finally, it is worth noting that studies on the effects of predators are particularly timely now, given the desire of certain conservation groups to reintroduce large predators into certain areas. The U.S. Fish and Wildlife Service would like to reintroduce the wolf into Yellowstone National Park and to stabilize its numbers in Montana and Minnesota, the only states other than Alaska to possess viable populations. Cattle ranchers are fearful that wolves would decimate their herds. There is also considerable argument as to whether wolves constitute a large predation pressure on other prey such as caribou herds and should be controlled in times of declining caribou numbers.

10

Herbivory

Plants appear to present a luscious green world of food to any organism versatile enough to attack and use it. Why couldn't more of this food source be exploited? After all, plants cannot even move to escape being eaten.

There are three possible reasons that more plant material is not eaten. First, natural enemies—predators and parasites—might keep *herbivores* below levels at which they could make full use of their resources. This idea is preferred by some ecologists, such as Lawton and Strong (1981) who studied insect herbivores. Second, herbivores may have evolved mechanisms of *self-regulation* to prevent the destruction of the host plant, perhaps ensuring food for future generations. Third, the plant world is not as helpless as it appears—the sea of green is in fact tinted with shades of noxious chemicals and armed with defensive spines and tough cuticles.

Plant Defenses

An array of unusual and powerful chemicals is present in plants, such as nicotine (an alkaloid) in tobacco, morphine and caffeine (also alkaloids), mustard oils, terpenoids (in peppermint and catnip), phenylpropanes (in cinnamon and cloves), and many others. A teaspoon of mustard should be enough to convince anyone of the potency of these chemicals. Such compounds are unusual—they are not part of the primary metabolic pathways of plants. They are therefore referred to as *secondary chemicals* and are thought to be synthesized specifically as a deterrent to herbivores.

Before this aspect of plant defense is discussed, it is important to note an alternative viewpoint—that secondary plant substances are merely the waste products of plant metabolism. It must be remembered that excretion is a necessary part of metabolism even for plants, but it must take a fundamentally different form (that is, storage) than it does in animals. Some authors argued that it is only coincidental that these waste metabolites have an effect on herbivores.

Ehrlich and Raven (1964) were the first to crystallize the notion that secondary plant compounds evolved specifically to thwart herbivores. They proposed that most secondary compounds are produced only at a metabolic cost to the plant, not as energy-free by-products. This is the prevalent view among plant-insect ecologists today. Rhoades (1979) has formulated a general defense theory based on the idea that such plant compounds are costly to produce:

1. Higher herbivory levels lead to more defenses.
2. Higher costs of defense lead to fewer defenses.
3. More defenses are allocated to the most valuable tissues.
4. Environmental stress may lessen the availability of energy for defensive mechanisms.
5. Defense mechanisms are reduced when enemies are absent and increased when plants are attacked. This is known as the *theory of induced defense*.

Types of Defensive Reactions

Defensive reactions in plants can be classified into several main types, of which the most commonly used are quantitative and qualitative.

Quantitative defenses are substances that gradually build up inside the herbivore as it eats and that prevent digestion of food. Examples are tannins and resins in leaves, which may occupy 60 percent of the dry weight of the leaf. In the case of tannin, the compound is not toxic in small doses, but it has cumulative effects. Tannins (the compounds in many leaves, such as tea, that give water a brown colour) act by binding with proteins in insect-herbivore guts. The more leaf the herbivores ingest, the more difficult it is for them to digest it. Feeny (1970) was the first to document such a defense (by oaks against externally feeding caterpillars). Zucker (1983) proposed that there are two main classes of tannins, each with differing biological functions. *Hydrolyzable tannins* inactivate the digestive enzymes of herbivores, especially insects, whereas *condensed tannins* are attached to the cellulose and fiber-bound proteins of cell walls, thereby defending plants against microbial and fungal attack. Some authors have found that other insect herbivores are not much affected by quantitative defenses, but the evidence that tannins affect vertebrate herbivory is stronger. Cooper and Owen-Smith (1985) showed that for browsing ruminants in Africa (kudus, impalas, and goats), palatability of 14 species of woody plants was clearly related to leaf contents of condensed tannins. The effect showed a distinct threshold; the browsers rejected all plants that contain more than 5 percent condensed tannins. Cooper and Owen-Smith suggest that the reason is that ruminants depend on microbial fermentation of plant cell walls for part of their energy needs. For grey squirrels, acorns with higher concentrations of tannins are less preferred than acorns with lower concentrations.

Qualitative defenses are, essentially, highly toxic substances, very small doses of which can kill herbivores. These compounds are present in leaves at low concentrations, like 1 to 2 percent of dry weight. Examples include alkaloids and cyanogenic compounds in leaves. Atropine, produced by deadly nightshade (*Atropa belladonna*), is a most potent poison. Of course, the plant must store many of these poisons in discrete glands or vacuoles or in latex or resin systems in order not to poison itself. Some compounds are stored as precursors and only become toxic when they are metabolized by the herbivore. For example, fluoroacetate found in certain Dichapetalaceae is metabolized by herbivores to fluorocitrate, a potent inhibitor of Krebs-cycle reactions.

A good review of theory and pattern in plant defense allocation is given by Zangerl and Bazzaz (1992). The two defense strategies, qualitative and quantitative, are correlated with plant "apparency". Apparent plants are long-lived and always apparent to the herbivores (for example, oak trees). Their defenses are thought to be mainly of the quantitative kind, effective against specialist herbivores with a long history of association with these *K*-selected plants. Unapparent plants are weeds, which are ephemeral and unavailable to herbivores for long periods. Their defenses are thought to be mainly qualitative, guarding against generalist enemies. Thus, trees nearly all contain digestibility-reducing compounds, and weeds contain the cheaper-to-make toxins. The value of these terms is of dubious value, however. An "unapparent" plant is not likely to be unapparent to the herbivores that specialize in finding it by chemical cues and eating it. Apparency probably best reflects the ability of human searchers to find plants. Nevertheless, the terms have spawned a great deal of research.

Cates and Orians (1975) set out to test the idea that early and late successional plants might be differentially palatable to generalist herbivores. They found, using a slug as a test animal, that early successional plants were preferred. Is this result so surprising, though, given that the slug is an early successional animal, presumably predisposed to feeding on early successional vegetation? Quite the reverse was found to he true when late successional herbivores were used. For instance, gypsy moths, *Lymantria dispar*, prefer the leaves of trees and other late successional plants to those of weeds and other perennials. Coley (1988) studied growth, herbivory, and defenses in 41 common tree species in a lowland Panamanian rain forest. Species with long-lived leaves had significantly higher concentrations of immobile defenses such as tannins and lignins. She also found that trees with lower growth rates generally suffered higher herbivory.

Other defense mechanisms consist of the following:

Mechanical defenses

Plant thorns and spines deter vertebrate herbivores, if not invertebrate ones. In Africa, in the presence of a large guild of vertebrate herbivores, much of the vegetation is thorny and spinose. Many neotropical plants are also armed in this fashion, for, although now absent, large browsers were abundant in this area until recently.

Failure to attract

Some plants may stop herbivory by failing to attract herbivores. They do so by lacking a certain chemical attractant that the herbivore uses as a cue.

Reproductive inhibition

Some plants, for example firs (*Abies* spp.), contain insect hormone derivatives that, if digested, prevent successful metamorphosis of insect juveniles into adults. In this way herbivory in the future is lessened by a decrease in the herbivore's reproductive output.

Masting

The synchronous production of progeny, seeds, in some years satiates herbivores, permitting some seed to survive. Nilsson and Wastljung (1987) compared seed predation on beeches (*Fagus sylvatica*). In mast years, 3.1 percent of seeds were destroyed by a boring moth; in nonmast years, this figure was 38 percent. Vertebrate predation of seeds was 5.7 percent in mast years but 12 percent in normal years.

Some plants defend themselves against herbivores by enlisting the help of other animals. Such a relationship can, of course, he seen as *mutualism*. A very common example is that plants attract ants by providing sugary nectar secreted from extra-floral nectary glands. African *Barteria* and neotropical *Cecropia* trees have hollow stems in which ants maintain populations. The ants are obligate occupiers of the trees; in return for food and shelter, they protect the trees from other herbivores and from encroaching vines, both of which they bite to death. Schupp (1986) demonstrated the benefits to juvenile *Cecropia* by removing ants. The experimental plants suffered more damage from nocturnal herbivorous Coleoptera than unmanipulated controls. The *Acacias* of Central America have hollow thorns, which provide homes for extremely aggressive *Pseudomyrmex* ants, which also kill other herbivores and chop away encroaching vegetation. In numerical terms, the effects of ants can be quite substantial. Schemske (1980) found that seed production of *Costus woodsonii* in Panama was reduced by 66 percent where ants were excluded. In England, Skinner and Whittaker (1981) used exclusion techniques to show that the wood ant, *Formica rufa*, was able to reduce herbivory from 8 percent of the leaves to 1 percent. Not all ant attendance is beneficial to trees, however, because many ants simply farm sap-sucking Homoptera and effectively protect their populations, a behaviour that is of little use to the plant.

Of course, some defensive schemes backfire on the plants. Some chemicals that are toxic to generalist insects actually increase the growth rates of adapted specialist insects, which can circumvent the defense or actually put it to good use in their own metabolic pathways. Danaid (monarch) butterflies are attracted to milkweed plants that contain cardiac glycosides. These substances are vertebrate heart poisons, and cattle will not eat the milkweeds, but monarch butterflies can assimilate these poisons and use them in their own bodies as a defense against their own predators, advertising their distastefulness with bright colours. With decreased rates of predation, monarch caterpillars may actually cause more herbivory than they would if they were acceptable to predators.

Induced Defenses

Plants do not necessarily keep their tissues permanently suffused with defensive and deadly chemicals. There is much evidence that chemicals arc produced only as they are needed. The initiation of herbivore attack is usually sufficient to start the metabolic pathways of defense grinding. This defensive tactic is known as *induced defense*. For example, Rhoades (1979) removed 50 percent of the leaves of ragwort, *Senecio jacobaea*, and detected a 45-percent increase in leaf alkaloids and N-oxidases in the undamaged leaves. Some of these effects can persist for long periods. Haukioja (1980) found that leaves of a birch tree were still poor-quality food for a moth three years after the tree had been damaged. This phenomenon may help to explain the severe oscillations of many forest insects. Sheep fertility was reduced six weeks after aphid attack an the alfalfa the sheep fed on had increased the production of its estrogen mimic, coumestrol.

Facultative defenses are not confined to chemical mechanisms. For example, the prickles on cattle-grazed *Rubus* plants were longer and sharper than those on ungrazed individuals nearby. Browsed *Acacia depranolobium* trees in Kenya have longer thorns than unbrowsed ones. Stinging nettles, *Urtica dioica*, exhibit increased density of stinging trichomes after herbivore damage. On holly trees, *Ilex aquifolium*, in Britain, lower leaves, subject to grazing, are heavily armed with spines; upper leaves, free from herbivory, are not. Interestingly enough, though the same phenomenon is apparent on American holly, *Ilex opaca*, Potter and Kimmerer (1988) regard it as an ontogenetic phenomenon rather than a defense against browsers. They suggest that the poor nutritional quality of holly and the high concentrations of saponins are more important as deterrents of vertebrate herbivores and that fibrous leaf edges deter invertebrate herbivores, which are unlikely to be affected by spines. Williams and Whitham (1986) have even argued that leaves infested with sessile insects (for example, gall makers and leaf miners) are abscised prematurely as a defense against herbivory; the insects are killed as the leaf senesces on the forest floor. Stiling and Simberloff (1989) examined such a situation on oak trees in northern Florida. Although it is true that leaves infested with leaf miners do abscise earlier than noninfested leaves and that premature abscission kills miners inside the leaf, leaf abscission is not likely to be a complex, induced defense. It is more likely that premature abscission is a simple wound response on the part of the tree to rid itself of damaged leaves, possibly to avoid infection. For abscission to work as an induced defense in this situation, miners would have to show a high fidelity to a host tree; otherwise the tree would simply be reinfected by miners from neighbouring trees. There is little fidelity in the leaf miner—oak tree system.

Though induced defenses undoubtedly occur in nature, their effectiveness as a deterrent is open to speculation and is probably less than that of permanent chemical defenses. Fowler and Lawton (1985) reviewed much of the work on induced defenses and concluded that as a result of such chemicals, only small changes, generally less than 10 percent, occurred in such things as larval development time or pupal weights. Other studies have shown that certain insects even benefit by feeding on previously damaged plants.

Mathematical Models

Crawley (1983) has provided a series of models of plant-herbivore interactions. The most simple of these assumes that there is an upper limit, a carrying capacity, K, for a population of plants. The rate of change of a population of plants is given by

$$\frac{dV}{dt} = A - B$$

where A is the gains, B is the losses, and V is plant abundance. Similarly, for the herbivore

$$\frac{dN}{dt} = C - D$$

where C and D are gains and losses and N is herbivore numbers.

It is assumed that, in the absence of herbivores, plant populations increase exponentially such that gains, A, $= aV$ where a is the plant's intrinsic rate of increase. Losses for plants, B, $= bNV$ where b is the feeding rate of herbivores. For the herbivores, $C = cNV$ where c describes the numerical response of herbivores, and $D = dN$ where d is herbivore death rate. The assumption is that, in the absence of plants, herbivores starve and their numbers decline exponentially. Assuming an upper carrying capacity, K, and following a basic Lotka-Volterra type of logistic model then $A = aV\ (K - V)/K$. The other elements, B, C, and D, remain the same. Therefore,

$$\frac{dV}{dt} = \frac{aV(K - V)}{K} - bNV$$

and

$$\frac{dN}{dt} = cNV - dN$$

At equilibrium both dV/dt and dN/dt are zero; there is no population change. The equilibrium plant density V^* and equilibrium herbivore abundance N^* can then be solved for because $A = B$ and $C = D$. Therefore,

$$V^* = \frac{d}{c}$$

and

$$N^* = \frac{a(K - (d / c))}{bK}$$

The effect of each of the parameters a, b, c, and d is best assessed by graphical techniques. Each parameter can be made to vary while

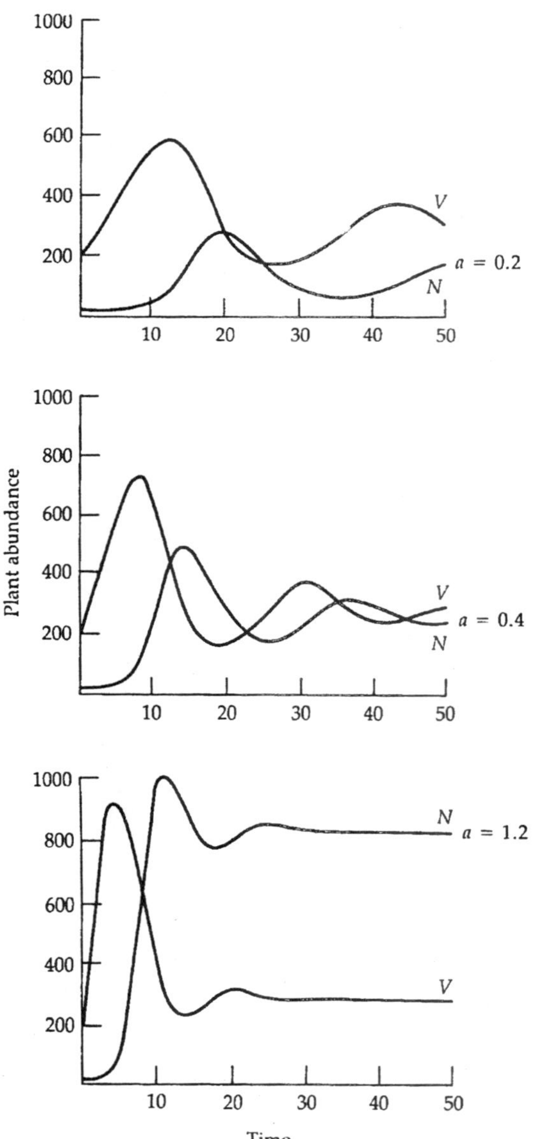

Fig. 10.1. The effect of the plant's intrinsic growth rate a on plant and herbivore abundance.

the others remain fixed. The effect of increasing the plant's intrinsic rate of growth, *a*, stabilizes and increases herbivore equilibrium density but has no effect on equilibrium plant abundance: essentially, the faster the plants grow, the faster the herbivores eat them up. When b, the feeding rate of herbivores, is increased, there is a lower herbivore equilibrium, but the size of the equilibrium plant population is unchanged. The more efficient an herbivore is at food gathering, the

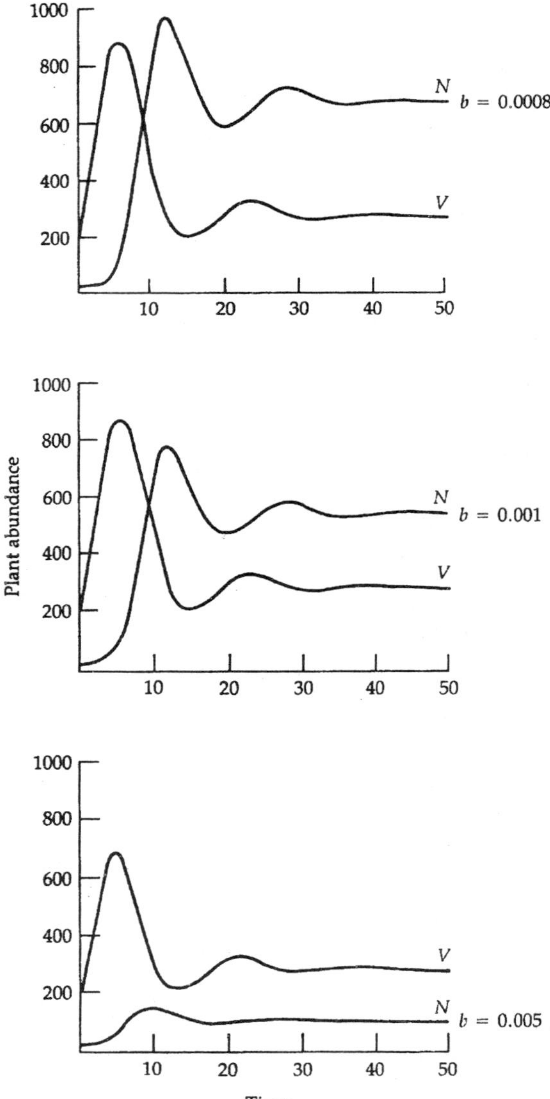

Fig. 10.2. Resource-limied plants—the effect of herbivore searching efficiency b.

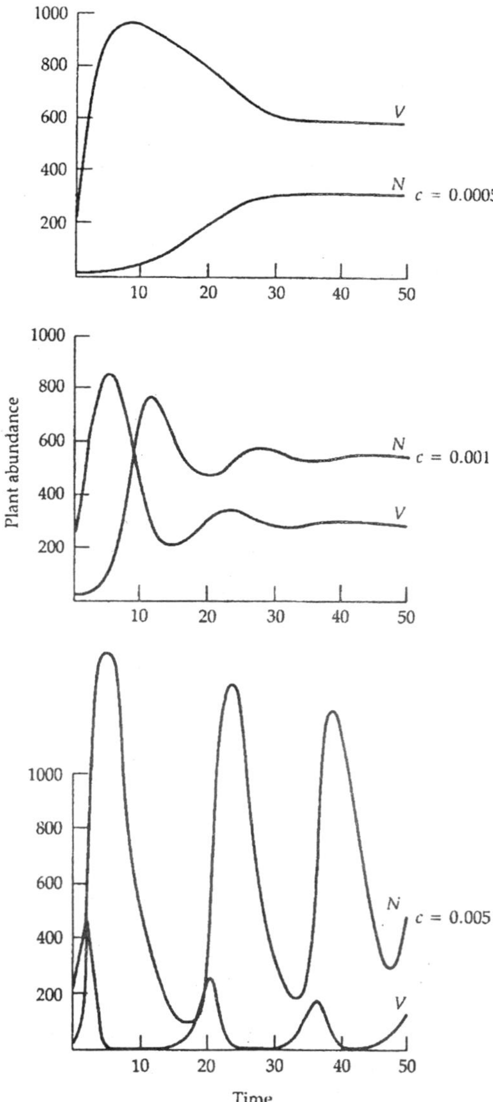

Fig. 10.3. The effect of herbivore growth efficiency c.

rarer the herbivore is because fewer herbivores can be supported by a given level of plant production. The numerical response of herbivores is described by c; the higher the value of c, the more herbivores can be borne per unit feeding. This is essentially the efficiency with which herbivores turn food into progeny. The higher this efficiency, the lower the equilibrium population of plants because an efficient herbivore turns all production into herbivores. Perhaps more important is the effect of c on stability. When plant equilibrium density is low, much below K (the carrying capacity), the population is under lax control and tends to increase exponentially when herbivore numbers decrease. Thus, the lower the equilibrium plant population, the lower the stability of plant and herbivore numbers. The effects of herbivore death rate, d, are exactly opposite to those of c. Increasing herbivore death rate increases *stability* because it leads to an increase in equilibrium plant numbers. Finally, plant carrying capacity, K, is critical to the stability of these models. If K is high, then populations may be under lax control even where equilibrium plant numbers are quite high. An increase in K could lead to an increase in the number of herbivores, too. However, equilibrium plant density V^* is determined solely by c and d. Thus increasing K reduces the relative level of plant equilibrium populations, and control of plant numbers is lax. This effect explains why high values of K decrease the stability of a system. This is what Rosenzweig (1971) called the "paradox of enrichment" and may account for the unstable population cycles of herbivorous insects in the vast boreal forest of the Northern Hemisphere, where the carrying capacity for evergreen trees is enormous.

Crawley has concluded that plants have a much more important impact on herbivore than herbivores have on the dynamics of plants. Mammals are especially prone to food limitation, though insect herbivores are probably more influenced by predation, parasitism, ani disease. Plant death rates themselves, Crawley suggests, arc largely determined by competition with other plants and by self-thinning.

Field Frequency of Herbivory

Despite the impressive array of defenses in their arsenals, plants do not have things all their own way in the plant-herbivore interaction. Herbivores can detoxify many poisons by four chemical pathways: oxidation, reduction, hydrolysis, and conjugation. Oxidation occurs in mammals in the liver and in insects in the midgut. It is brought about by a group of enzymes known as mixed-function oxidases (MFOs). Conjugation, often the critical step in detoxification, involves the uniting of two harmful elements into one inactive and readily excreted product. Given that herbivores can circumvent plant defenses in certain situations, what is their measured effect on plant populations in the field? There are numerous studies showing how herbivores reduce plant growth, flowering, reproduction, and survival. There is even good

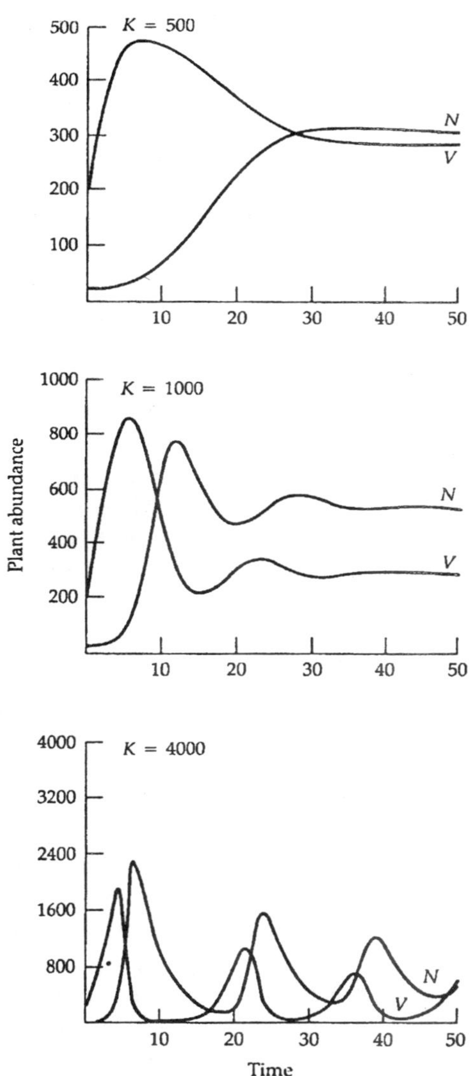

Fig. 10.4. The paradox of enrichment.

evidence that long-term suppression of insect herbivores increases the production and growth of roots. However, on average, no more than 10 percent of net primary productivity seems to be taken by herbivores and about 90 percent by decomposers, in most natural systems. In a review of 93 cases of herbivory in terrestrial systems, an average of 7 percent was found to be consumed. It must be remembered, of course, that such figures mask large and important variations. For example, the larch budmoth may take less than 2 percent of the net production of forest trees in some years but 100 percent in others. Also, workers interested in herbivory would probably not choose to study a plant that suffers very little herbivory; thus, even 7 percent could be an overestimate.

Sometimes, damage from insect feeding causes wilting or allows disease to cause loss of more plant biomass. Grasshoppers feeding on needlerush, *Juncus roemerianus*, in salt marshes feed in the middles of the tall, narrow leaves so that even though only the middle of the leaf is actually digested, the top half is cut off and added to the litter layer. This author has observed that regions of *Quercus geminata* leaves that are distal to *Stilbosis* leaf mines often turn brown and senesce, even though the miners do not damage them directly.

Ultimately, the best way to estimate the effects of herbivory on plant populations is to remove the herbivores and examine subsequent growth and reproductive output. Some of the best evidence on the impact of herbivores on plants comes from the biological control of weeds. Following its importation from the Americas in 1839, the prickly pear cactus, *Opuntia stricta*, became a serious pest in Australia, occupying by 1925 more than 240,000 km^2 of once-valuable rangeland. After some initial imports of insects that failed to control the growth of the cactus, *Cactoblastis cactorum*, a moth, was introduced from South America in 1925. By 1932, the original stands of prickly pear had collapsed under the onslaught of the moth larvae. Despite a small resurgence of prickly pear in 1932-1933, *Cactoblastis* has devastated prickly pear populations ever since, and the cactus is now confined to isolated areas. Similar sucess stories were reported in South Africa, Hawaii, and the Caribbean. Unfortunately, the moth has now invaded south Florida from the Caribbean and is damaging some rare and endemic cacti there. This is important because it is one of the relatively few examples of biological control that has unintended side effects.

There have been several other successes in the biological control of weeds by natural enemies, and these tend to overshadow the probably more numerous failures. Klamath weed (*Hypericum perforatum*), a pest of pastureland in California, was controlled by two French beetles. Floating fern, *Salvinia molesta*, choked a lake in Australia and was controlled by the weevil *Cyrtobagus Salvinae*, introduced from Brazil, where the fern is native. Alligatorweed was controlled in Florida's rivers by the so-called alligatorweed beetle, *Agasicles hygrophila*, from

South America, and hopes are high that water hyacinth can be controlled biologically, too. On the other side of the coin, large numbers of insects have been introduced to control *Lantana camara*, an introduced weed in Hawaii. Very few have had any impact on the growth of the plant, though its spread might have been slowed.

In successful cases of biological control of weeds, all the plants have been alien, and most were perennials. No native weeds and very few annual weeds have yet been controlled by insects. Annuals can produce huge amounts of seed even when heavily infested by insects. No seed-eating insect has yet controlled a weed plant. The weevil *Apion ulicis* was introduced into New Zealand to control gorse, *Ulex europaeus*, and has become one of the most abundant insects in New Zealand. Unfortunately, although the insect eats up to 95 percent of the gorse seed produced every year, there has been no appreciable effect on plant numbers.

In a natural setting, removal of herbivores from their host plants has been done less commonly and with mixed results. Karban (1982) used this technique to demonstrate reduction in growth of apple trees due to insect infestation, and he detected a similar effect in scrub oak, *Quercus ilicifolia*. Australian *Eucalyptus* trees with insects removed were almost 100 percent taller than control trees after three years of this treatment. Increased growth in herbivore removal experiments has also been shown for sand-dune willow trees. Meyer (1993) was able to show, by selective removals, that xylem-feeding spittlebugs had more of an effect on goldenrod (*Solidago atissima*) growth than either leaf-chewing beetles or phloem-feeding aphids. However, we are not yet in a position to know which type of herbivores have the greatest effects on which type of plants. One of the strongest effects was shown by Gibbens et al. (1993) who excluded rabbits from browsing rangeland plants in New Mexico for more than 50 years. By the end of this period, the basal area of some plants was 30-fold greater in the exclusion treatments than in the controls.

The most serious effects of herbivory may be on reproductive output. Crawley (1985) removed herbivores from oaks in Britain by spraying techniques. Though unsprayed trees lost only 8 to 12 percent of their leaf area, the sprayed trees consistently produced from 2.5 to 4-5 times the number of seeds produced by unsprayed plants. Waloff and Richards (1977) found almost three times more seed on broom bushes sprayed for insect control than on unsprayed ones. Once again, however, results tend to be mixed. Karban (1985) could find no effect of periodical cicada nymphs on acorn production by *Q. ilicifolia* in New York State. In instances where an exotic herbivore is introduced in the absence of its enemies, the results are much more dramatic. Bermuda cedar, *Juniperus bermudiana*, was virtually wiped out by an introduced scale insect, *Lepidosaphes newsteadi*. Young trees in forestry plantations can also suffer large mortalities when they are girdled by introduced goats, rabbits, sheep, or squirrels. In times of abnormally high densities, other animals can have these same effects. For example, in 1958 only 24 percent of mature trees surveyed in an African *Terminalia glaucescens* woodland were dead, but in 1967 after elephant densities were boosted by immigration beyond the carrying capacity of the region, almost 96 percent of the trees were dead. Crawley (1983) has provided many other examples of the impact of herbivory.

In most cases, herbivores cause subtle alterations of growth rates of stems and roots, rather than outright death of the plant. Flower, seed, and fruit production can also be influenced, though it is likely that predators of fallen seed and fruit are more important in a scheme of plant fitness. In this respect, herbivores can he seen as successful *parasites* because they do not kill their hosts—they merely reduce the growth rate. In an agricultural setting, of course, there are many estimates of losses of crops to herbivores. Damage is often severe enough not only to justify control but to cause economic hardship to agriculturalists. Even though plants are rarely killed outright, the effects of even minor damage on crop yields can be substantial. It is clear, however, that control measures are often initiated when populations of pests are so small that significant damage is unlikely to occur.

Beneficial Herbivory?

Some authors have argued that herbivory can be beneficial to plants. The rationale is that plants are stimulated to regrow after damage—they end up overcompensating, growing even more than they would have had they not been damaged. The result is often more seed production from more vegetative plant parts. Simberloff, Brown, and Lowrie (1978) noted that the action of isopod and other invertebrate root borers of mangroves tended to initiate new prop roots at the point of attack. More prop roots meant greater stability of mangroves against wave and storm action, so root herbivory could in fact be beneficial. However, in a review of the 20 papers most commonly cited as evidence for beneficial herbivory, Belsky (1986) found fault with the logic, experimental design, or statistics of nearly all of them. Even newer papers that purport to support the beneficial-herbivory theory

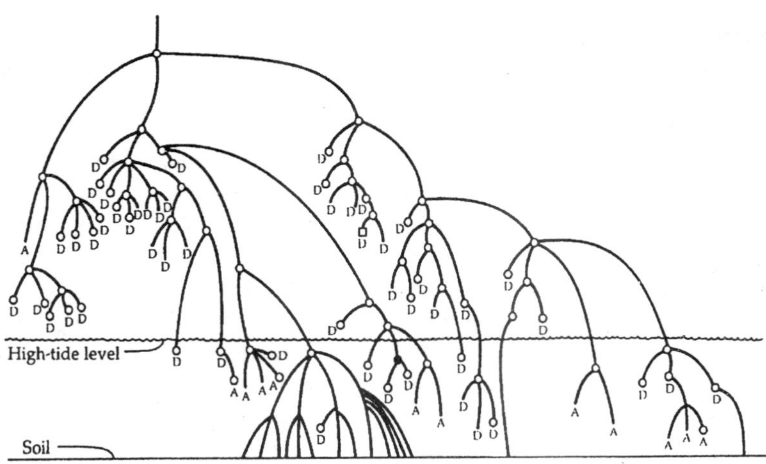

Fig. 10.5. Branching pattern for a single Rhizophora mangle root.

are usually fraught with methodological or technical errors. In addition, Strauss (1988) has pointed out that very carefully designed experiments involving measurements of plant size before and after herbivory are needed because herbivores themselves naturally choose larger plants, which might be expected to show more growth than would stunted plants, even after herbivory. Because of the economic damage supposedly done by herbivores, especially insects, it is not trivial to assess whether or not defoliation is beneficial. Mattson and Addy (1975) have tried to model forest growth with and without insect herbivores. They examined two situations. In the first, aspen was defoliated by forest tent caterpillars. Forest tent caterpillars begin to infest a forest slowly, reach a peak in numbers, remain there for three or four years, and then subside. Stemwood production in the years of peak infestation is much reduced, but foliage production increases to compensate for insect defoliation. Within roughly ten years after the infestation, the biomass production was identical to that of unaffected forests. In the short term, the caterpillars reduced wood production, but in the long term they had no major effect.

In a second example, balsam fir was defoliated by spruce budworms. These larvae actually kill mature trees aged 55 to 60 years, but they leave young trees largely alone. The saplings grow quickly after their parents are killed, and a resurgence of the forest, from its juveniles or saplings, occurs. The end result is the same as in the previous case: in the short term, there is a considerable effect on wood production, but there is not in the long term. Production rates in the young forest remain elevated above that of the mature forest for 15 years because in a mature woodland, most trees have passed their rapid-growth phase. The role of foresters in this cycle is not at all clear. Perhaps an effective action would be to harvest those trees in the center of budworm outbreaks because they would be killed in any case and because many larvae would be removed with them.

In herbivory on grasses, the placement of the meristem is important. In most species, the meristem is down low, safe from herbivores and protected in the basal leaf sheaths. The few species that produce elongated vegetative shoots are very vulnerable to grazing. Because growth is very slow from axillary buds, which must take over after the meristem is eaten, such plants tends to be outcompeted by species whose meristems survive. The inflorescences of most species, however, must extend upward into the air to ensure pollination of the flowers. Because certain species die only after flowering, heavy herbivory or constant clipping by a lawnmower can, by preventing flowering, virtually ensure their immortality. Grasses in sports stadiums may effectively live forever.

Effects of Plants on Herbivores

There is much evidence that herbivores themselves select the plants that are the most nutritionally adequate in terms of nitrogen content of the tissue or amino-acid concentration of the sap. Iason, Duck, and Clutton-Brock (1986) showed how red deer fed preferentially on grasses defecated upon by herring gulls (*Larus argentatus*). Where the number of gull drop- pings increased, so did the vegetation nitrogen content. For birds, Watson, Moss, and Parr (1984) showed how food enrichment affects numbers and spacing behaviour of red grouse. In some cases, however, such correlations between host-plant quality and herbivore density are present, but observed population patterns of herbivores are more dependent on other phenomena, such as predation or parasitism. There seems to be no easy way to predict when herbivore densities are controlled by the quality of their hosts. When food quality declines, many herbivores, especially vertebrates, respond simply by feeding at a higher rate or for a longer time. However, this will often result in increased exposure time to predators and parasites and higher enemy-induced mortality rates. On the reverse side, deaths of herbivores due to depletion of food plants are witnessed very infrequently, perhaps because herbivores can leave an area of poor food availability, but they cannot easily escape the weather or an outbreak of disease. The relatively few examples of mass starvation due to overexploitation of plants come mainly from studies of insects that habitually undergo periodic outbreaks or from cycles of Arctic rodents.

Herbivores are not only influenced by plant quality; they are affected by host defenses, many of which are genetically inherited. Herbivore densities can also he affected by the weather, as noted by Davidson and Andrewartha's (1948a,b) studies on thrips in Australia. Sometimes, there is an interaction between host genotype and environment such that at some sites plants are resistant to herbivores, but at other sites they are susceptible. Presumably the abiotic environment, soil type, water levels, or nutrient levels influence whether or not plants can maintain their resistance to herbivores. Which of these factors is most important? Rick Karban (1992), in an excellent review of the effect of plant variation on herbivorous insects, showed that more than 80 percent of studies that had used experiments to test for the presence of plant genetic variation on insect herbivore densities found it. This is a high percentage. However, Karban also reviewed studies that compared the effects of plant genotype to other factors and found that in 17/30 cases (56.7 percent), plant genotype explained less variation in herbivore numbers than did other factors, such as yearly variability. For example, Stiling and Rossi (1995) showed that environmental conditions affected the population densities of all eight major herbivores on coastal plants in Florida more than did plant genotype. This kind of result reminds us of the ideas of Davidson, Andrewartha, and Birch who suggested that environmental factors such as the weather may be of paramount importance in determining herbivore abundance.

11

EVOLUTIONARY ADAPTATION

The study of ecology encompasses many points of view. One of these holds that ecological relationships manifest interactions between the gene pools of populations and selective factors in the environment. The form and function of organisms are adaptations with genetic bases; understanding their evolution requires a knowledge of the environment, the organism, and the genetic potential of the gene pool. The environment embodies the selective pressures that establish fitness differences between individuals with different genotypes. Evolutionary responses of the phenotype to these pressures depend on functional interrelationships that limit form and function to combinations of traits that work, in the sense that they obey physical laws. Responses also depend upon the availability of genetic variation in the population, upon which selection acts. Moreover, genetic variation itself makes advantageous certain types of adaptations, principally of the breeding structure of the population, that optimally manage the expression of genetic variation in the progeny of each individual.

Mutual evolutionary responses of interacting populations, affirm that a consideration of evolution is central to the study of ecological systems. Indeed, from its beginning ecology has formed links to three related biological disciplines: genetics, which studies the mechanism of heredity; evolution, which includes the study of change in the genetic makeup of populations; and development, which represents the realization of the genetic blueprint in the form of the individual. A distinct subdiscipline, called *evolutionary ecology*, concentrates especially on the interpretation of form and function in animals and plants as adaptations to their environments.

Evolutionary ecology has five major programs of inquiry. The first is to understand the mechanisms of evolutionary change from a genetical standpoint, including limits imposed by the availability of genetic variation. This is discussed in detail in texts on population genetics and evolution and will be touched upon only briefly here. Second, genetic variation in populations confronts individuals with the problem of choosing the genotypes of their mates so as to optimize the genotypes of their offspring. This problem may be solved through modification of the breeding system and pattern of mate choice, as we shall see below.

The third program, to which we shall devote most of our attention, is the interpretation of form and function in the context of adaptation. This program, often called the adaptationist paradigm and what Eric L. Charnov (1982) calls "selection thinking," presumes that evolution can achieve that combination of traits best-suited to any particular environment and permissible within the bounds set by physical laws. The major goals of this program are to learn how physical limits on form and function prescribe the possible phenotypes among which the environment selects; to determine appropriate measures of fitness, especially with respect to traits that govern the interactions of individuals within a population; and to understand where the adaptationist paradigm can be applied validly and where it cannot.

The fourth program elaborates the last point: it is to determine the degree and mechanisms of matching of phenotypes to environments. Matching can occur both by selection of the phenotype by the environment and selection of the environment by the phenotype. The issue is the degree to which response to the environment is a property of the individual or of the gene pool of the population. Furthermore, organisms retain tangible evidence of their evolutionary history in the form of traits that are shared with related species regardless of the environment. For example, plants in the rose family have flowers with five-part symmetry; those in the lily family are endowed with a three-part or six-part symmetry. Whatever the adaptive significance of flower symmetry, it is lost in the dim evolutionary history of these groups, and does not stand at issue in their present-day evolutionary response to the environment, as do the size, colour, and form of individual flower parts.

Finally, the fifth program of evolutionary ecology is to determine the extent to which properties of larger ecological systems—communities and ecosystems—depend upon evolutionary relationships among their parts.

Evolutionary Response

Much of the variation among organisms within a population has a genetic basis. In some cases, the differences between individuals express variation at a single genetic locus. The principal colour groups-of eyes (blue or brown) in humans and typical (salt-and-pepper) and melanic (black) individuals of the peppered moth result from different forms of the same genes (alleles). In other cases, traits come under the influence of many loci whose effects may be additive, complementary, or modifying. Evolutionary change reflects the substitution of new alleles for old ones at individual genetic loci. When selection is applied to some traits, such as the amount of black pigment in the wing scales of moths, a single gene locus or a few gene loci are affected. Selection upon other traits, such as body shape or patterns of social behaviour, may affect numerous genes responsible for producing the trait, many of which in turn influence other characteristics of the phenotype.

The evolutionary mechanics of selection and genetic responses are the subject of population genetics. A primary task of population geneticists since the late 1920s has been to develop quantitative predictions of changes in gene frequencies in response to selection. In such simple cases as selection upon single genetic loci with simple genetic dominance between alleles, these models describe mathematically the intuitive result that the allele whose bearers leave the most descendants eventually predominates in the gene pool. The equations of population genetics also allow one to predict rates of change in gene frequency, hence how rapidly a population can respond genetically to a change in the environment.

The time required for a dominant gene to replace its recessive allele depends on its frequency at the beginning and end of the substitution process and the strength of selection. In the case of the replacement of the typical form of the peppered moth by the *carbonaria* form in polluted woods of England, this substitution is known to have taken about a century. From the results of H. B. D, Kettlewell's experiments on the peppered moth, one can estimate that the fitness of the allele for typical colouration was only 47 per cent that of the *carbonaria* allele; hence the fitness differential, or strength of selection, was 0.53. Plugging this value into equations predicting the time required for allele substitution, a change in frequency of the carbonaria allele from 0.05 to 0.95 would have required 47 generations, while a change from 0.01 to 0.99 would have required 204 generations. The peppered moth has 1 generation each year, and so the population genetics equations appear to be consistent with the observed time course of the gene substitution, assuming an initial frequency prior to the industrial revolution of about 1 per cent. Had selection been only a tenth as strong, the same change in frequency would have required 10 times as long.

The same equations may be turned around to estimate fitness differentials from changes in gene frequency. For example, since the introduction of pollution control programs in England, the frequency of the *carbonaria* allele has decreased at a rate consistent with a 12 per cent selective disadvantage. Records of trapped red and silver foxes kept by the Moravian mission posts in Labrador for a hundred years, 1834-1933, show that the proportion of silver foxes in the catch declined from about 0.15 to 0.05. Because silver coat is a recessive phenotype, the frequency of the gene must have decreased from 0.39 to 0.22 over a period of 100 years, a change that would have required a fitness differential of 0.035, or 3.5 per cent. It does not seem unreasonable that a greater demand for silver fox furs might have led trappers to cause an annual mortality of silver foxes 3 per cent in excess of that for red foxes.

Polygenic Basis

The relationships between organisms and their environments often depend upon modifications of continually varying traits, such as lengths of appendages, body size and shape, thickness of hair or cuticle, and continuous gradations of behaviour. Because variation in these characters depends on the contributions of many genetic loci, their response to selection cannot be analyzed by the simple population genetics models that have been applied to single-gene traits such as melanism in moths. Animal and plant breeders interested in such traits as milk production, oil content of seeds, and rate of egg production have developed a mathematical treatment of continuously varying traits and their responses to selection, known as quantitative genetics. The theory of quantitative genetics rests on the assumption that variation within a population results from the additive contributions of many genes with similar effect. Thus the length of an appendage may come under the influence of a dozen genetic loci each of which may cause a small increase or decrease in length relative to the population average, depending on the allele. Individuals with a net excess of length-incrementing alleles at these 12 loci would have appendages longer than the average.

Although quantitative genetics greatly simplifies our concept of gene action, its models have proven successful in predicting the results of selective breeding programs. To the extent that artificial selection mimics selection in nature, quantitative genetics can help us to understand the evolutionary responses of natural populations.

The heart of quantitative genetics is the variance of a trait within a population. Variance is a statistical measure of variation; specifically, it is the average of the squared deviations of individuals from the mean of the population. Hence the variance (*V*; also *Var* and s^2) in a trait *X* (a measurement) in a population of *n* individuals is

$$V = \frac{1}{n}\sum_{i=1}^{n}(X_i - \overline{X})^2 \qquad ...(1)$$

where X_i is the value for each individual *i* (i = 1 to *n*), and $\overline{X}$ is the mean value of *X* in the population.

Continuously variable traits frequently exhibit a bell-shaped distribution of values in natural populations, with most individuals clustered near the mean and with the frequency diminishing at extreme values. Each individual's value of a particular trait (the phenotypic value) is determined by deviations from the population mean caused by genetic and environmental influences. Because both sources of deviation enter into the calculation of variance for all values in the population, we may speak of phenotypic variance (V_P) as having two components, one attributable to genetic constitution (V_G) and one resulting from environmental factors (V_E). The two components added together equal the total phenotypic variance, or $V_P = V_G + V_E$.

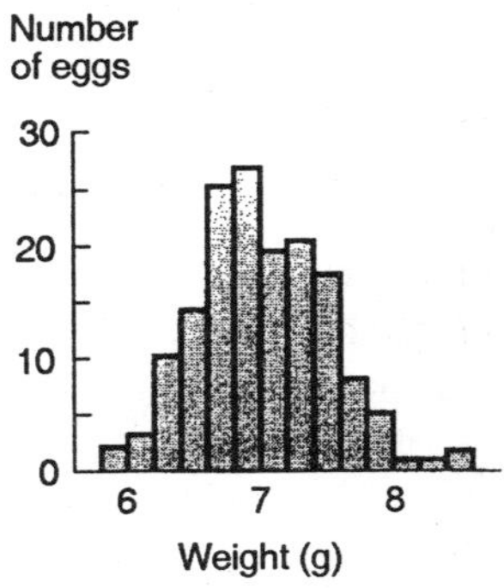

Fig. 11.1. Frequency distribution of the weights of eggs of the European starling Sturnus vulgaris.

The genotypic variance can be subdivided further: V_A is the additive variance determined by the expression of alleles in homozygous form; V_D is the dominance variance determined by the interaction of alleles in heterozygous form; and V_I is the interaction variance, comprising the influences of different genes on the expression of alleles at a particular locus. In general, $V_G = V_A + V_D + V_I$, although variance may be further incremented by the correlation between particular genotypes and particular environments and other, usually minor factors.

The principal task of quantitative genetics has been to estimate the magnitude of the several components of phenotypic variance. This is made necessary by the fact that response to selection derives only from the additive genetic component of variance, because only V_A reflects the genetic diversity of the population — that is, the different alleles that replace, and are replaced by, others during evolutionary change. Phenotypic variance can be partitioned into its several components by statistical analyses of the results of breeding programs designed for the purpose. Let it suffice to say that these analyses utilize correlations of phenotypic values between close relatives, usually between parents and their offspring or between siblings.

The proportion of the phenotypic variance due to additive genetic factors is often expressed as their ratio, called the *heritability* (h^2) of a trait: $h^2 = V_A/V_P$. Most studies of heritability involve traits of commercial value in livestock, poultry, and crops, for which representative values of h^2. The data indicate that sizes have higher heritabilities (0.50-0.70), hence less environmental influence, than weights (0.20-0.35). Among traits related to production and fecundity, those creating the greater drain on energy and nutrients have the lower heritabilities. Thus the percentage of butterfat in milk is under strong genetic control ($h^2 = 0.60$), while total milk production has a low heritability (0.30); variation in egg size in chickens has a large additive genetic component ($h^2 = 0.60$), while rate of egg production has a lower heritability (0.30). Any character that requires a large commitment of resources must be sensitive to environmental variation in those resources. The heritabilities of fecundity and life history characteristics are generally low (0.05 to 0.50). Heritability estimates accumulating for traits of individuals in wild populations resemble those of domestic animals and crops.

Artificial Selection

Artificial selection usually is accomplished with large fitness differentials applied over short periods; in this way, it caricatures the less intense selection that occurs in natural populations in that selection is exaggerated and some results would have no chance of surviving in the wild. Despite these differences, selective breeding programs suggest the kinds of evolutionary response that one might observe under natural conditions.

The change in a quantitative trait resulting from a single generation of selection (*R*, for response) depends on the deviation of selected individuals from the mean value of the population (*S*, for selection differential) and the heritability of the trait, according to the relationship

$$R = h^2S \quad ...(2)$$

For example, if h^2 were 0.5 and males and females 10 size units larger than the population average were bred together, their progeny would be 5 size units overall above the average of the unselected population. The greater the heritability of the trait, the more rapidly it can respond to selection.

Values of *R* and *S* are conveniently expressed as multiples of the standard deviation of measurements within the population, the standard deviation (*SD* or *s*) being the square root of the variance. Each value expressed in standard deviation units corresponds to a particular percentile rank in the population. Zero SD units is the population mean; when values are symmetrically distributed about the mean, half the individuals lie above that value and half below. When values have a normal distribution, 31 per cent of individuals lie above +0.5 SD and 31 per cent lie below -0.5 SD from the mean; 16 per cent have values more extreme than +1.0 SD or -1 SD; 7 per cent exceed 1.5 SD; and only 2.3 per cent exceed 2.0 SD in each direction from the mean. As a result, the more intense is selection (the larger the value

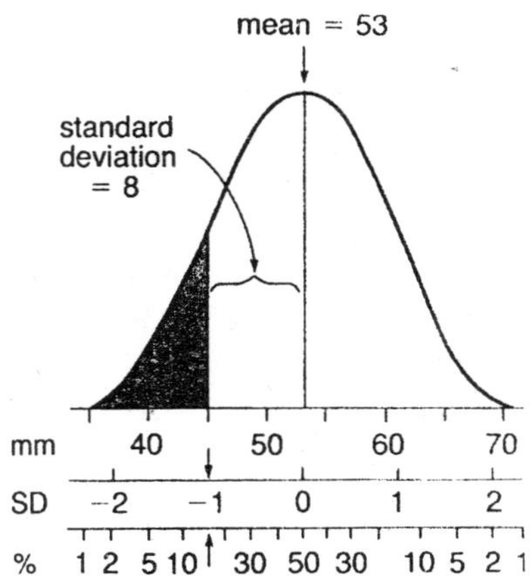

Fig. 11.2. Schematic diagram of variation in a hypothetical trait showing the relationship between the measurement scale, the standard deviation, and the proportion of the population with phenotypic values more extreme than a particular value.

of *S*), the smaller the number of individuals selected and the smaller the number of resulting progeny for the next generation of selection. When selection is too strong the population dwindles, eventually to extinction. Even in artificial selection programs, the strength of selection is limited by the size of the stock population and the reproductive rate of selected individuals, which must be at least as large as the number of individuals eliminated by selection each generation.

The relationship between selection intensity (per cent of individuals selected), phenotypic selection differential (*S*), and response (*R*) is illustrated by a program of selection for rate of egg laying in the flour beetle *Tribolium castaneum*. In the stock population, the number of eggs laid from 7 to 11 days after adult emergence (the phenotypic trait investigated) had a mean of 19.0, a standard deviation of 11.8, and a heritability (h^2) of 0.30. One control line and five lines with different levels of selection (ranging from 50 to 5 per cent) were established. Knowing the variability, heritability, and selection differential, one can estimate the initial response of the population to selection. For example, in the C line a selection intensity of 20 per cent corresponds to a selection differential of 1.4 SD, or 16.5 eggs (1.4 × 11.8). With a heritability of 0.30, the response to selection should be about 5.0 eggs per generation ($R = h^2S = 0.30 \times 16.5$). The observed response fell somewhat short of this prediction (about 3.0 eggs per generation), probably because the estimate of heritability included maternal and dominance effects as well as additive genetic

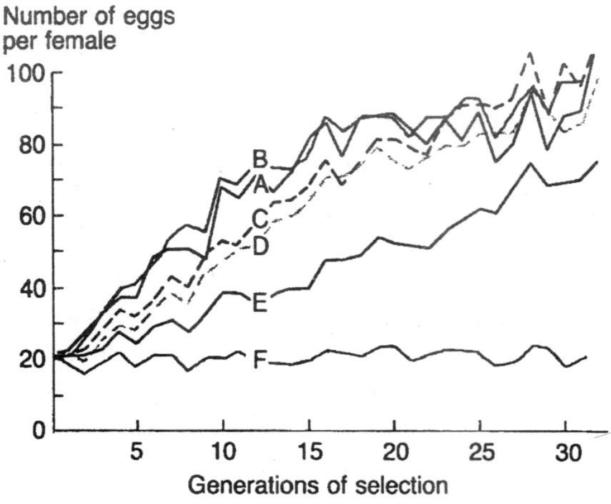

Fig. 11.3. Change in rate of egg laying in Tribolium castaneum lines exposed to different levels of selection, increasing from F to A.

variation. But the beetle population did behave as predicted in that the rate of response varied in direct proportion to the intensity of selection.

The slowed response is the result of two factors: erosion of genetic variation by selection and the application of opposing selection by correlated changes in other traits. Selection works only when individuals vary genetically within a population. When all unfit alleles are removed by selection, evolution pauses until new mutations or gene combinations appear.

The maximum long-term response to selection in the absence of new genetic variation can be estimated from properties of the probability distributions governing quantitative traits. Suppose that a measurement is influenced equally by n genetic loci each having two alleles: (-) and (+). An individual homozygous for (-) alleles at each of the n loci exhibits the minimum measurement. Each (+) allele in the genotype adds a single unit increment to the measurement, so that the maximum possible is $2n$ units greater than the minimum (there are two copies of each gene in diploid organisms). Suppose the frequency in the population of the (+) allele is the same at each locus and is p; then the mean value of the measurement in the population is $2np$ greater than the minimum possible. If the alleles segregated randomly, probability theory tells us that the variance in number of (+) alleles among individuals would be $2npq$, hence the standard deviation would be $\sqrt{2npq}$. How much of a selection response can be expected when the (+) alleles are strongly selected? Without the addition of new genetic variation, the maximum possible measurement is $2n$ above the minimum. Thus the maximum response (maximum phenotype - average phenotype) is $2n - 2np$, which may be expressed as $2n(1 - p)$ or $2nq$. In terms of numbers of standard deviations, the magnitude of this response is $2nq / \sqrt{2nqp}$, which may be rearranged to give $\sqrt{2nq / p}$. For example, when a trait is controlled by 8 loci at which the frequency of (+) alleles is 0.5, the maximum phenotypic response to selection is $(2 \times 8 \times 0.5/0.5) = 4$ standard deviations above the mean. Somewhat surprisingly, therefore, selection can produce phenotypes well beyond the extreme value observed in an unselected population. Probability theory reminds us that the proportion of individuals receiving 16 (+) alleles at 8 loci just by chance, when the frequency of (+) is 0.5, is less than 1 in 66,000 (that is, 0.5^{16}); 12 or more out of 16 (+) alleles (1.4 SD units above the mean) occur in less than 4 per cent of individuals. So we see that the short-term evolutionary potential of quantitatively varying traits may extend far beyond the distribution of

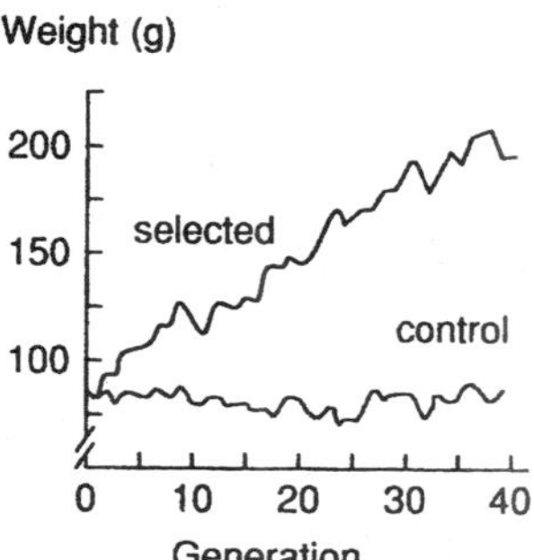

Fig. 11.4. Four-week body weight of selected and control lines of Japanese quail.

phenotypes in a population, providing that the multilocus, additive model of quantitative variation is reasonable.

Laboratory studies show that the large gains predicted by quantitative genetics models can, in fact, be realized. Mature body weights of unselected Japanese quail average about 91 g with a standard deviation of 8 g. After 40 generations of selection for high body weight, the population average in one study increased to 200 g, or almost 14 SD units above the mean of the unselected population. Realized heritabilities ($h^2 = R/S$) decreased from 0.30-0.45 in generations 1 to 10, to 0.15-0.20 in generations 11 to 30, and 0.05-0.10 in generations 31 to 40.

What is often more difficult to explain than the response of a trait to selection is the leveling off the response, often after only modest progress. This usually is not due to exhaustion of genetic variation for the trait, because reverse selection (back toward the mean of the unselected population) typically produces an immediate response, which can result only from genetic variation remaining in the population. Furthermore, when selection is merely relaxed, so that selection coefficients are zero, a selected trait will sometimes return toward the preselection measurement, apparently by itself. The most reasonable explanation for these results is that selection applied to one trait causes changes in other traits that affect the fitness of the organism. Increase in rate of egg laying, for example, may cause physiological or morphological changes that reduce viability and thus oppose the artificial selection regime.

Correlated Responses

Evolutionary responses often include characters other than the one selected Both development and function integrate the parts of organisms, bringing about an interdependence of phenotypic traits, particularly among those involving size or rate of growth and production. Animal and plant breeders have estimated genetic correlations between traits in many domestic species. For example, in poultry the genetic correlation between body weight and egg weight is 0.5 0; between body weight and egg production it is -0.16. Therefore, one cannot apply selection to body weight without also obtaining a relatively rapid increase in egg size and a slower, but steady decrease in rate of laying. Simultaneous selection of large egg size and small body size goes against the grain of genetic correlation, and usually is unsuccessful.

Many characters, such as body and egg size, are linked in developmentally, genetically, or functionally related groups that tend to respond to selection in concert. For example, Larry Leamy (1977) determined genetic correlations among skeletal measurements of mice and identified four clusters of traits highly integrated genetically among themselves, but relatively independent of each other: (1) skull length; (2) skull width, body weight, and tail length; (3) skull width (providing a link to group 2), scapula length, and other measurements associated with the pectoral girdle; and (4) limb bones and total body length. In mice, therefore, selection for body weight produces responses in tail length and the proportions of the skull, as well as in body weight itself.

When Peter Dawson (1966) selected for fast and slow larval development in *Tribolium castaneum* and *T. confusum*, he observed a number of correlated responses that decreased fitness regardless of the direction of selection. Selection for rapid development resulted in decreased size and, in *T. castaneum*, decreased larval survival and an increase in incidence of adult abnormalities. Selection for slow development resulted in increased adult weight, increased frequency of adult abnormalities, and decreased fertility of females.

Dawson assessed the fitness of each of the selected strains of flour beetles by placing them in competition with unselected stocks of the other species (1967). Both the "fast" and "slow" lines suffered decreased fitness during early stages of selection, suggesting that development rate is optimized with respect to its effect on fitness. (As Dawson continued his selection experiments, the fitness of one of the selected lines began to improve, apparently because of increased cannibalism by the selected larvae. Evolution frequently springs such surprises on us.)

Results similar to Dawson's were obtained by M. W. Verghese and A. W. Nordskog (1968), who examined correlated responses in reproductive fitness in selected lines of chickens. Selection for both increase and decrease in body weight and egg weight caused a decline in reproductive fitness, as indicated by rate of egg production, hatch rate, and survival of chicks to 9 months. Regardless of the character or direction of selection, fitness in the selected lines varied from 54 to 85 per cent of the nonselected line. These and other experiments emphasize the fact that natural populations are balanced genetically, and their adaptations are both well-tuned to the environment and finely adjusted to each other.

Russell Lande (1979) used the concept of genetic correlation to examine the short-term response of many characters to simultaneous selection. Genetic correlation between two traits [r_G (XY)] measures the degree to which genetic variation in one is related to genetic variation in the other. One also may calculate the environmental correlation (r_E) and the total phenotypic correlation (r_P) between two traits. Separation of these components of covariation requires statistical analysis of particular breeding programs, as does the separation of genetic components of variance.

Correlation is calculated from the covariance between two traits, estimated by an equation analogous to that for variance,

$$Cov(XY) = \frac{1}{n}\sum_{i=1}^{n}(X_i - \overline{X})(Y_i - \overline{Y}) \qquad \ldots(3)$$

The square of the correlation (r^2) is equal to $Cov(XY)/[Var(X)Var(Y)]$. Now, whereas the response of trait X to selection directly upon itself is $R(X) = h^2(X)S(X)$, the correlated response of trait Y to selection on trait X is

$$R(Y) = h(X)h(Y)r_G(XY)S(X)\frac{V_P(X)}{V_P(Y)} \qquad \ldots(4)$$

where V_P is the phenotypic variance and h is the square root of heritability. From equations (3) and (4), one can see that the ratio of the response in Y to that in X, resulting from selection on X, is

$$\frac{R(Y)}{R(X)} = r_G(XY)\frac{h(Y)V_P(Y)}{h(X)V_P(X)} \qquad \ldots(5)$$

Lande illustrated the implications of correlated response by substituting the following values for brain and body weight of mice: r_G(brain-body) = 0,68, h(brain) = 0.8, h(body) = .66, V_P[log(brain)] = 0.058, and V_P[log(body)] = 0.145. For these values, $R(Y)/R(X)$ =

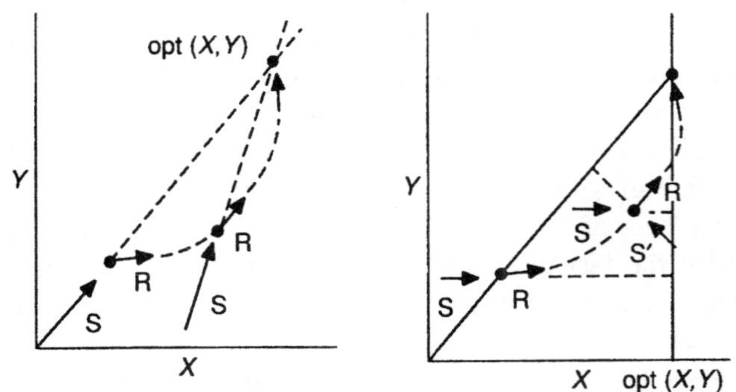

Fig. 11.5. Left: Response (R) to selection (S) for optimum phenotypic values of a pair of traits when genetic correlations deflect the responses trajectory. Right: Response to selection on a single trait X when there is an optimum ratio between X and Y.

0.36. Because measurements *X* and *Y* were log-transformed values, the ratio $R(Y)/R(X)$ expresses the allometric relationship between the selected and unselected populations. Lande used matrix algebra to generalize his equations for many gene loci, and emphasized that some patterns of phenotypic correlations between two traits could result fortuitously from selection on only one of the traits, or even on a third trait genetically correlated to the first two.

While genetic correlation may steer short-term responses to selection, both the matrix of genetic correlations relating different traits and the vector of selection coefficients change as the phenotype responds. In the first case, selection pushes both *X* and *Y* toward some optimum value, but genetic correlation dictates that the initial response deviates from the shortest phenotypic route to the goal. But as the position of the population changes in phenotypic space, the direction of the selection vector also shifts, pushing the population in a curved, perhaps even spiral course toward the goal. Achievement of the optimum is delayed, but not prevented, so long as the genetic correlation is not perfect. In the second case, selection is applied on one trait only (*X*), but there is an optimum ratio between *X* and *Y* (perhaps an optimum shape). The selection response causes the population to deviate from the optimum ratio, which establishes secondary selection on *Y*, leading in a curved path to both the optimum value of *X* and the optimum ratio of *X* to *Y*.

Genetic Variation

The gene pool of a population represents a balance between selection, which tends to reduce genetic variation, and several processes that increase genetic variation. Most genetic variation in a local population arises from mutation, immigration of alleles from other areas having different selective environments, temporally varying selective factors, and changes in gene frequency from generation to generation arising purely by chance in small populations (genetic drift). Given that such variation exists in all populations, we may ask: How much variation is good for individuals? And what strategies can individuals adopt to obtain the best genetic constitution for their offspring?

The answer to the first question depends upon the situation. When the individual's environment is highly predictable from generation to generation, and the individual is well-adapted, imposition of genetic variation reduces fitness. In such a case, individuals should reproduce asexually, as by parthenogenesis in animals and vegetative reproduction in plants, producing offspring genetically identical to the parent. Although most organisms or their close relatives can reproduce sexually, many abandon the practice seasonally, some mix both sexual and asexual reproduction, while others commit their futures wholly to asexual modes of procreation.

Much has been written about the evolutionary consequences of sexual reproduction both for the fitness of the individual within populations and for the evolutionary future of the population as a whole. Most evolutionists agree that sex is somehow necessary as a source of the new gene combinations that build the genetic foundation of long-term evolutionary change. We can infer this because most higher taxa of organisms (genera, families) contain sexual forms, and few strictly asexual groups are thought to have had long evolutionary histories. Yet there is a tremendous fitness incentive for the individual to give up sex. Each offspring of a female bears one set of her genes and one set of those of her mate. When females reproduce parthogenetically, each offspring contains two sets of the genes of the parent. So, unless a mate is required to help rear offspring—certainly not the case in most organisms—the fitness of a genotype that destines an individual to be parthenogenetic has twice the fitness of the genotype of a sexual individual, all other things being equal.

Sex remains a mystery to biologists. Most attempts to explain how sexual reproduction can be maintained in the face of an overwhelming short-term advantage to asexual reproduction have proposed that the compensating advantage to sex resides either with increase in the genetic variation among progeny or reduction of deleterious genetic variation. Variable offspring may confer a fitness advantage in a highly unpredictable environment where variation assures that at least some progeny will be well-suited to whatever environmental conditions should come their way. In strongly competitive environments, variation among progeny increases the chance that at least some will have the extremely high fitness necessary to persist.

Alternative explanations for sex argue that genetic recombination made possible by sexual reproduction is necessary to eliminate deleterious mutations from the germ line. In an asexual clone, deleterious mutations can be eliminated only by selection among progeny. When the entire mutation rate per genome exceeds the number of selective deaths per generation, mutations will accumulate and reduce the vitality of the clone. In contrast, individuals in sexual populations continually exchange genetic material between family lines. Because the entire population is joined into a common gene pool, selection picks and chooses among continually reshuffled genotypes. The deleterious mutation that crops up in one line can be exchanged for a beneficial allele through recombination.

Whatever the raison d'etre of sex, it provides three potential benefits: long-term evolutionary flexibility of the population, medium-term elimination of deleterious mutations from the family line, and short-term production of variation among the progeny of the individual.

Breeding Systems

Accepting as given that most populations reproduce sexually and that most gene pools contain variation, some of it deleterious, ecological geneticists are now beginning to address the effects of different sexual strategies on fitness. A major problem is the optimal degree of outcrossing for a population. Everyone knows that close inbreeding is bad. Brother-sister matings and, where possible (especially in plants), selfing may result in the expression of deleterious recessive genes. The mechanism for this is simple. Suppose an individual is heterozygous for a rare, deleterious, recessive gene (the gene pool is full of them and most individuals have some). If it were to mate with an individual from the general population, which probably does not have the same rare allele, half of their progeny would be heterozygous, like the one parent, and half would be homozygous for the common form of the gene, like the other parent. None of the progeny would be disadvantaged by the union. If, however, the individual selfed, one-quarter of its offspring would be homozygous for the deleterious allele and would suffer loss of fitness as a result. Matings between close relatives produce the same result, only less frequently.

Most species employ mechanisms, including dispersal of progeny, recognition of close relatives, and negative assortative mating, to reduce the occurrence of inbreeding. Hermaphroditic species of plants, in which individuals bear both male and female sexual organs, have additional mechanisms to prevent selfing, including self-incompatibility, temporal separation of male and female function, and elaborate flower structures designed to make self-fertilization difficult.

While close inbreeding generally creates problems, it may confer benefits as well. In particular, selfers can guarantee fertilization of their flowers in habitats lacking suitable pollinators or where individuals are widely spaced. Many weedy species that colonize isolated patches of disturbed habitat (for example, dandelions) are selfers. One assumes that most deleterious variation was weeded out of such populations as it was exposed in homozygous individuals during the transition between outcrossing and selfing.

Outcrossing at great distance may also reduce fitness when populations of plants include spatially defined ecotypic variation over small scales of distance, particularly in complex, heterogeneous environments. In such cases, local adaptation to particular habitat patches enhances fitness and receiving pollen from individuals adapted to different habitat conditions may reduce the fitness of progeny that become established near the female parent.

Several studies have reported an optimal outcrossing distance in populations of plants. Nearby individuals are likely to be close relatives, which raises the specter of inbreeding. Distant individuals are likely to be adapted to different conditions. M. V. Price and N. M. Waser (1979) fertilized flowers of the larkspur *Delphinium nelsoni* in central Colorado with pollen obtained from the same individual and from individuals located at distances of 1, 10, 100, and 1000 meters. Their results showed that the number of seeds set per flower was greatest when the pollen source came from a distance of 10 meters, and was least for selfed pollen and that obtained 1000 meters distant. Furthermore when these seeds were planted, survivorship to 1 and 2 years greatly favoured matings across the intermediate distance of 10 meters.

Although Price and Waser's study indicated an optimum outcrossing distance, larkspurs cannot exploit this advantage. The plants are pollinated primarily by bees and hummingbirds, which tend to visit nearby flowers in succession. As a result, coloured dye particles placed along with natural pollen on the male anthers were recovered primarily on flowers within 2 meters from its source. Perhaps no adaptation of flower structure could modify the behaviour of pollinators (which have little interest in the plant's fitness) to achieve the optimal outcrossing distance.

Another possibility for managing genetic variation is the selective abortion of developing ovules on the basis of the genotype of the embryo. Most plants produce many more flowers than they can mature as fruits; flowers are relatively cheap, fruits expensive. Excess fertilized ovules are reduced in part by predation or other extrinsic damage and in part by programmed abortion. Abscission of flowers and fruits can be highly selective with respect to number of ovules fertilized per flower or amount of herbivore damage. These phenomena have been reviewed by Andrew Stephenson (1981), who cited evidence that some species more likely abort fruit when flowers are self-fertilized than when they are outcrossed.

Noting that plants appear able to "recognize" the father of particular offspring, Daniel H. Janzen (1977) suggested that the abortion of developing ovules within fruits might be the result of a mechanism by which the female parent can select among its offspring according to their genotypes. So while plants may be unable to control the behaviour of their pollinators, they may through overproduction of flowers and ovules within flowers exercise control over the genotypes of their progeny. At present, this is an untested hypothesis with exciting potential. Animals can choose mates more actively than plants and there is abundant evidence of mate selection based upon kinship and genotype, often with a rare mating advantage.

Environmental Adaptation

Organisms are well adapted to their environments. This assumption has provided a powerful tool for conceptualizing the organism-environment interaction and understanding design limitations placed upon the responses of organisms to environmental change. This research program also has led to the discovery of new components of the fitness of organisms and has emphasized the important roles of interactions between individuals—predators, competitors, mutualists, society members, mates, parents, and offspring—in directing the course of adaptation. Therefore, even if the assumption of adaptation is not fully correct, it has expanded our ecological concept tremendously.

The primary question addressed by evolutionary ecology is: How do the adaptations of organisms reflect their environment? To answer this question requires an understanding of selective factors in the environment and the evolutionary responsiveness of the phenotype. To have scientific validity, an idea concerning the phenotype-environment relationship must include a suitable criterion for fitness of phenotypes, the genetic basis of phenotypic variation, and a model to link aspects of form and function that determine fitness to each other and to conditions of the environment.

One class of models of adaptation, that of phenotypic optimization, does not incorporate genetic variation explicitly, but rather assumes that phenotypic variation has a parallel genetic basis and that selection of optimum phenotypes brings about appropriate genetic change. For example, particular conditions may dictate an optimum allocation of resources between producing reproductive structures and continuing to grow; a certain genotype presumably produces that optimum phenotype and it will be selected irrespective of the particular genetic basis of the trait. The core of an optimization model is the relationship between phenotype and fitness. Each such relationship is unique for a particular environment. To visualize the relationship, one must either measure the fitnesses of a range of phenotypes directly or devise a realistic model that predicts the fitnesses of nonexistent phenotypes. The degree to which observed phenotypes match the predictions of the model measures the validity of the mechanics embodied in that model, although we must keep in mind that alternative models may make similar predictions.

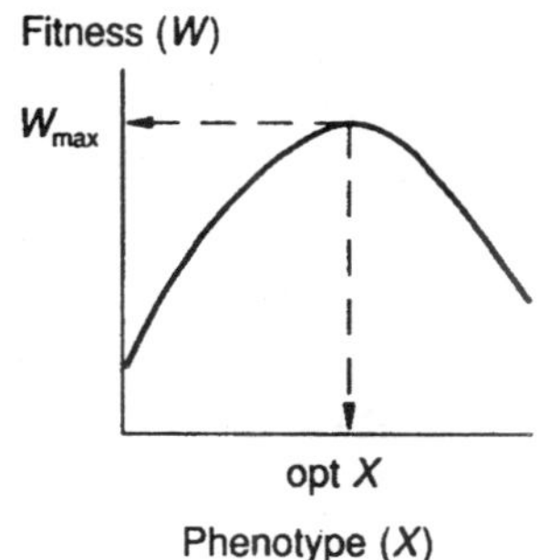

Fig. 11.6. Relationship between fitness and phenotypic value when there is a single optimum phenotype.

Evolutionary Strategy

A second approach to understanding adaptation, particularly useful in the case of discretely varying phenotypes and when phenotypes interact with one another, is that of determining the Evolutionarily Stable Strategy. An Evolutionarily Stable Strategy, or ESS, is that phenotype

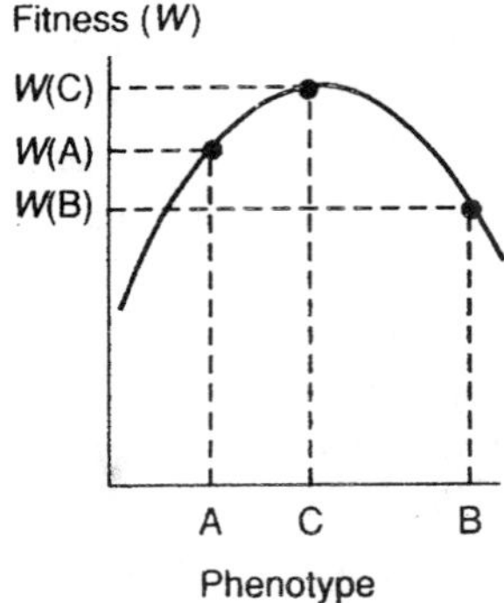

Fig. 11.7. Phenotype C is an evolutionarily stable strategy because no other phenotype can invade a population of C individuals.

or combination of phenotypes, which when constituting a population make it impossible for individuals with alternative phenotypes to invade the population. As John Maynard Smith (1982) puts it, an ESS is "a strategy such that, if all the members of a population adopt it, no mutant strategy can invade." Referring back to the relationship between phenotype and fitness, let us suppose that only two phenotypes, A and B, occur. A population consisting only of B individuals can be invaded by A individuals—that is, the phenotype will increase when rare—owing to the superior fitness of A. A population consisting only of A individuals resists invasion by the B phenotype; hence A is the ESS. Considering the range of all possible phenotypes, it is clear that only C, which confers the greatest fitness, will resist invasion by all other phenotypes. Therefore, a phenotype identified by the maximum-fitness criterion is also an Evolutionarily Stable Strategy.

The utility of ESS thinking becomes more apparent when phenotypes interact with one another and the fitness of each depends on the proportions of other phenotypes in the population. Hence the concept of the Evolutionarily Stable Strategy has found broad application in the study of social behaviour and mating systems, as we shall see a few chapters hence.

Phenotype Interaction

Regardless of the fitness criterion adopted, and irrespective of whether genotypic or phenotypic models are employed, the principal challenge to the evolutionary ecologist is to understand how small changes in phenotype affect fitness. To accomplish this goal, ecologists must understand the interrelationship of the many characteristics of the organism and the fullness of its interaction with all aspects of its environment. At any given time, our maturing concept of the organism-environment complex can be summarized in the form of models, whose consistency with observations on natural systems is continually scrutinized by further research, followed by reformulation of ideas when inconsistencies arise.

I shall illustrate this process with an example from my own work on the life histories of birds. In 1968 David Lack, the prominent English ecologist, suggested that growth rates of birds optimally balance two environmental factors: predation and other sources of mortality, which select rapidly growing individuals that pass through vulnerable developmental stages quickly; and food supply, which selects more slowly growing individuals that demand less food and thereby allow their parents to provide for more offspring. From this simple idea. Lack suggested that growth rate should vary in direct proportion to vulnerability of nestlings to mortality factors. But observations on birds revealed, to the contrary, that growth rate is relatively insensitive to variation in mortality rate, suggesting that Lack's hypothesis was not sufficient.

In an attempt to resolve this problem, I developed a model to describe the influence of growth rate on fitness through the consequences of growth rate for number of chicks produced per brood and number of surviving broods per season. The details of such a model require

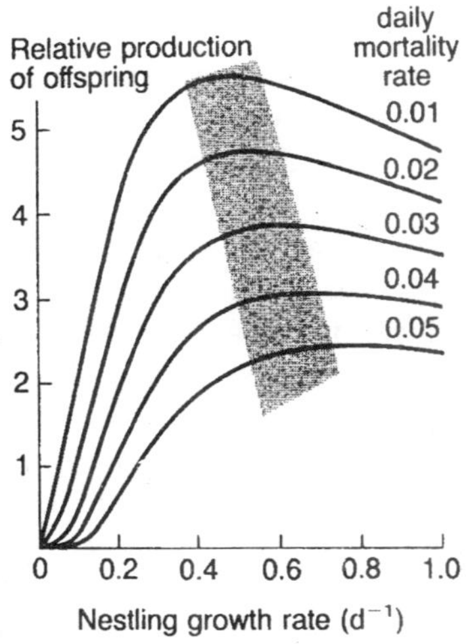

Fig. 11.8. Relative fitness as a function of nestling growth rate in a model relating the two through the influence of growth rate on brood size, number of broods per season, and brood survival for brood mortality rates from 0.01 to 0.05 day^{-1}.

knowledge of the food requirements of growing chicks and the initiation of new nesting attempts following both success and failure of the previous attempt. A set of expressions for each of these effects resulted in equations relating an index of fitness (number of young reared per season) to growth rate. The curves defined by these equations have maxima at intermediate growth rates close to rates measured in nature. Furthermore, when nest mortality rate is varied over a wide range in the model, the position of the optimum growth rate changes little, contrary to Lack's expectation but consistent with observation. This is related to the fact that birds rapidly replace destroyed clutches and so the impact of mortality on number of young reared per season is correspondingly diminished. The consistency of the model's predictions with observation suggests that Lack's concept of how growth rate influences fitness is essentially correct and that factors other than the ones included in the model are not important to the evolutionary optimization of growth rate.

Adaptationist Program

My confidence in understanding growth rates in birds rests firmly on the assumption that the organism-environment relationship reflects solely the fitness relationships of phenotypes with selective factors in their surroundings. This general view has been challenged on a number of grounds. One of these is that many processes oppose the perfection of adaptation: lack of suitable genetic variation for response to the environment; continuous production of less fit phenotypes by mutation and immigration; changes in environment that leave previously well-adapted phenotypes behind. In addition, some phenotypes of potentially great fitness are not possible because of limits imposed by physical laws. Some things, such as rates of physical diffusion, are beyond the realm of biological influence. Evolution also drags along the baggage of past adaptation, which may contain no particular relevance for the present. Body plans of major taxa have different number of limbs: four in terrestrial vertebrates, six in insects, eight in spiders, ten in crabs, and so on. The number can be changed by reduction of appendages (snakes have zero limbs), or the modification of other parts, such as antennae or mouth parts, to function partly as limbs, but the basic number of limbs characteristic of each major group has no particular meaning, other than historical, and may restrict future evolutionary potential. Finally, some adaptations have fortuitous consequences for developmentally or genetically linked features of the phenotype. Lande's observation that selection of body size alone automatically results in a particular increase in brain size, owing to their genetic intercorrelation, makes this point.

Some difficulties with adaptationist thinking were eloquently argued by Harvard evolutionary biologists Stephen J. Gould and Richard Lewontin (1979) in a paper entitled "The spandrels of San Marco and the Panglossian paradigm: A critique of the adaptationist programme." Spandrel refers to the space, generally triangular, between an arch supporting a ceiling or other horizontal structure and the ceiling itself. In many cathedrals, these were decorated with painted scenes. Gould and Lewontin's point was that the spandrel arises as a consequence of architecture, but is not a key feature of building design, as are arches and the horizontal members they support. The analogy to organism architecture cautions us that some structures may be fortuitous or secondary consequences of other, strongly selected adaptations. As Gould and Lewontin put it, "One must not confuse the fact that a structure is used in some way . . . with the primary evolutionary reason for its existence." They summarized their reservations about the adaptationist program by questioning the ubiquity of its basic assumption:

> This [adaptationist] programme regards natural selection as so powerful and the constraints upon it so few that direct production of adaptation through its operation becomes the primary cause of nearly all organic form, function, and behaviour. . . . We would not object so strenuously to the adaptationist programme if its invocation, in any particular case, could lead in principle to its rejection for want of evidence.

In this passage Gould and Lewontin suggest that evolutionary ecologists should test the basic assumption of adaptation as well as determine the validity of adaptationist explanations for patterns in nature.

Taxonomical Adaptation

Evolutionists divide the characteristics of organisms into one set that reveals the phylogenetic history of a group and another set that responds easily to selection and reflects the contemporary environment. Characters do not, of course, fall discretely into taxonomically revealing and ecologically revealing sets. Rather they are arranged along a continuum from evolutionarily stable to labile. Different characters are useful in distinguishing different levels of taxonomic groups. Among birds, for example, orders are distinguished primarily by skeletal features, such as the structure of the palate and arrangement of bones in the skull, whereas families are distinguished by variations in the beak, pattern of scales on the tarsus (lower leg), and number and relative length of the primary (flight) feathers. The lower taxonomic categories—genera and species—are often based upon small differences in measurements and ratios of measurements, as well as plumage colouration and song.

The differences that distinguish mammals from reptiles, truly a drastic reorganization of certain parts of the body plan, physiological processes, and patterns of reproduction, evolved over tens of millions of years through intermediate stages that enjoyed varying levels of success, judging from their abundance in the fossil record. Eventually, however, evolution arrived upon a particularly providential set of characteristics, those shared by all modern mammals, and the group underwent an explosion of evolutionary diversification between 70 and 60 million years ago. Regardless of subsequent modifications that now distinguish bears, rabbits, mice, seals, and wildebeests, all mammals retain the fundamental class characteristics, including warm-bloodedness and nursing of the young. At each lower taxonomic level, other traits have been set aside by evolution.

How do characters at each level in this hierarchy match up with the ecological distributions of organisms? Which ones may be thought of as ecological characters in the sense of the adaptationist program, and which are too remote to infer the ecological mold into which they were cast? The matching of organisms to their environments comes about in two ways: by the organism's choosing its environment to match its evolved features, and by selective modification of the gene pool of a population. Larger taxonomic groups appear to be widespread with respect to climate and other aspects of the physical environment, but often specialized with respect to diet. Thus the insect order Homoptera (leafhoppers, cicadas) may be found wherever vascular plants occur but, because their mouth parts are specialized for sucking plant juices, their local ecological roles are narrowly prescribed. All the evolutionary modifications of this group have taken place within an ecological context established by the ordinal characters. As Gould (1982) has put it, "current utility permits no necessary conclusion about historical origin. Structures now indispensible for survival may have arisen for other reasons and been 'coopted' by functional shift for their new role."

Smaller taxonomic groups are often distinguished by differences in body size, habitat, microhabitat, and selection of diet according to prey or host species, rather than manner of feeding. We may presume that the modifications responsible for such diversification are evolutionarily more malleable and, therefore, more amenable to study by evolutionary ecologists. As a general rule of thumb, adaptationist thinking may provide a useful tool for understanding aspects of the organism-environment interaction based on characters that distinguish close relatives.

Large Systems

The functioning of biological communities and ecosystems is determined by the collective adaptations of all their constituent species. To a large degree these species constitute important aspects of each other's selective environments and so each has evolved with respect to the evolution of others in the system. This mutual accommodation of species to each other is broadly referred to as coevolution.

Ecologists have often wondered about the importance of coevolution to the maintenance of community function, and whether there is a criterion for fitness at the level of the system as well as that of the individual genotype. One approach to this problem would be to perform an experiment—impossible to pull off other than with laboratory microcosms—in which ecosystems are constituted with species appropriate to a natural system, that is, representing all the ecological roles found in a natural system, but obtained from different places and thereby having independent evolutionary histories. If such artificially concocted systems functioned less well than corresponding natural systems, one could conclude that "coevolved" properties conferred unique qualities to large ecological systems.

The degree to which communities and ecosystems have an evolved genetic integrity has not been resolved by ecologists, and may not be for decades, if ever. Smaller parts of this larger question, concerning the role of adaptation in suiting the organism to its environment and the mutual coevolution of populations, have, however, contributed importantly to our concepts and understanding of ecology, as we shall see in the chapters that follow.

12

Environmental Variability

The world is anything but constant and uniform. A single genotype cannot be best suited to all the conditions encountered by each individual. These facts pose a fundamental and difficult challenge to understanding adaptation, raising several related issues to prominence in the ecological literature of the past three decades. One is the definition of fitness under variable conditions. That is, how does one best integrate the varied success of a genotype under different conditions into a single measure of fitness? A second issue concerns whether environmental heterogeneity facilitates coexistence of more than one genotype in a population. Interest in this problem was greatly stimulated by discovery in the 1960s of high levels of genetic variation in most populations. Third, heterogeneous environments present individuals with choices concerning habitat use, prey selection, mate selection, and so on. The behaviours that govern these choices are, to a large extent, under genetic control and therefore are evolved adaptations. To comprehend such adaptations, one must understand the consequences of rules adopted for making choices.

Heterogeneous Environment

Arguing over the maintenance of genetic variation within populations has been a traditional pasttime of population geneticists for decades. Until the mid-1960s, when biochemical techniques uncovered vast amounts of previously unknown genetic variation, obvious polymorphisms of structure and colour were thought to arise from the superior fitness of heterozygotes, as in the maintenance of the sickle-cell gene in human populations by the resistance of heterozygotes to malaria. But experiments with flies and other organisms had also shown that constant laboratory conditions revealed no such heterosis; rarely did two or more morphs or genotypes coexist in population cages. This led A. B. da Cunha and Theodosus Dobzhansky (1954), Richard Lewontin (1958), and others to suggest that polymorphism persisted only when different genotypes were selectively favoured in different parts of the environment or at different times.

Starting with these general concepts, Richard Levins (1962) set out to "attempt to explore systematically the relationship between environmental heterogeneity and fitness in populations." His theoretical approach took a giant step beyond contemporary bounds of empirical data and experimental result.

The heart of Levins's theory was fitness set analysis. A fitness set portrays phenotypes on a graph whose axes are components of fitness. One then superimposes a fitness criterion upon this graph to identify that phenotype having the greatest fitness. For simplicity. Levins restricted his analysis to the case of a population exposed to two different environments (1 and 2) for which the components of fitness are the respective fitnesses of each phenotype in each of the two environments. When all possible phenotypes are positioned on a graph according to their fitnesses in environment 1 (W_1) and environment 2 (W_2), the solid figure outlined is the fitness set. Only those phenotypes lying on the periphery of the figure are of interest because at least one of these will have a fitness superior to any phenotype within. But which point on the periphery represents the greatest fitness? When conditions are constant, and individuals experience only environment 1 or only environment 2, the selected phenotype will be the one having the greatest fitness in that environment.

To determine the most fit phenotype in a heterogeneous environment Levins devised the "adaptive function," a mathematical statement

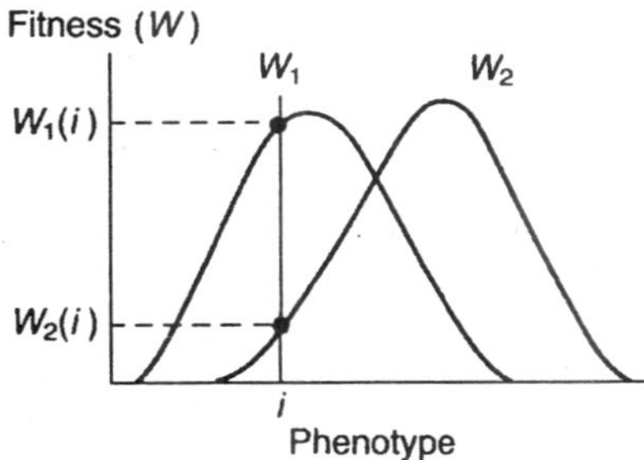

Fig. 12.1. Relationship between phenotype and fitness in two different environments (1 and 2). The fitnesses of a single phenotype (i) in the two environments (W_1 and W_2) are indicated.

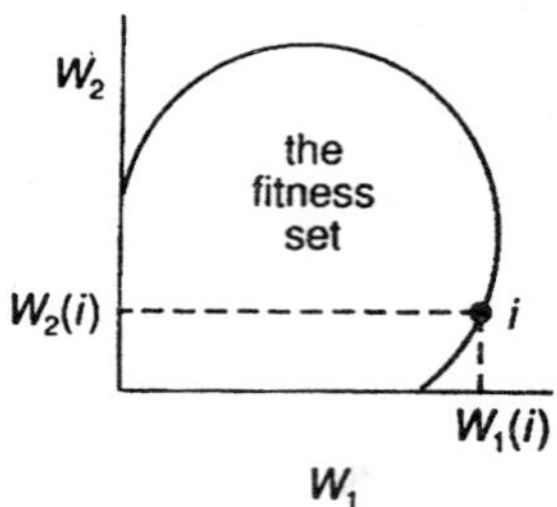

Fig. 12.2. The fitness set portrays the relationship between the fitnesses of all possible phenotypes in environments 1 and 2.

combining fitnesses under each of the conditions (for example, 1 and 2) into a measure of the overall fitness of a phenotype in a heterogeneous environment. When the environment varies spatially and organisms encounter habitat patches of type i in direct proportion to their frequency (p_i), the overall fitness of the phenotype (W) is the average of the fitnesses in each habitat patch weighted by their frequency. Thus, $W = p_1W_1 + p_2W_2$. By rearranging this expression, we obtain $W_2 = (W/p_2) - (p_1/p_2)W_1$, which is the equation of a straight line on a graph whose axes are W_1 and W_2, the same axes used to define the fitness set. This equation defines a family of lines having slope — p_1/p_2 and intercept, or distance from the origin of the graph, determined by the value W—the overall fitness. Each line represents the combinations of W_1 and W_2 resulting in the same overall fitness. These lines of equal fitness—or adaptive functions, as Levins called them—superimposed upon the fitness set identify the phenotype with

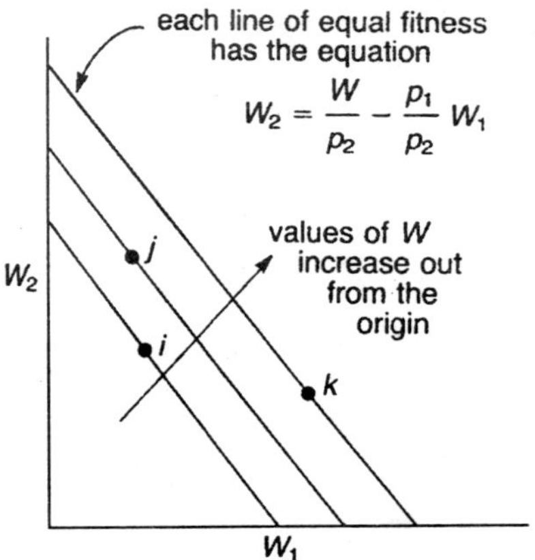

Fig. 12.3. Adaptive functions are lines of equal fitness (W) whose positions on the graph are determined by the relative proportions of environments 1 and 2.

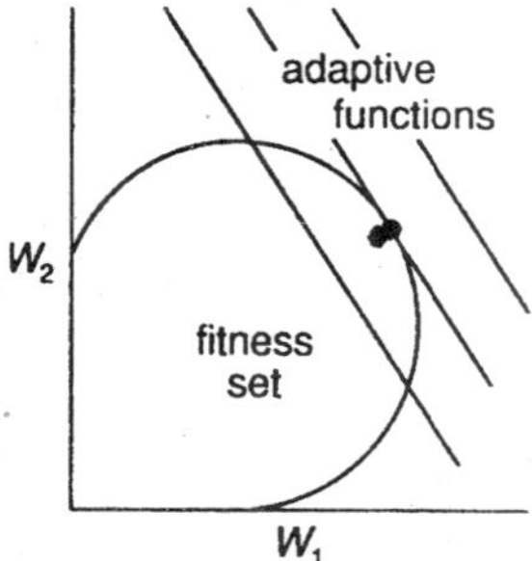

Fig. 12.4. The superimposition of adaptive functions on the fitness set reveals the most fit phenotype as that touched by the adaptive function of highest value.

the highest overall fitness as the point on the periphery of the fitness set touched by the adaptive function lying farthest from the origin of the graph, hence representing the highest value of W among all phenotypes. This adaptive function is a tangent to the fitness set.

As the ratio of habitat patches varies from place to place, the slope of the adaptive function ($-p_1/p_2$) varies and the tangent adaptive function touches the fitness set at different points. From this result, Levins suggested that along a dine of frequency in habitat patches one should expect to see a cline in selected phenotype.

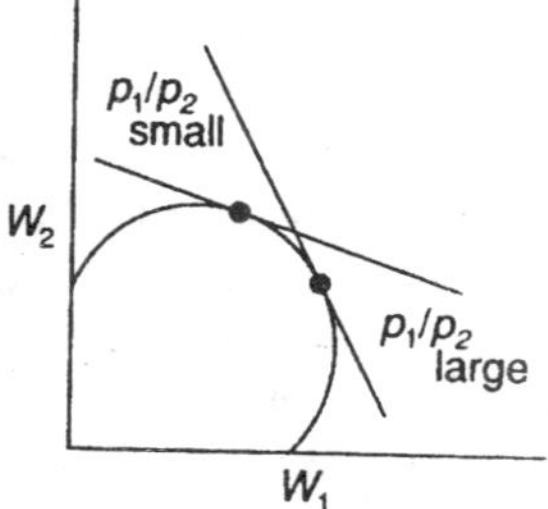

Fig. 12.5. As the slope of the adaptive function changes with changes in the proportion of environments 1 and 2, the most fit phenotype also changes.

The fitness was based on habitat patches sufficiently similar that the fitnesses of each phenotype in the two habitats did not differ greatly. Levins also explored the situation in which the habitat patches present markedly different selective conditions. In such cases, the fitness set has a depression in its center, which results in a different outcome of selection. Linear adaptive functions superimposed on this fitness set show that phenotypes lying along the inward bulging part of the fitness set's periphery cannot be selected. As the ratio of habitat patches changes, the phenotype with maximum fitness shifts abruptly from one

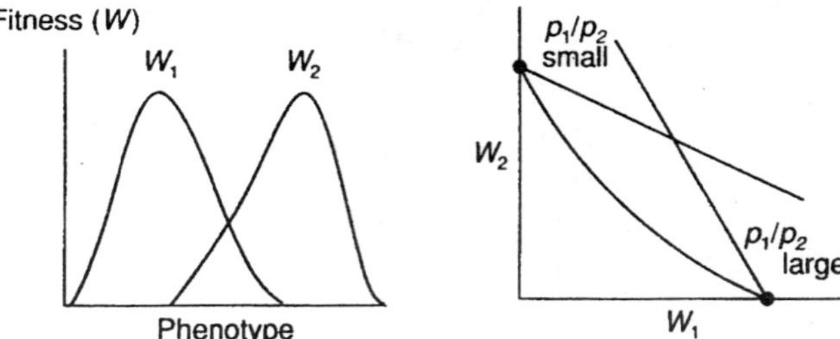

Fig. 12.6. When environments 1 and 2 differ greatly, the fitness set is concave and changes in the proportion of environments cause a stepwise shift in optimum phenotype.

lobe of the fitness set to the other, skipping over intermediates. Intuitively, when habitats differ greatly, phenotypes intermediate between those superior in each type of habitat are not well suited to either. This result gave the adage "jack of all trades, master of none" considerable currency in ecological thinking. It also led to an interesting prediction: along a gradual cline of frequency of habitat patches, one could expect to see an abrupt change in phenotype.

When the environment varies temporally, so that habitat patches are encountered in a time sequence, the adaptive function may be more appropriately calculated as a geometric mean, rather than an arithmetic mean. This is because fitnesses represent factorial increases in the numbers of phenotypes from generation to generation. When population growth rate varies over generations, the long-term expectation of population size is directly proportional to the product of the population growth rates experienced during each time segment. Therefore, in a temporally varying environment, the average fitness of a phenotype is $W = W_1^{p_1} W_2^{p_2}$, which can be rearranged to give the adaptive function $W_2 = (W/W_1^{p_1})^{1/p_2}$.

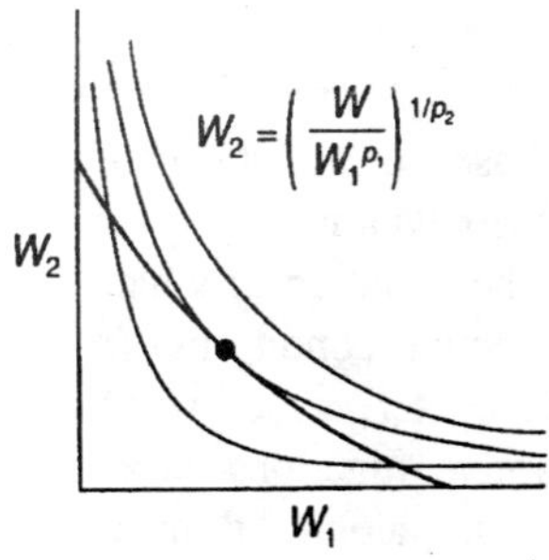

Fig. 12.7. When environments are encountered in sequence the adaptive functions are hyperbolas and may favour an intermediate phenotype even for a concave fitness set.

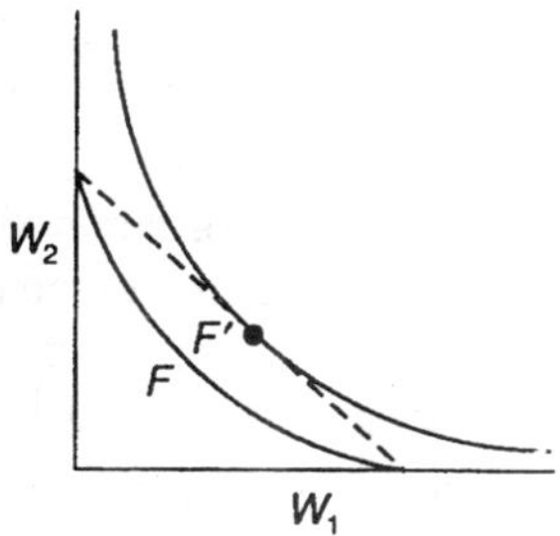

Fig. 12.8. Levins's concept of the extended fitness set (F′) as a mixture of different proportions of extreme phenotypes, which he thought could explain the maintenance of polymorphisms in populations.

At this point in the development of his theory. Levins abandoned logic and asserted the existence of an extended fitness set (F'), which he defined as "the set of all points in the fitness space that represent possible populations, that is, all possible mixtures of the phenotypes represented in the set F". According to Levins, selection favoured a mixture of genotypes represented by the tangent of the hyperbolic adaptive function on the extended fitness set. Thus he believed that he had achieved an explanation for polymorphism in populations, and further predicted that along a cline of frequency of temporally heterogeneous conditions, one should observe a cline in the frequency of two phenotypes (a morph-ratio cline).

Levins erred in thinking that in a temporally varying environment mixtures of two morphs assume the arithmetic mean of their component fitnesses. In fact, the "fitness" (that is, population growth rate) of a mixture of phenotypes is the average of the geometric mean of the component fitnesses of each phenotype. Hence Levine's extended fitness set does not exist. The absurdity of the concept can be appreciated most readily when the fitness set bows inward from one axis to the other. A mixture of two phenotypes, each well suited to a different type of condition but having zero fitness in the alternative condition, has zero fitness overall: neither of the phenotypes can persist in a temporally varying environment. Levins made this mistake because he ignored the mechanics of selection in his model.

Although he claimed to have done so. Levins failed to solve the problem of genetic polymorphism in the absence of heterozygote advantage. Other hypotheses similarly rely on temporal variation in the environment, but have proper genetic bases and explicitly incorporate the mechanics of selection. But in spite of the shortcomings of Levins's analysis it left a legacy of concepts, explored initially by Levins's collaborator Robert MacArthur. Among these is the concept of environmental grain, which we touched upon earlier in this book.

MacArthur and Levins (1964) were interested in the outcome of competition between populations, or of competition between genotypes within populations, in heterogeneous environments. From the beginning, it was clear that the outcome would depend on the degree to which organisms could choose among the different habitat patches:

> To make these ideas more precise, we first consider an imaginary habitat in which there is a scattering of uniform units or grains of resource 1 and another scattering of uniform grains of resource 2. In such an environment we can distinguish as "fine-grained" an individual or species which utilizes both resources in the proportion in which they can occur. (If the actual grain size of the resources were so fine the species could not discriminate and select, then the species would have to be "fine-grained," hence the terminology.) An individual or a species will be called "coarse-grained" if it discriminates and selected only grains of one of the resources.

Levins's earlier treatment of fitness in a heterogeneous environment assumed that patches of habitat or resource were fine-grained; that is, individuals confronted habitat patches in direct proportion to their frequency. When patches are coarse-grained, the individual may use both types of patches or specialize on one or the other. But when an individual can choose, what is the best strategy to follow? MacArthur and Levins argued that "normally, coarse-grained utilization [i.e., specialization] will be expected only where the time and energy lost due to neglecting the other possible resource is slight compared with the benefits of specialization."

Theory of Optimal Foraging

When an animal moves through a habitat in search of food, it sequentially encounters potential prey. With each encounter, the organism is confronted with the choice of pursuing and eating the prey, which requires time and the expenditure of energy, and results in either the acquisition of food energy or the passing up of potential prey in favour of continuing to search. Intuitively, when a predator meets potential prey infrequently compared to the time it requires to subdue and consume them, it should eat all prey types. But when a predator encounters prey often, it should pass by less desirable types because it will soon find better ones.

MacArthur and Eric Pianka (1966) formalized these arguments in a graphical model, which employed a cost-benefit analysis to determine the number of kinds of prey that a predator should include in its diet. They assumed that the best diet selection minimized the average time required to find and consume an individual prey. To simplify the model, we shall consider only the case in which prey species have identical nutritive value but vary in abundance and predator escape tactics. Increasing diet breadth by adding a new kind of prey affects the rate of consumption of prey in two ways: first, a broader diet avails the predator of more potential prey individuals; opposing this benefit, however, the average ease of pursuit, capture, and consumption decreases, provided that the predator always includes prey species in its diet in descending order of suitability. A predator should broaden its diet until the decrease in average quality of its prey more than offsets any decrease in searching time due to greater abundance of prey included within the diet.

The average search time (T_s) and pursuit and handling time (T_p) per prey are graphed as a function of diet breadth, with prey ranked according to their suitability. The optimum diet breadth is that which results in the lowest sum of search and pursuit time ($T_s + T_p$). The slope of the search-time curve becomes less steep with increasing diet breadth because each new prey species adds proportionately less to the total diet. The slope of the pursuit-time curve increases because as the suitability of the prey decreases the capture and handling time increase more rapidly. Because of the shapes of the T_s and T_p curves, the sum of the two usually assumes a U-shaped curve, the lowest point of which defines the optimum diet breadth.

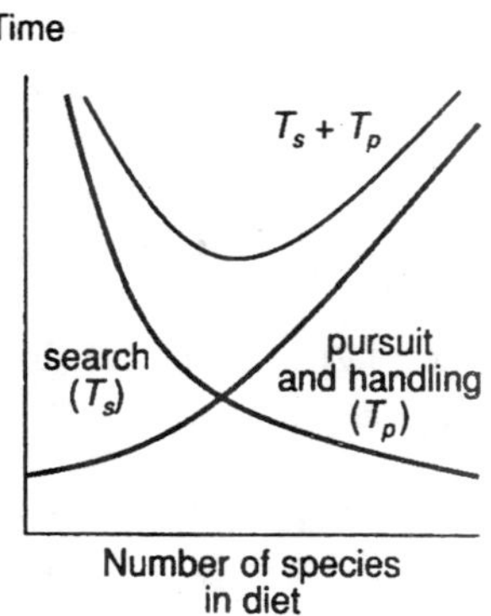

Fig. 12.9. A graphical model of the optimization of diet breadth by a predator.

A change in the overall abundance of prey, resulting from a change in the productivity of the habitat or the number of competing species,

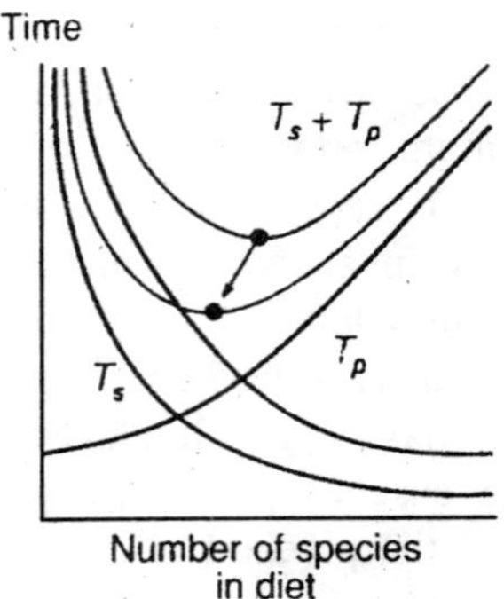

Fig. 12.10. When availability of prey increases, due either to increased habitat production or reduced competition, searching time (T_s) decreases and the optimum diet breadth narrows.

will change search time, but not pursuit and handling time; only the quantity of prey vary, not their quality. When productivity increases, the T_s curve decreases and the optimum diet breadth shifts to the left. Thus increased productivity favours specialization. When competition increases, prey become more scarce and the T_s curve increases, shifting the optimum to the right and favouring increased generalization.

How does prey diversity affect diet breadth? When the number of potential prey species increases without increasing their overall variety or total production, the T_s curve increases because each species becomes proportionately less abundant, the T_p curve decreases somewhat because the number of easily caught prey species increases, and optimum diet breadth shifts to the right. But although the diet includes more species of prey, the proportion of potential prey eaten and the variety of prey, assessed by their predator-escape characteristics, do not change. When the number of potential prey species increases by increasing the variety of species (adding species more difficult to capture), only the search time curve increases (each prey becomes less abundant), and although diet breadth increases slightly, proportion of total prey eaten as well as rate of prey capture decreases. In general, then, increased production leads to diet specialization and increased rate of consumption of prey. Increased competition is equivalent to reduced production and has an opposite effect. Increased variety of prey species may lead to a somewhat broader food niche, but reduced rate of prey consumption, all other things being equal.

Profitability of Individual Prey

Optimization of choice by predators has been explored theoretically by a number of authors. These treatments agree with MacArthur and

Pianka's (1966) prediction that increased production of resources favours increased diet specialization. Mathematically, the problem of optimal diet has been characterized by the "classical" model of prey choice. The exposition here follows that of John Krebs and N. B. Davies (1981), who, for the sake of simplicity, considered a predator hunting two types of prey (1 and 2, again). The pertinent characteristics of the prey are encounter rate (λ), reward (E), handling time (h), and profitability (E/h). A predator searches until it encounters the first prey (i). It then pursues, subdues, and consumes that prey with handling time h_i. During this handling period, the predator cannot search and therefore passes up any other prey that it might have encountered had it continued searching. When prey i is consumed, the predator once again starts searching.

When two prey are sought, the total encounter rate of prey is the sum of the encounter rates of each individually ($\lambda_1 + \lambda_2$). The total number of prey encountered in a series of search intervals of total time T_s is equal to $T_s(\lambda_1 + \lambda_2)$, and the expected reward obtained from these prey is

$$E = T_S(\lambda_1 E_1 + \lambda_2 E_2)$$

The total handling time associated with search time T_s is $T_s(\lambda_1 h_1 + \lambda_2 h_2)$, and therefore the total time required to obtain reward E is

$$T = T_S + T_S(\lambda_1 h_1 + \lambda_2 h_2)$$

the sum of the search and the handling times. Now, the amount of reward received per unit of time when the diet includes prey items 1 and 2 is therefore

$$\frac{E(1,2)}{T(1,2)} = \frac{\lambda_1 E_1 + \lambda_2 E_2}{1 + \lambda_1 h_1 + \lambda_2 h_2}$$

The reward per unit time for selecting only one of the two prey types is, similarly,

$$\frac{E(1)}{T(1)} = \frac{\lambda_1 E_1}{1 + \lambda_1 h_1}$$

Supposing that prey type 1 provides more reward per unit time than type 2 [that is, $E(1)/T(1) > E(2)/T(2)$], the predator should specialize on prey type 1 rather than include both 1 and 2 in its diet when $E(1)/T(1) > E(1,2)/T(1,2)$, or

$$\frac{1}{\lambda_1} < \left(\frac{E_1}{E_2} h_1 - h_1\right)$$

In words, the more specialized diet results in a higher feeding rate when the time to encounter the next item of type 1 ($1/\lambda_1$) is generally less than the difference in handling times ($h_2 - h_1$). Because h_2 is adjusted by the ratio of rewards from items of type 1 and type 2, the greater the reward from prey type 1 relative to that of type 2, the more likely specialization will be favoured.

Krebs et al. (1977) attempted to test the "classical" foraging model by putting great tits (*Parus major*) through an ingenious prey choice situation. Pieces of meal worms of two sizes were placed on a conveyor belt that passed under an opening, allowing the caged subject access to the "prey." Reward (E) was related to the size of the prey, which was under the experimenters' control, handling time (h) was measured, and encounter rate (λ_1) could be varied. According to the theory, when the benefit of selecting the single, most profitable prey type [$E(1)/T(1) - E(1,2)/T(1,2)$] exceeded zero, the subjects should have switched from picking prey items off the belt at random to selecting only the larger pieces. The results of the experiment conformed generally to prediction but, contrary to expectation, prey type 1 was never selected to the exclusion of prey type 2.

Krebs and R. H. McCleery (1984) discussed many possible explanations for deviations of the experimental results from theory, but their bottom line was that the theory tremendously oversimplifies both the food resource and the behavioural response of an astute predator to that resource. Certainly errors in discriminating prey types, long-term learning of preferences for one type or another, short-term learning

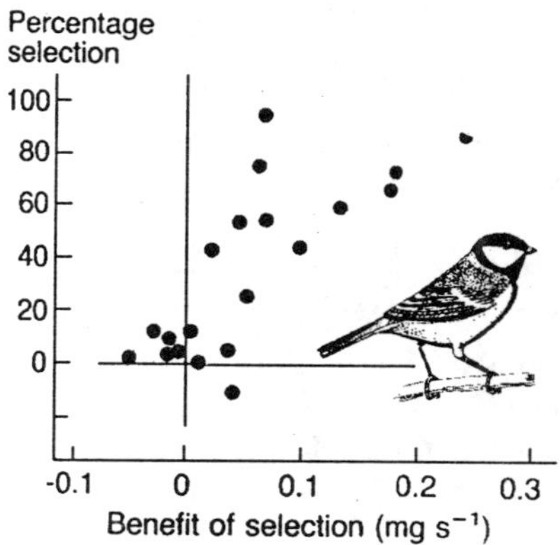

Fig. 12.11. A test of the "classical" optimal diet model with great tits showing the relationship between preference for a more profitable prey and the benefit of specializing on that prey.

during runs of one or the other prey type, simultaneous encounter of more than one prey, and non-substitutability of prey because each provides a different essential requirement add complications that are difficult to incorporate into simple models. Further considerations include the reduction of risks associated with variable prey availability and spatial distributions of resources with respect to nesting, perching, or roosting sites.

Optimal Patch

Most prey occur in patches of high abundance separated by unsuitable habitat. Predators must travel between patches, and so they are confronted with choices concerning prey use within patches and must decide when to leave a patch to search elsewhere. As in optimal prey choice, discussed above, this problem has received considerable attention from theoreticians and has been investigated experimentally in some laboratory systems. Most models of the patch-choice problem derive from the idea that the quality of a patch decreases as the predator captures prey and reduces the level of resource within. Eventually, the predator receives greater reward by moving to a new patch than by continuing to search in a depleted one. The amount of time in a patch before leaving is called the giving-up time.

As a predator initially searches within a patch, it encounters and consumes food at a high rate. As it depletes the resource, its return per unit of time decreases and the total gain from the patch begins to level off. The rate of gain decreases to zero as the resource is exhausted. During a period of search (T_s), the average rate of gain from a patch [$F(T_s)$ is the amount of food consumed [$P(T_s)$] divided by the sum of the search time and the time required to travel between patches (T_t). Thus $F(T_s) = P(T_s)/ (T_s + T_t)$. Assuming that T_t is fixed

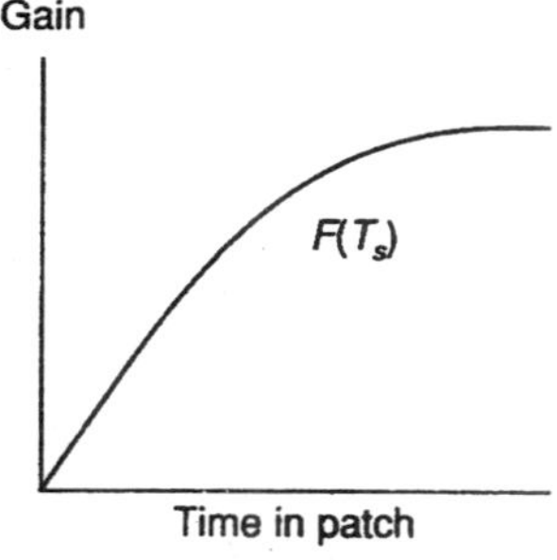

Fig. 12.12. Cumulative amount of resource consumed as a function of amount of time in a patch of resource.

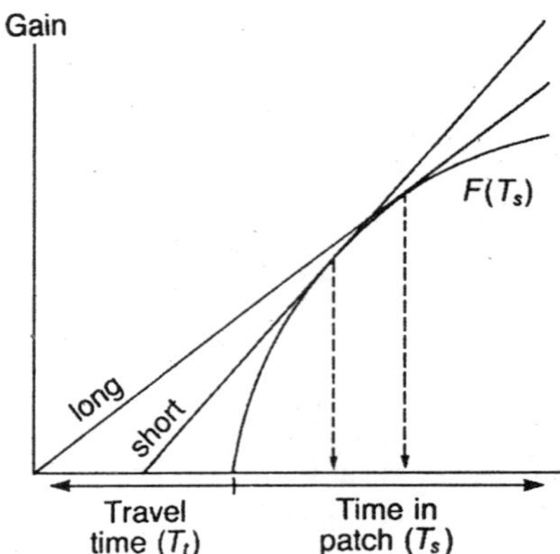

Fig. 12.13. Graphical portrayal of the influence of travel time between patches on optimal giving-up time for a consumer within a patch.

according to the spatial distribution of patches, the optimum giving up time (T_s^*) can be found by taking the derivative of this expression and solving to find the maximum (that is, $dF/dT_s = 0$). A major prediction of this model is that optimal giving-up time decreases as travel time between patches decreases. Intuition tells us that when a new patch with high resource level can be reached quickly, one cannot gain by staying long in a patch of diminishing quality.

The graphical result can be given some quantitative substance by specifying the predation function $P(T_s)$. Let N be the total quantity of resources in a patch; a is a constant representing predation efficiency. Now, when the predation function is, for example, $P(T_s) = aNT_s/(1 + aT_s)$, the shape of which one can solve for $T_s^* = \sqrt{(T_t / a)}$. This result leads to three conclusions: optimal giving-up time (T_s^*) is (1) independent of the total amount of resource per patch (N); (2) decreases as the rate of prey depletion (predation efficiency, a) increases; and (3) increases as T_t increases.

Tests of the theory of optimal giving-up time have been hard to come by. They rely on matching the behaviour of a predator to quantitative predictions obtained from specific formulations of the model. In order to make the comparison, one must be able to measure the predation function $P(T_s)$ and the travel time between patches (T_t) directly. This is possible only in contrived laboratory situations, among which parasitoid-host systems and birds feeding on mealworms have proved especially suitable. Although results often conform to the predictions of models, predators in such situations may be performing more like small biological analogues of simplistic models than revealing the richness of their behaviour in the natural world. Model systems do, however, provide insights into the behavioural capacities of organisms (perception, learning, memory, flexibility) relevant to making choices among the variety of opportunities presented by the environment. Perhaps it is naive to think that behaviour patterns, evolved in richly complex natural settings, will function "optimally" in simplified, unnatural laboratory experiments. But it is too early to disavow this approach as misleading, particularly inasmuch as it has already greatly enlarged the ecologist's concept of the organism-environment relationship.

Free Distribution

Models of optimal choice discussed to this point address the behaviour of a single individual confronted with environmental heterogeneity. Frequently, however, many individuals face the same options and optimal behaviour may be influenced by the decisions of other members of the population. For example, a patch becomes less attractive to newcomers as the number of predators exploiting it increases. Such a patch likely has fewer remaining resources relative to undiscovered patches; furthermore, competing individuals may precipitate costly behavioural conflicts. Presented with many patches of varying intrinsic quality, consumers having complete knowledge and freedom of choice should occupy or exploit patches in direct proportion to their quality. Those with the highest levels of resources should attract the most consumers, those with the lowest levels the fewest.

Each consumer bases its choice of a patch on criteria that maximize its own rate of gain of resources. Imagine two patches, one having more resources than the second. At first, consumers will choose the intrinsically better patch. But as the population of consumers builds up in that patch, its apparent quality decreases, owing to depletion of resources and antagonistic interactions, until the second patch becomes the better choice. At this point, additional consumers choose the second and first patch alternately as the quality of both continues to decrease. As a result, each individual in the population will exploit a patch of equal realized quality, regardless of the variation in intrinsic patch quality in the absence of consumers. This is called the "ideal free distribution," a term used by Stephen Fretwell and H. L. Lucas (1970) to describe the filling of different habitat patches by territorial species of birds.

The existence of ideal free distributions of individuals in nature is suggested by several lines of evidence, that for habitat selection by birds being among the most compelling. In migratory species, individuals arrive on the breeding grounds over a period of several weeks. Early arrivals generally fill certain habitats that confer high breeding success before newcomers begin to establish territories in poorer habitats. When individuals are removed from "good" habitats, their places are quickly taken by others moving in from "poor" habitats. When population densities are low, perhaps following a particularly severe winter, occupancy of poor habitats decreases more than that of good habitats. Usually, however, breeding success in poor habitats does not match that in good habitats, thus contravening one expectation of an ideal free distribution. This discrepancy may reflect unequal competitive ability within a population based on age or social dominance; dominant individuals may be able to monopolize resources by defending large territories and set an upper limit to population density below the density attained under an ideal free distribution.

The concept of the ideal free distribution has been applied to several phenomena, perhaps most successfully by G. A. Parker (1974, 1978), who examined the number of male dungflies (*Scatophaga*) that compete for mates on cowpats (also known as meadow muffins and cow pies) of different size, hence quality and attractiveness to females. The distribution of males among these patches of "habitat" approximates an ideal free distribution, in which males gain similar numbers of matings regardless of patch quality.

The achievement of an ideal free distribution has also been investigated in the laboratory, where the quality of patches can be controlled. M. Milinski (1979) conducted experiments in which six stickleback fish were provided with food (water fleas) at different rates at opposite ends of an aquarium. Each end could be regarded as a patch, and the system had the following conditions conducive to establishing an ideal free distribution: (1) the two patches differed in profitability; (2) profitability decreased as the number of fish using a patch increased; (3) the fish were free to move between patches, hence they had free choice.

Hungry fish were placed in the aquarium about 3 hours before the start of the experiment. During trials, the numbers of fish in each half of the tank were recorded at the end of each 20-second interval. Prior to adding any food to the tank the fish were distributed equally between the two halves. In one experiment, water fleas were added at a rate of 30 per minute to one end of the tank and 6 per minute to the other, a ratio of 5 to 1. Within 5 minutes, the fish had distributed themselves between the two halves in the same ratio as predicted for an ideal free distribution. In a second experiment, water fleas were provided at rates of 30 and 15 per minute, a 2 to 1 ratio. Again the distribution of fish followed suit, and when the better and poorer patches were reversed in the tank, the fish reversed their distribution within about 5 minutes. Behavioural mechanisms used to achieve an ideal free distribution were not determined, but cues for behavioural choices must incorporate both the rate of food provisioning and the number of competitors within the patch. Such experiments demonstrate considerable sensitivity of organisms to conditions of their environment as well as behavioural flexibility in making choices.

13

Sex and Adaptation

The study of natural selection and adaptation is made all the more interesting by sex—by the fact that most animals and plants reproduce sexually, that is. We measure fitness in terms of number of descendants, and because each sexually produced descendant arises from the union of one female and one male gamete, individuals can gain fitness through either male or female function. Many populations, including that of humans, consist of a mixture of individuals that are either male or female. Others consist of a single type of individual that develops both male and female function. In both cases, individuals must divide their resources between male and female functions. With regard to unisexual individuals we may ask, What is the optimum ratio of males to females in their progeny? With regard to bisexual individuals, What is the optimum allocation of resources between male and female functions?

The presence of male and female functions within a single population gives rise to three additional considerations. The first is choosing among potential mates having different genotypes, made necessary by genetic variation within the population. The second is the mating system; that is, the number of other individuals that one mates with, and how one organizes these arrangements and manages them behaviourally. The third is the selection of traits in one sex by choices exercised by the second, referred to as sexual selection. In this chapter, we shall examine these consequences of sexuality for selection and adaptation beginning with the conditions under which sexual function should be combined in each organism or separated between males and females.

At the outset, we shall accept sex and the distinct qualities of maleness and femaleness as general properties of present-day life. The origin of sexual reproduction, whose initial purpose presumably was associated with advantages conferred by genetic recombination, has been lost in the dim evolutionary past. Sex came along early in the evolution of life, judging from its near ubiquity among eukaryotic organisms. The maintenance of sexual reproduction in many groups of organisms, as opposed to adopting parthenogenetic reproduction as some animals and plants have, is also something of a mystery, with no single hypothesis being particularly compelling.

Sex is the union of two haploid gametes to form a diploid cell. No law of nature states that the gametes must be different (that is, male and female), and indeed in such single-celled organisms as yeast, protozoa, and some green algae the gametes are identical (a condition known as isogamy) and "sexes" are distinguished only as genetically defined mating types. The evolution of maleness and femaleness—basically the divergence of big (female) and little (male) gametes—followed upon the initial establishment of sexual reproduction. Conditions for the evolution of anisogamy are discussed by Jeffrey A. Parker et al. (1972) and Parker (1978). Judging by the absence of isogamy in multicellular organisms, and the extreme specialization of the female gamete, which provides nourishment for the developing embryo, and the male gamete—a set of chromosomes with a way of getting around—anisogamy must be powerfully selected. But it is at this point, with the assumptions of sex and male/female function behind us, that we shall begin our inquiry.

Sexual Separation

As with many issues, botanists and zoologists have divergent views about sex. This is not to say that one or the other has more fun, but just that their perspectives differ. In most species of plants, individuals have both male and female function, leading plant ecologists to ask. What conditions favour separation of the sexes? In most animal species, male and female functions are separated between individuals, and animal ecologists ask. What conditions favour the joining of both sexual functions in a single individual? These questions are, of course, opposite sides of a common coin. Before attempting to answer them, however, some definitions.

The condition of separate male and female individuals is referred to as gonochoristic when applied to animals, from the Greek *gonos* pertaining to procreation (for example, gonad), and *choris*, apart or separate. Botanists refer to the same condition as dioecious, from the Greek *di-* (two) and *oikos* (dwelling).

When both sexes are together, the individual is referred to as an hermaphrodite, after Hermaphroditus, son of Hermes and Aphrodite, who while bathing became joined in one body with a nymph. (The circumstances of this happy union are not important here.) Hermaphrodites may be simultaneous, as in the case of many snails and most worms, or they may be sequential (that is, first one sex, then the other). When male function is followed by female function, as in some molluscs and echinoderms, the individual is said to be protandrous (Greek *andros*, male). When female function comes first, as in some fishes, the individual is protogynous (Greek *gyne*, woman). Botanists apply an additional term, monoecious, to the condition of an individual plant bearing separate male and female flowers, rather than so-called perfect flowers with both male and female parts. While perfect-flowered hermaphrodites are the rule among plants (72 per cent of species, by one estimate), almost all imaginable combinations of sexual patterns are known, including populations with hermaphrodites and either male or female plants, populations with male, female, and monoecious plants, and hermaphrodite individuals with perfect flowers and which also bear either male or female flowers.

Situations favouring hermaphroditism have been investigated theoretically by specifying the conditions under which an hermaphrodite strategy can invade a population consisting of separate males and females, and, conversely, when a population consisting of hermaphrodites can resist invasion by either males, females, or both. The approach, that of determining the *Evolutionarily Stable Strategy* (ESS), was outlined by Eric A. Charnov et al. (1976) and elaborated by Charnov (1982). We shall consider only the simplest situation here, that of simultaneous hermaphroditism. Suppose that males, females, and self-incompatible hermaphrodites occur in a population with frequencies m, f, and h, respectively ($m + f + h = 1$). Each male contributes to future generations through N matings, and each female through n seeds. Hermaphrodites contribute through both male and female functions, at a level aN and bn, respectively. The proportion of male contribution to progeny from each male is $W(m) = 1/(m + ah)$, and the proportion of female contribution from each female is $W(f) = 1/(f + bh)$. For each hermaphrodite, the combined contribution through male and female function is $W(h) = a/(m + ah) + b/(f + bh)$.

An hermaphrodite can invade a population of males and females only when hermaphrodites are more fit than either males or females; because males and females have equal fitness in an outcrossing

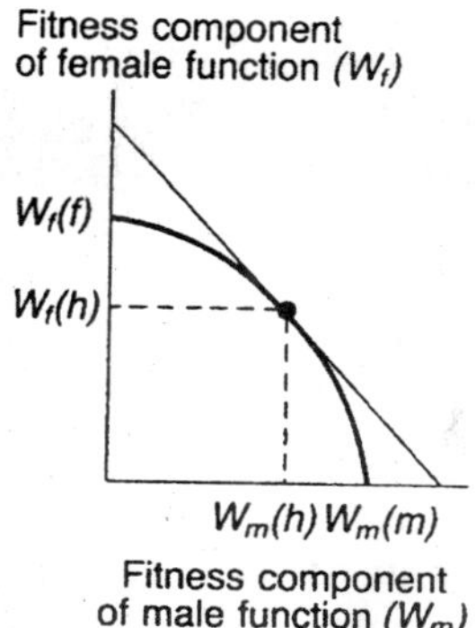

Fig. 13.1. Convex fitness set for allocation to male function between 0 and 1.

population $W(h)$ must exceed $W(f)$. A little algebra gives the necessary inequality $a(f + bh) + b(f + ah) > m + ah$. When hermaphrodites are rare and h is therefore close to 0, and when males and females are equally abundant in the population ($m = f = 0.5$), the inequality reduces to $a + b > 1$. This inequality can be shown to be a sufficient condition to prevent the invasion of a population of hermaphrodites by either males or females. Therefore, hermaphroditism is an ESS when the sum of the combined male and female contributions exceeds the sum of the contributions from each sex separately.

Charnov's analysis showed that when a female can achieve a certain amount of male function by giving up a smaller amount of her female function, she (subsequently it or they) should do so. Similarly, males should add female function when it does not cut deeply into their male productivity. For the flowers of female plants to produce a little bit of pollen, or increase the rate of pollen transfer, seemingly few resources would have to be transferred from female to male function.

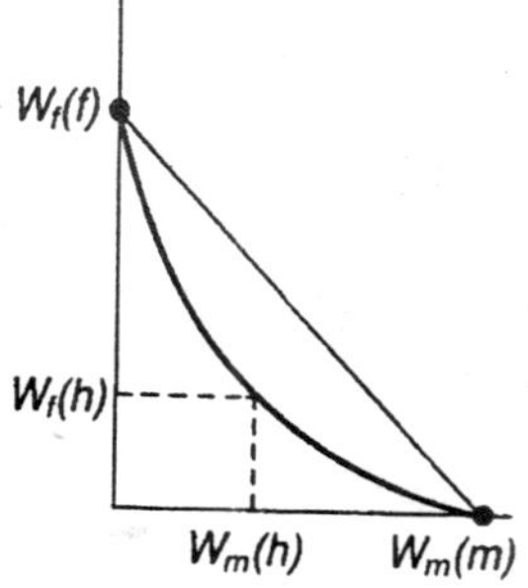

Fig. 13.2. When the fitness set for allocation between male and female function is concave, extreme phenotype are selected and hermaphrodites are excluded from the population.

And so the strategy should be adopted frequently, as it is among plants. For hermaphrodites to be excluded from a population, the fitness set must bulge inward so that gains in adding function of one sex are more than offset by losses in function of the other. This may occur when the establishment of new sexual function in an individual requires a substantial fixed cost before *any* gametes can be produced. Sexual function requires gonads, ducts, and other structures for transmitting gametes, secondary sexual characteristics for attracting mates and for competition among individuals within sexes. In many animals, where maleness requires specializations for mate attraction and antagonistic interaction with other males, or where femaleness requires specializations for egg production or brood care, fixed costs may be high and hermaphroditism disadvantaged compared to sexual specialization. In fact, hermaphroditism is rare among active animal species and those with brood care; it is particularly common among sedentary aquatic species that shed gametes into the water.

Some populations include hermaphrodites and individuals of one sex, usually females. Fitness set analysis shows that this situation may arise when female function is gained at large cost in male function. For females to invade a population of hermaphrodites, $W(f)$ must exceed $W(h)$ when $m = f = 0$ and $h = 1$. The condition for this is $b < 0.5$. It can be shown that the fitness of both females and hermaphrodites is frequency dependent and that an ESS is achieved when the proportion of females is $f = h(1 - b)$.

ESS and fitness set analysis elucidate general guidelines for thinking about sex allocation, but the method of inquiry is so new that little has yet been done to measure the shape of fitness sets relating male and female function, establish values of a and b, and verify basic

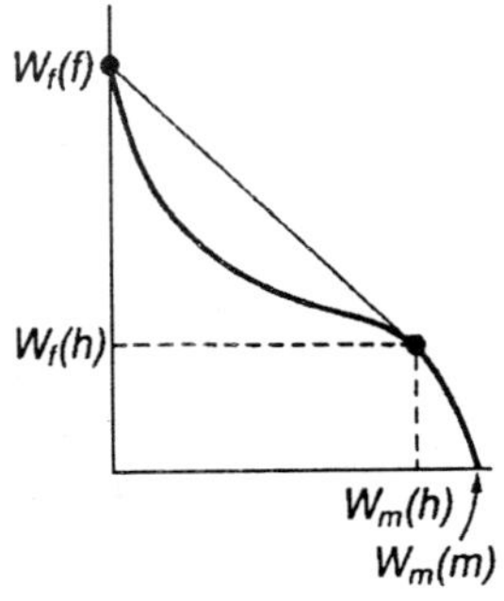

Fig. 13.3. Fitness set representation of the conditions required to maintain a mixed population of females and hermaphrodites.

assumptions of the theory, such as degree of self-compatibility and level of inbreeding, in nature. So, why are so many species of plants hermaphroditic? Charnov (1982) outlined eight possibilities that are associated either with self-fertilization or the interaction of male and female function in the same plants:

> (1) Hermaphroditism allows self-fertilization which is adjusted to an intermediate optimum. (2) It allows facultative selfing. (3) Pollen is a primary attractant for pollinating agents; a female would simply not be visited. (4) Pollen limits seed set and the visit of a pollinating agent achieves gains for both genders. Both genders share the flower resource, or cost of attraction, which makes the fitness set convex [bulge outward]. (5) Temporal displacement of availability of resource for the two sex functions means that female resource is simply not available for male function. (6) Seed set is resource limited; pollen (the male gain curve) fitness saturates because of limitations in mate availability . . . or saturation of pollen vectors. (7) Seed set is resource limited, but limited fruit dispersal results in sib competition and thus a law of diminishing returns through seeds. (8) Male and female function may depend upon (be limited by) different resources (e.g., protein versus carbohydrate).

In sum, low overlap in resource requirements of male and female function, and "cost-sharing" in which sexual organs, such as flowers, can serve both male and female function, tend to produce an outward-bulging fitness set and favour hermaphroditism.

Sequential hermaphroditism reflects changing costs and benefits of male and female sexual function as an organism grows. In some marine gastropods having internal fertilization, such as the slipper shell *Crepidula*, insemination requires the production of only small amounts of sperm. Hence male function requires few resources and has little effect on somatic growth. As a consequence, many such species are protandrous hermaphrodites, being male when small and female when they are large and can produce correspondingly large clutches of eggs. When male competition for mates is important, large size can be an advantage. This apparently has led to protogynous hermaphroditism in some species of fish that inhabit reefs, where males compete among themselves for breeding territories.

Optimal Sex Ratio

Two facts are remarkable about sex ratio. First, so many populations have very nearly equal numbers of males and females. Second, there are so many exceptions to this rule. Because of its prevalence, symmetry, and application to the human population, the even, or 1: 1, sex ratio is considered the standard condition, deviations being special cases.

R.A. Fisher (1930) summed up the theoretical basis of sex-ratio theory when he observed that every product of a sexual union has exactly one mother and one father. One consequence of this truism is that individuals of the rarer sex in the population will enjoy greater fitness because they compete with fewer others of the same sex for matings. For example, when a population of 5 males and 10 females produces 100 offspring, each male contributes 20 sets of genes but each female contributes only 10 sets of genes. With such a skewed sex ratio, any parental genotype giving rise to a larger proportion of male offspring would be favoured and the frequency of males in the population would increase. Similarly, if females were the rarer sex, genotypes that increased the proportion of female progeny would be favoured, and the frequency of females would increase in the population. Fitnesses are balanced, and there is no selective pressure to alter the sex ratio, when males and females occur with equal frequency, in which case individuals of both sexes contribute equally to future generations and the frequencies of males and females among the progeny of an individual are of no consequence to its fitness.

With respect to selection upon the sex ratio, the relevant measure of fitness is the number of grandprogeny, not progeny. This is because a gene that influences the sex ratio expresses itself in the proportion of males or females among the progeny, not the total number of progeny, of the individual bearing that gene; its consequences are not felt until the progeny have passed their genes onto the following generation.

Mathematical development of this theory of sex ratio also provides a quantitative basis for understanding selection of deviations from the even sex ratio. Consider a population in which the frequencies of males and females among zygotes (the primary sex ratio) are M and F, respectively ($M + F = 1$). Because all individuals have one father and one mother, the fitness contribution of an individual through its male offspring is proportional to $1/M$ and its fitness contribution through female offspring is proportional to $1/F$. Now suppose that a mutant appears whose offspring are male and female in the frequencies m and f ($m + f = 1$). Provided that mutant and typical individuals produce the same number of offspring, the fitness of the mutant is proportional to $(m/M) + (f/F)$; scaled in the same way, the fitness of the typical genotype is proportional to $(M/M) + (F/F) = 2$. Now, the difference in fitness (D) between the mutant and typical individual is $D(m,f) = (m/M) + (f/F) - 2$, which, with a little rearranging, is

$$D = \frac{(m-M)(F-M)}{MF} \qquad \text{...(1)}$$

We see that D is equal to 0 (evolutionary equilibrium) either when $m = M$, which is uninteresting because there is no genetic variation in fitness, or when $M = F$ (an even sex ratio). The condition $M = F$ represents a stable equilibrium; when the frequency of males exceeds that of females in the population ($F - M < 0$), a mutant that produces a lower frequency of male offspring than the population average ($m - M < 0$) will increase [$D(m,f) > 0$]. The survival of the zygote to reproductive maturity does not bear on the evolution of sex ratio in this case because the fitness value of the zygote is unaffected. That is, if fewer males survive than females, each has an enhanced fitness value as an adult, but fewer make it to that stage. The total contribution of each sex to future populations is identical.

Sex Ratio

Under some circumstances an individual male or female offspring requires more resources to produce than an individual of the other sex. This may pertain especially when sexual dimorphism in size results in different parental investment in individual offspring of each sex. Suppose that males cost some fraction (a) of females to produce such that $F + aM = 1$ and $F = 1 - aM$. Now equation (1) becomes $D(m,f) = (m - M)(F - aM)/MF$ and $D(m,f) = 0$ when $F = aM$, or $F/M = a$. Thus if males are the more expensive sex ($a > 1$), the equilibrium sex ratio will favour females. Furthermore, the total expenditure on the sexes will be equal, as indicated by the expression $F = aM$.

Differential Mortality

Suppose that males survive less well than females, by fraction s, to the termination of parental care at time T. If resources not delivered to deceased males could instead be invested in females, then equilibrium sex ratio should be adjusted in favour of the more poorly surviving sex. When such transfer of resources, referred to as compensation, applies, $F + sM = 1$ and $D(m,f) = 0$ when $F = sM$, or $F/M = s$. Therefore, when males survive less well than females ($s < 1$), the equilibrium sex ratio favours males. Furthermore, at the termination of parental care (time *T*), the sex ratio should be even, owing to the greater mortality of the more prevalently produced sex.

Differential costs and differential early survival of male and female offspring should lead to deviations in sex ratios of populations from 1 :1. Yet surprisingly few data for vertebrates suggest either substantial deviations in the sex ratio or genetic variation among individuals in the sex ratio of their offspring. Indeed, the situation is so uniform that George Williams (1979) considered the 1:1 sex ratio a trivial consequence of the sex-determining mechanism. In most vertebrates, one pair of chromosomes (the sex chromosomes) for which the population is polymorphic (for example, X and Y) controls an individual's sex. When an individual is homozygous for one of the chromosomes it is either male or female, depending on the type of organism; when heterozygous, it is the other sex. Thus, in humans, XX homozygotes are female and XY heterozygotes are male. Because matings occur between XX and XY individuals, the genotypes of the offspring are half XX and half XY—half male and half female. The heterogametic sex may be either male (mammals) or female (birds and butterflies), but the sex-determining mechanism will produce equal numbers of males and females in either case.

Charnov (1982) objected to Williams's "nonselection" thinking in this case on two grounds. First, the occurrence of sex chromosomes may signal the predominant selective advantage of the even sex ratio. That is, rather than sex ratio being the inevitable consequence of a particular sex-determining mechanism, the mechanism itself may have evolved to its prevalance because it produced the most fit ratio of the sexes among progeny. After all, animals employ many other sex-determining mechanisms, including temperature-sensitive sex ratios in many reptiles. Moreover, even in the case of XY sex determination, meiotic drive and sperm competition could produce unbalanced sex ratios, should they be selected.

Charnov's second objection to Williams' s thinking is that the existence of sex chromosomes may make it difficult to evolve toward some alternative system of sex determination. Thus even when selection favours an unbalanced sex ratio, there may be no genetic variation in sex ratio to work on. This does not invalidate selection thinking about sex ratio, but rather illustrates how the genetic system might limit evolution.

Breeding Condition

In many species, competition among individuals of one sex (usually males) for matings leads to tremendous variance in fitness contributions among individuals. Presumably, where competition is keen, some males achieve many matings, others none. In harem-forming species, for example, a few males control most of the females and others lower in social dominance have little access to mates. Generally contests among males are won by the largest individuals. Robert Trivers and D. E. Willard (1973) suggest that if the condition of male progeny is directly influenced by the condition of the mother, then females in good condition should produce male offspring, which will grow to large size and fare well in male-male competition for mates; females in poor condition should place their investment in female offspring, which will mate successfully regardless of the parental care they are given. Certainly in mammals, in which the female cares directly for her offspring through the periods of gestation and lactation, the condition of the mother will likely influence that of her offspring.

Experimental confirmation of Trivers and Willard's idea has emerged from an experimental laboratory study on reproducing female wood rats (*Neotoma floridana*). In this species, as in most mammals, females normally invest equally in male and female offspring. But when Polly Ann McClure restricted food drastically during the first 3 weeks of the lactation period to below the maintenance level of a nonreproductive female, male offspring were selectively starved (mothers actively rejected their attempts to nurse) and the sex ratio at 3 weeks was altered to about 0.5 males for each female.

Facultative Control

The hymenoptera (bees, ants, and wasps) have an unusual sex-determining mechanism by which fertilized eggs produce females and unfertilized eggs produce males. As a result, females are diploid and males are haploid, giving rise to the term "haplodiploid" sex-determining mechanism. Reproductive females can control the sex of their offspring simply by fertilizing eggs or not. The haplodiploid system raises questions about its evolutionary origin and adaptive significance, but also offers the hope of testing certain aspects of sex-ratio theory because sex ratio is under the direct control of the female, presumably through mechanisms that have genetic variation.

Haplodiploidy has evolved many times from more usual diploid sex-determining systems. Two steps are required. First, genes must arise in the maternal genome that enable the female to suppress the fertilization of eggs; these eggs also must spontaneously initiate development without being fertilized and develop into males. Second, genes must arise that suppress the development of maleness in diploid individuals. The first step confers tremendous advantage to the genes responsible for it. The haploid males contain only genes from the mother, including the genes responsible for haploidy. These are transmitted to the next generation at twice the frequency of maternal genes transmitted through sexually produced offspring, in which half of the genome comes from the father. Therefore, if uniparental males make at least half the fitness contribution as biparental males, then genes favouring haplodiploidy can invade a population. Why isn't the uniparental sex female? Presumably this would lead to strict parthenogenesis with haploid clones of asexually reproducing females. In spite of initial advantages in the transmission of genes to future generations, such clones might die out owing to the lack of sexual reproduction and the reduced genetic potential of haploids compared to diploids (heterozygotes are not possible).

Many wasps are parasitoids on other insects or complete their larval development within the fruits of certain plants. For some of these species, hosts are so scarce and mates so difficult to find that females mate on the host on which they grew up before dispersing to find new hosts on which to lay their own eggs. When a host is parasitized by a single female wasp, females within each brood are limited to mating with their brothers. Under this circumstance, because the number of grandprogeny of a female is directly proportional to the number of female offspring that disperse to search out new hosts, male offspring have reduced fitness value. One might therefore expect a reduced proportion of males in each brood. This is, when males compete only with their brothers for the opportunity to mate, one male offspring is as good as several from the standpoint of their mother, Thus it is not surprising that most species of parasitoid wasps have sex ratios skewed greatly in favour of females. Several produce only one male per brood. Such males are often wingless and, in extreme cases, males fertilize females as larvae within the host, or, as in viviparous pyemotid mites, within the mother. In the last case, males become sexually functional as larvae and never develop into adult forms. These observations confirm the idea that when local mate competition occurs only among sibs, the fitness value of males is greatly reduced and, where possible, mothers limit the number of males among their progeny. In normally outcrossing haplodiploid species, a sex ratio closer to 1:1 is the rule.

In many parasitoid species, hosts are superparasitized, meaning that more than one female may lay her eggs in the same host. Also, in most cases, ovipositing wasps can use chemical cues to determine whether eggs have been previously laid by another female or not. When a female attacks a "virgin" host, and if few hosts are superparasitized, her daughters will likely engage in sib-mating because their brothers frequently will be the only males around. When a female attacks a host that has been parasitized recently, her sons can compete to mate with the daughters of the first parasitoid female, and their value will be considerably enhanced. One might expect superparasite females to produce more male offspring per brood than the first female to lay on a particular host. And this is, in fact, frequently the case. Moreover, the proportion of sons in the progeny of the second female ought to increase as the ratio of the number of offspring of the first female to the number of her own offspring increases, because opportunities for outcrossing increase in direct proportion to this ratio.

Jack Werren (1980) established conditions for testing this idea in laboratory experiments with the wasp *Nasonia*, which parasitizes larvae of the fly *Sarcophaga*. The results conformed to the predictions of William D. Hamilton's quantitative model of male and female contributions to fitness. The progeny of primary parasites were 9 per cent male. The proportion of males produced by secondarily parasitic

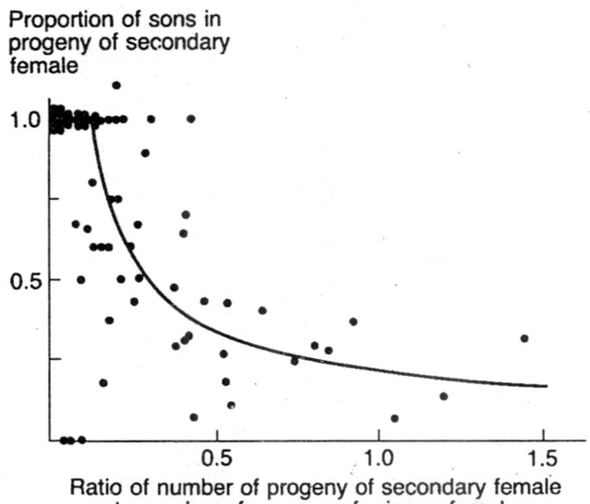

Fig. 13.4. Relationship between the sex ratio in progeny of secondary females of the parasitoid Nasonia and the probability that their offspring will mate with the offspring of different females.

females varied from a similarly low level when their own offspring predominated in the host to 100 per cent males when their offspring were few compared to those of the primary parasites.

Haplodiploid wasps also provide an opportunity to test Trivers and Willard's (1973) idea about controlling the sex of offspring in relation to their expected fitness contributions in a variable environment. In many parasitoids, larger hosts result in the production of larger female offspring with greater fecundity than the smaller females that emerge from smaller hosts. Males presumably are less disadvantaged by developing to small size within a small host because male sexual function is less dependent on body size than is female sexual function. In his 1979 paper on sexual selection, Charnov quoted the results of an experiment by the Russian Ivan Chewyreuv, published in 1913, on the ichneumonid wasp *Pimpla instigator*, which lays a single parasitic egg on the pupae of moths and butterflies. When Chewyreuv presented females with large hosts, all of 23 offspring produced were female; when he presented them with small hosts, 38 out of 47 offspring were male. Large and small hosts could be alternated, in which case the female would either fertilize the egg or not as appropriate. In these laboratory experiments, pupae of different species of moth were used to present hosts of different sizes. But Chewyreuv also showed that when the wasp *Exenterus* parasitized cocoons of the moth *Lophyrus* in nature, the larger host pupae tended to produce a preponderance of females. In *Lophyrus*, males are only half the size of females. Seventy-nine per cent of the wasps emerging from the larger, female pupae were females, compared to only 47 per cent of those emerging from the smaller, male pupae.

Werren's and Chewyreuv's results show that when females can control the sex of their progeny, they do produce more of the sex with the greater fitness contribution. This contribution depends on the expected fecundity of progeny of each sex, which depends in turn on resources for egg production in female wasps and availability of potential outcrossed matings for males.

Mating Systems

The theory of sex allocation discussed above deals with the distribution of resources by a parent to male and female offspring. Adults also are selected to obtain as many matings of as high quality as possible. It is a basic asymmetry of life that the fitness contribution of a female is limited by her ability to make eggs and otherwise provide for her offspring, while the fitness contribution of a male is usually determined by the number of matings that he can procure. Because the female gamete is larger than the male gamete, it requires more resource to produce and a female's fecundity is likely to be limited by her ability to gather resources to make eggs. When males contribute no resources to their mate, as in some species they can do by defending territories for the females to feed in or by feeding the female directly, or when males do not care for their young, a male can increase its own fecundity only by mating with additional females.

When a male mates with as many females as he can locate and properly persuade, and provides his offspring nothing more than a set of genes, he is said to be promiscuous. Among animal taxa as a whole, promiscuous mating is by far the commonest system. Promiscuity generally precludes a lasting pair bond. When adopted by a population, it also tends to increase variance in mating success among males, some individuals obtaining perhaps dozens of matings while others get none. Over the population as a whole, the number of matings must average out to one per period of female receptivity.

When males can contribute to the fecundity of their mate by procuring resources for the female or caring for the offspring directly, pair bonds may outlast the copulatory rapture of promiscuous species. As males increasingly contribute to the realized fecundity of a single mating, mating systems progress from promiscuity (least care), through polygamy and serial polygamy, to strict monogamy (most care, relative to the female). Polygamy describes the situation in which a single individual of one sex forms long-term pair bonds with more than one individual of the opposite sex. Normally males are mated to more than one female, in which case the system is referred to as polygyny (literally, many females). Polygyny may be expressed through the defense of several females against the mating attempts of other males (a harem), or the defense of territory or nesting sites to which more than one female are attracted to breed as well as mate. Thus polygyny may arise through the ability of a male to control matings by defending females, in which the contribution of the male to its progeny may be primarily genetic, or through the ability of a male to control or provide resources necessary to the female to reproduce.

In serial polygamy, an individual of one sex forms a durable bond with one individual of the opposite sex, but eventually abandons its mate, leaving it to care for their offspring, while it goes off to seek a new mate. There is never more than one pair bond formed at a time in such systems, but one sex always takes the primary responsibility for caring for the offspring of a mating. Most commonly it is the male that abandons the female to seek greener pastures (serial polygyny), but the reverse situation (serial polyandry) is also known.

Monogamy refers to the formation of a pair bond between one male and one female that often persists through the period required to rear the offspring of a mating, and which may last until one of the pair dies. Monogamy arises primarily in situations in which males can make a large contribution to the number and survival of offspring. Hence it is most common in species with prolonged dependence of offspring, in which both sexes can provide for the young. Monogamy is not common in mammals because providing milk is a specialized task of the female. But it is common among birds, especially those in which parents feed their offspring, a task of which both sexes are equally capable.

Habitat and Diet

Because the habits of birds have been so keenly scrutinized by biologists, and because birds encompass the full range of mating systems and express these diverse interactions between the sexes in a spectacular variety of structures and behaviours, it is not surprising that studies of birds have provided the seed from which theory concerning mating systems has grown. An early and crucial step toward interpreting mating systems as evolved adaptations was recognizing associations between mating systems of populations and their habitats, habits, and diets.

The English behaviourist John Crook (1964, 1965) was among the first to recognize that mating system and ecology went hand in hand. Surveying species of African weaverbirds, he noted that monogamy predominated among insectivorous species inhabiting forest and savanna, whereas polygyny was the rule among seed-eating species of open savannas and grasslands. Furthermore, most of the monogamous species bred as solitary pairs on widely dispersed territories while polygynous species were invariably colonial. In a broader study of North American songbirds, Jered Verner and Mary Willson (1966, 1969) found a similarly strong association of polygyny with grassland, prairie, and marsh habitats. But simple correlations between variables cannot explain how these associations arose. Empirical patterns, such as those noted by Crook and Verner and Willson, may, however, suggest hypotheses or models that embody a plausible mechanism relating selective factors in the environment to evolved behaviours.

Verner (1964) was the first to propose a model for the origin of polygynous mating systems. His reasoning was summarized by Verner and Willson (1966):

> It seems clear that polygyny would be advantageous to a male whenever the total number of successful offspring (those that survive to reproduce) from all his females exceeds the number that would be reared from one nest if he mated monogamously. Among females also, selection will favour those leaving the greatest number of successful offspring, whether these are reared in monogamous or polygamous associations. Consequently, in those species in which females select mates from among available males, there will be selection against polygyny unless this results in greater reproductive success for the females as well as for the males. Verner (1964) contends that polygyny can be advantageous for females if, within the limited area from which a female is likely to select a mate, the difference between two males' territories is sufficient that a female is able to rear more offspring on the better territory, by herself, than she could rear on the poorer one even with full assistance from the male. This difference between territories can be regarded as a "polygyny threshold," since it is likely that polygyny will be favoured by natural selection whenever the difference exceeds a certain level. The parameters of a male's territory that might operate in this regard include all requisites of successful breeding (food, cover, space, nest sites, etc.).

Gordon Orians, a behavioural ecologist at the University of Washington, with whom both Verner and Willson worked as students, devised a simple graphical representation of the process of mate selection according to the Verner-Willson polygyny threshold model. The model contrasts the fitness of a female mated to a monogamous male with that of a female mated to a polygynous male as a function of the quality of the male's territory. Several predictions can be made from this model. First, polygyny will occur only when the quality of territories varies among males so much that some females will have higher fitness mated to a polygynous male on a territory of high quality than they would mated monogamously to a male (and receiving his undivided attention) on a territory of poor quality. This is Verner's polygyny threshold of territory quality. Second, if females can assess the quality of territories, the first individuals to pair should mate monogamously, latecomers choosing mates with territories of progressively lower quality until the polygyny threshold is reached. At this point, females should be ambivalent about pairing up with unmated or previously mated males.

According to the polygyny threshold hypothesis, polygynous males should occupy territories, or otherwise defend resources, of intrinsically

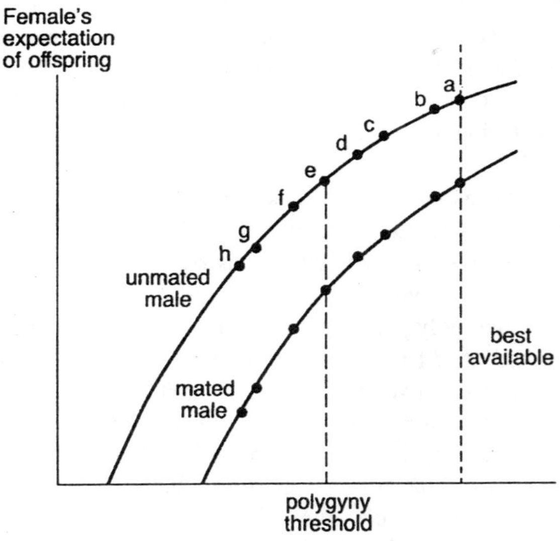

Fig. 13.5. The polygyny threshold of Verner and Wilson.

higher quality than those of monogamous or unmated males. Several studies have attempted to measure territory quality and evaluate the polygyny threshold model, the most direct being those of Sarah Lenington (1980) on red-winged blackbirds and W. K. Pleszczynska (1978) on lark buntings. Red-winged blackbirds frequently breed in cattail marshes where, in the early spring, males establish territories, among which females choose to breed. Males accept as many females into their territories as will mate with them; 2 or 3 are common, but up to 12 have been observed. Because the sex ratio of blackbirds is even, many males obtain no mates. Females of polygynous males defend small territories of their own against other females within the male's space.

To a female, the quality of a territory depends on characteristics that contribute to the number of offspring she can rear: those that influence the amount of food she can gather for her brood (males do not feed the young) and that determine the safety of her nest from storms and predators. In her study, Lenington was able to relate both components of a female's productivity to physical characteristics of the territory: number of young fledged per successful nest (an index to feeding conditions) was correlated with territory size and the amount of edge of cattail habitat; safety of the nest was related to the density of the cattails, thus the effectiveness of the vegetation as cover.

Using these attributes of territories, Lenington ranked males according to the number of offspring that a single female could anticipate. Although Lenington expected the first females to arrive on the breeding grounds to choose males with territories of the highest "quality," observations at a marsh near Princeton, New Jersey, failed to show any particular order with respect to her criteria. But when only food-related aspects of territory quality were considered, and the quality of potential nest sites was ignored, the order of choice corresponded much more closely to territory quality. A second prediction, that secondary females should choose only the highest quality territories, still was not supported. A third prediction, that the reproductive success of females on a territory of a given quality should be inversely related to number of females, was, however, supported.

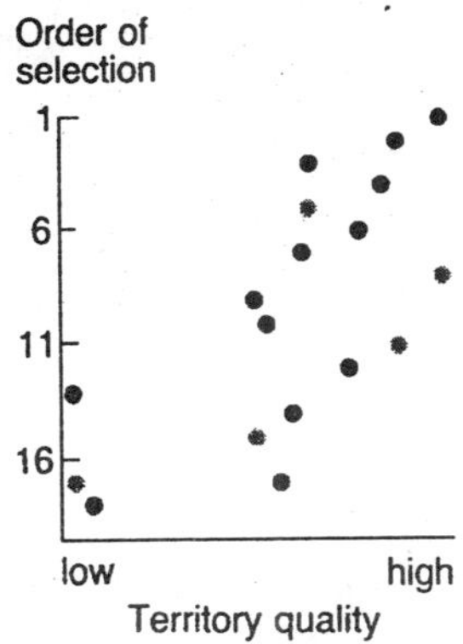

Fig. 13.6. Order of selection of males as a function of territory quality by primary females (black symbols) and secondary females (gray symbols) in a marsh-nesting.

Pleszczynska (1978) studied a small songbird, the lark bunting, nesting in a 4-hectare alfalfa field in South Dakota. The buntings nest somewhat colonially in high densities. Males are polygynous. Broods of primary females receive care from their fathers; those of secondary females do not. Because most food for the offspring is gathered from areas outside the territory, the quality of a territory is related to characteristics of potential nest sites, primarily the density of vegetation. Indeed, vegetation cover (assessed by light levels at nest sites) was strongly correlated with nest success.

As in red-winged blackbirds, male lark buntings establish territories before females arrive from the wintering grounds. In Pleszczynska's study area, females settled into the alfalfa field over a 9-day period in the spring. The order in which they settled territories of different "quality" corresponded almost precisely to Orians's model. Females arriving over the first 4 days mated monogamously with males on high-quality territories; those settling on days 5 and 6 tended to become secondary females; those arriving on days 7 through 9 mated usually monogamously with males on territories of lower quality.

According to Orians's model of mate selection, once the polygyny threshold has been reached the choice between becoming a secondary female on a territory of higher intrinsic quality or a primary female on a territory of lower intrinsic quality is ambivalent. And, indeed, Pleszczynska found that primary and secondary females that choose mates on the same days had nearly identical nesting success, about 0.6 young fledged per egg laid.

Pleszczynska investigated the role of vegetation cover in nest success experimentally by placing plastic leaves over nest sites to simulate denser cover. Over 3 years of such manipulations, nest success increased to between 65 and 76 per cent from control levels of between 42 and 52 per cent. Confident that female lark buntings could judge nesting success, therefore territory quality, by vegetation cover, Pleszczynska predicted the eventual mating status of males in locations in Colorado and in North and South Dakota by this criterion, proving to be correct in 52 out of 58 cases. Finally, Pleszczynska and R. Hansell (1980) experimentally altered territory quality in the alfalfa patch by either stripping alfalfa plants of leaves in areas of high cover or adding plastic leaves to alfalfa plants where cover was sparse. The distribution of females settling into the patch was consistent with the altered territory values, even to the point of creating some cases of trigamy where plastic leaves were added to enhance territory quality. These experiments lend considerable force to the idea that polygyny can arise where females perceive large differences in quality between males, which is most likely to occur in heterogeneous habitats.

Promiscuous Mating System

Although promiscuous mating is the rule in the animal and plant kingdoms, it surfaces rarely and sporadically among birds, being predominant only among grouse and a few groups of tropical, frugivorous species—the manakins, cotingas, and birds-of-paradise. Representatives of all these groups have communal mating areas, called leks, in which congregated males perform elaborate displays to attract females to mate with them. Because such mating systems are so unusual among birds, evolutionary ecologists have paid considerable attention to the surfacing of promiscuity on the calm sea of avian monogamy.

The males of most species of birds contribute substantially to the fecundity of their mates (and therefore to their own), either through defending territories and their included resources or through directly providing for the young. The key to promiscuity has, therefore, been perceived as the emancipation of the male from parental investment without substantially reducing the fecundity of his mates. This may be possible in one of several situations: first, the young of precocial birds feed themselves and so fecundity may be limited by factors other than the provisioning of food by the parents. Females of such species, including the promiscuously mating grouse and their relatives, may be able to rear nearly as many young by themselves as they could with the help of males. Second, in mammals, the female is specialized to nurse the young, and potential male contributions are limited to defending resources, protecting the litter, and providing food directly to the female, none of which elevates the potential contribution of the male to near that of the female. Of course, this explanation begs the question of biparental nursing in mammals. Third, resources may be conspicuous and of fixed quantity, in which case a female alone may be able to gather as much as a male and female working together. This explanation may apply to fruit-eating species.

Emancipation of Parents

The evolution of promiscuous mating from a monogamous system depends on the emancipation of the male from caring for the offspring directly or indirectly. In a few birds, and more commonly among other organisms, such as fish, it is the female that abandons the eggs or young to the care of the male parent. The choice between caring for the young or deserting them resides with each individual parent, but the consequences depend in large part on the choice made by its mate. Frequently one parent may abstain from caring for the offspring with little reduction in fitness so long as the other parent remains faithful to the clutch or brood; but when both parents abandon their progeny few or none survive. It is generally assumed that by reducing care for one set of offspring, a parent can go on to lay more eggs or obtain more matings. The evolutionarily stable strategy for the two parents can be investigated by game theory analysis.

Each parent may either care for the offspring of a mating, which Smith refers to as guarding, or may desert them. The male may obtain an additional mating with probability p if he guards the brood and P if he deserts. It is assumed that $P > p$. The number of eggs that a female can produce is V if she deserts her clutch and v if she remains to guard ($V > v$). The probability of survival of the eggs is S_0 when neither parent guards, S_1 when one parent guards, and S_2 when both parents guard.

Both parents guarding is an Evolutionarily Stable Strategy when the strategy of desertion by neither males nor females can invade the population. That is, biparental care is an ESS when $vS_2 > VS_1$ for females and $S_2(1 + p) > S_1(1 + P)$ for males. In words, these conditions require that remaining to care for the offspring increases fitness more than abandoning them increases the number of matings for both parents; that is $S_2/S_1 > V/v$ and $(1 + P)/(1 + p)$. The ESS conditions for all four combinations of behaviour reveal that when biparental care is an ESS, uniparental care cannot be; also, when biparental abandonment is an ESS, uniparental abandonment (== uniparental care) cannot be. Both biparental care and biparental abandonment may be ESSs under the special conditions $S_2/S_1 > V/v > S_1/S_0$ and $S_2/S_1 > (1 + P)/(1 + p) > S_1/S_0$. Because neither type of uniparental care is an ESS under these conditions, the evolutionary step between biparental and uniparental care may be impossible to achieve, even if the alternative strategy conferred greater fitness on both males and females. When these inequalities are reversed, both female and male uniparental care can be Evolutionarily Stable Strategies and the outcome of evolution is ambivalent.

Unambiguous female uniparental care occurs when the ESS for the female is to guard no matter what the male does—that is, $vS_2 > VS_1$ and $vS_1 > VS_0$—and for the male to desert so long as the female cares, $S_1(1 + P) > S_2(1 + p)$. These inequalities can be rearranged to give the conditions S_2/S_1 and $S_1/S_0 > V/v$ for females and $S_2/S_1 < (1 + P)/(1 + p)$ for males. Quite simply, the female's gain must be greater through caring for offspring than with leaving eggs either unattended or in the care of a male; the male must be better off deserting when his mate remains with the brood. The conditions for male uniparental care are the reverse.

In birds, the ratio S_1/S_0 is almost always very large; in no species do both parents abandon the eggs, although the newly hatched chicks of megapodes are free of all parental care. Female uniparental care is quite common, being the situation in promiscuous mating systems, but male uniparental care is known only in rheas, the mallee fowl (a megapode), and a few polyandrous shorebirds, in which the female lays a set of eggs that are incubated by her first mate while she lays a second set of eggs fertilized by a second male. The predominance of female uniparental care is undoubtedly due to the fact that $(1 + P)/(1 + p)$ is usually much larger than V/v; when either sex abandons the clutch, males can gain additional matings more rapidly than females can produce more eggs.

Under the ambivalent conditions where either parent's abandoning the clutch is an ESS, which one stays and which leaves depends on the relationship between fertilization and egg-laying. When fertilization is internal, the male's mating function is completed before the eggs are laid and the male is free to leave the female—holding her bag of eggs, so to speak. When fertilization is external, the eggs are laid first and the male must subsequently fertilize them. Smith (1978) summarized data of C. N. Breder and D. E. Rosen (1966) showing a strong relationship between external fertilization and male uniparental care in fish.

A final consideration pertinent to male uniparental care is the certainty of paternity. When fertilization is external and eggs are laid in a nest prepared and guarded by the male, the male can be certain that the offspring are his own progeny and that subsequent care is likely to enhance the survival of his own genes. When fertilization is internal and females mate at random within the population, males cannot know for certain that they have fertilized eggs laid by a particular female, unless they guard the female throughout her receptive period.

Selection of Mating

There is more than one way to skin a cat—or for a male to obtain matings. If the fitness of two alternative tactics increases as their frequency in the population decreases, such frequency-dependent selection may lead to stable polymorphism, or mixed Evolutionarily Stable Strategy. At the evolutionary equilibrium, selective pressures favouring one tactic and the other exactly balance and individuals expressing either one have the same fitness. Several cases of alternative mating tactics within the same population seem to exist. Although the degree to which individuals (usually males) differ genetically often is not known, K. D. Kallman et al. (1973) have described a sex-linked genetic locus in platyfish *Xiphophorus maculatus* that controls the age (hence size) of maturation of males. This undoubtedly influences mating tactics, as shown in a similar case of the bluegill sunfish *Lepomis macrochirus* by Gross and Charnov (1980). Two situations with more extensive behavioural observations involve dispersal to find mates in fig wasps and an unusual lek-forming bird, the ruff.

Fig wasps lay their eggs in the flowers of figs. This is a mutualism in which the adult female effects fertilization of the flowers, but the larvae consume some of the fig's developing seeds. William Hamilton (1979) has described the mating behaviour of 18 species of fig wasps associated with 2 species of figs in tropical American forests. In the most abundant species, several females frequently lay their eggs in a single fig flower. Males developing from these broods are both flightless and have enormous mandibles and heads (for a wasp) with which they fight other males for matings within the same fruit. Evidently the usual presence of unrelated females in the same fruit selects strongly for males to compete among themselves to mate locally. Females, of course, must fly to find flowers in which to lay their eggs.

In the rarest species, males tend to be winged and to disperse to find mates, perhaps to avoid sister mating. It may be less difficult for such species to find mates than the males of parasitoid species because females are attracted to rather obvious resources, the fig flowers. Some of the less common species are, however, polymorphic in that some of the males developing within broods have wings and disperse to find mates while others are wingless and remain within the fruit of their birth to mate.

The ruff (*Philomachus pugnax*, literally a quarrelsome lover of combat) derives its English name from the distinctive collar of elongated feathers ringing the necks of the males; it derives its scientific name from the intense contests among males on their leks. This system of mating behaviour, described by A. 1. Hogan-Warburg (1966) and I. G. Van Rhijn (1973), is unusual among the group of wading birds to which the ruff belongs and is further distinguished by polymorphism in appearance and behaviour among the males. Males are of two types: independents, which establish territories on the mating arena, and satellites, which do not attempt to establish their own courting spot but are tolerated by independent males. In appearance, independent males have a predominantly dark plumage, with much variation among individuals; satellite males are predominantly white. Males having intermediate plumage may be either satellites or independents. The polymorphism almost certainly is genetic, although there is no direct evidence; the behaviour and plumage may be pleiotropically linked through the effects of a few genes on the levels of certain hormones that affect both traits. Observations show that both independent and satellite males mate with females attracted to the lek. Independent males may tolerate satellites on their territories to enhance the attractiveness of their mating court to females.

In each case of dimorphism in appearance and mating behaviour of males, the two forms would appear to balance conflicting selective pressures. Independent ruffs may obtain more matings than satellite males, but they assume greater risk of injury through combat and of being taken by predators. W. Cade (1979) described a situation in field crickets (*Gryllus*) in which calling males are more likely to attract females than are silent males, but they are also more likely to be parasitized by a fly that is attracted to the mating call.

Sexual Selection

Sexual selection is the situation in which one sex determines the fitness of traits expressed in the other by exercising choice in mating. The usual result of sexual selection is strong sexual dimorphism, especially of ornamentation, colouration, and courtship behaviour. Darwin, in his book *Descent of Man and Selection in Relation to Sex* (1871), was the first to propose that sexual dimorphism could be explained by selection applied differentially to one sex. Evolutionary ecologists pretty much agree that sexual dimorphism can arise in three different ways. First, the different sexual roles of males and females may place each in a different relationship to the environment, causing differential selection and response. For example, because females produce large gametes fecundity often is directly related to body size; this may provide a basis for the larger size of females in many species. In addition, the special nutritional requirements for egg production, and for protecting the eggs and young, which usually falls upon the female, may lead to females using the environment differently than males, thereby bringing different selective factors upon themselves. Simply having to find suitable nest sites may require females to utilize different habitats than males during the nesting season.

Second, sexual dimorphism may also arise through contests between males for opportunities to mate with females. Such contests may select elaborate weapons for combat, such as the antlers of deer and the horns of the mountain sheep. Although some authors include male-male competition in their definitions of sexual selection, it is more properly distinguished as intrasexual selection. Because evolution following upon intrasexual competition presents no conceptual difficulties, in spite of its prevalence it will be given little attention here.

Third, sexual dimorphism may arise through intersexual selection; that is, the direct exercise of choice among individuals of the opposite sex based upon their appearance and behaviour. With few exceptions, females do the choosing, and males respond with magnificent displays of vainglorious courtship ritual. Why females choose, and males compete among themselves for the opportunity to mate, derives from the general asymmetry of parental investment that defines the male and female conditions. Males enhance their fecundity in direct proportion to the number of matings they obtain; females are limited in number of offspring by the number of eggs they can produce, but they stand to gain in quality of offspring by choosing to mate with males bearing superior genotypes.

The goal of female choice presumably is to recognize and favour males with particular genetic constitutions. The dilemma presented by sexual dimorphism is that many of the traits selected, such as bright colouration, long feathers, and elaborate courtship behaviour, would seem to put males at great risk. How can females select traits that appear to reduce the fitness of males among their progeny? Darwin was aware of this problem, but not until the mid-1970s did the beginnings of a resolution appear.

R. A. Fisher (1930) was the first to provide a detailed explanation for the sexual selection of male adornment, by a mechanism he termed "runaway sexual selection." The elements of Fisher's model progressed logically as follows: (1) variation among males in a fitness-related trait made some individuals more desirable as mates than others; (2) females that perceived this difference among males and selected mates accordingly had higher fitness than nonselective females; (3) persistent female choice for extreme values of the male trait under selection leads to continued male response and eventually to the bizarre courtship antics of, say, the birds-of-paradise, to pick a conspicuous example. Fisher's model includes the origin of both female choice and male sexual-selected traits as adaptive modifications. Also, the "runaway" behaviour of the model can be achieved only if female choice is based on comparison of a trait among males, rather than upon some absolute ideal. In the latter case, selection would stop when male adaptations coincided with the ideal. In the former case, sources of new variation would always provide a superior male by comparison with others in the population, just as, in artificial selection programs, breeders continually up the ante even as the population pays through selective deaths to stay in the game.

The difficulty with Fisher's model has persistently been that female choice must initially be based on traits that intrinsically confer advantage to males in the absence of female choice. It is difficult to imagine how some sexually selected traits could have originated in this way, considering the ridiculous extremes to which they have been carried. Models designed to explore the process of sexual selection in more detail have been presented by Trivers (1972), Halliday (1978), and, incorporating specific genetic bases for selected traits, by O'Donald (1980), Lande (1980, 1981), Kirkpatrick (1982, 1987), and Ten Cate and Bateson (1988). These theoretical investigations generally confirm that runaway sexual selection is feasible under a nonrestrictive range of conditions, but they fail to address the origin of female choice. Once the system gets going, male advantage can be sustained when female choice more than compensates any increased encumbrance endured by favoured males.

Female choice is certainly a fact of life, experienced at some level by most males. Richard D. Howard (1978) demonstrated that female bullfrogs strongly preferred to mate with larger males. Male bullfrogs are territorial, and females may perceive that the larger males are able to secure larger territories in male-male competition, or perhaps the overall genetic quality of an individual is reflected in its body size.

A particularly compelling demonstration of female choice was the experimental study of Malte Andersson (1982) on tail length of male long-tailed widowbirds (*Euplectes progne*). This polygynous species inhabits open grasslands of central Africa. Females, about the size of a sparrow, are mottled brown, short-tailed, altogether ordinary in appearance. The males are black, except for a red shoulder patch, and sport a half-meter-long tail conspicuously displayed in courtship flights. Males may attract a half-dozen females to nest in their territories, but they provide no care for their offspring. Tremendous variance in male reproductive success provides the classic conditions for sexual selection. Andersson's experiment was simple and straightforward. He cut the tail feathers of some males to shorten them, and glued the clipped feathers onto the ends of other males' tails to lengthen them, and observed the subsequent success of males in attracting females to their territories. Controls were unclipped males and males whose tails were cut and glued back into place.

Andersson found that length of tail had no effect on a male's ability to maintain a territory, and fitness apparently is not influenced by the effect of tail length on contests between males. But, strikingly, males with experimentally elongated tails attracted significantly more mates than those with shortened or unaltered tails. This result strongly

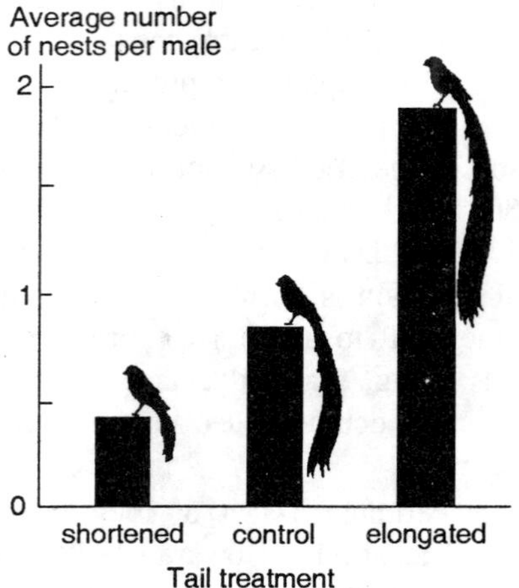

Fig. 13.7. Relative reproductive success of male long-tailed widowbirds with artificially shortened and lengthened tails.

suggests choice of mate on the basis of tail length and that preference is relative; that is, selection persists regardless of response. Even though male widowbirds have gone a long way to accommodate female preference, longer tails still look better to females, whether they are in the best interests of the male or not. Andersson's study was too brief to determine whether an extra-long tail reduced a male's survival.

Even though the basis for runaway selection in female choice has been demonstrated, its origin is still shrouded in mystery. One possibility is that outlandish traits had their origin in more conservative, intrinsically adaptive beginnings, but that establishment of strong female choice took hold of the processes and predominated environmental selective factors. Another, intriguing possibility is the "handicap principle" suggested by the Israeli behavioural ecologist Amotz Zahavi (1975). He viewed male secondary sexual characteristics as handicaps: that a male can survive while bearing such a handicap signals a female that he has an otherwise superior genotype. It may sound crazy, but if you wanted to demonstrate your strength to someone, you might carry around a large set of weights. A weaker individual couldn't do it, so there is little chance of a lesser genotype falsely advertising strength. According to Zahavi, the greater the handicap borne, the greater the ability of the individual to offset the handicap by other virtues.

You may wonder why a female should choose males that cancel out their superior constitutions by taking on burdens. The strong man may be superior intrinsically, but can he dance while holding a set of weights? This difficulty has stimulated a flood of comment and countercomment on the idea of the handicap principle without much agreement. But there are several possible bases for resolution. One is that the superior traits of males are passed on to both male and female offspring, but their handicaps are expressed only in males. So by choosing the handicapped male, a female may produce superior daughters. If true, this would reinforce the restriction of male purpose to providing genes, and make proper choice of genotype by females all the more important. It also would establish a conflict between satisfying female choice and being able to succeed in other phases of life, including male-male contest.

Another plausible scenario for the origin of female choice is that male handicaps may arise from normal, environmentally promoted variation within the population. Vagaries of upbringing may cause some males to have, for example, longer tails than others. If long tails arc a handicap in some way, only males with otherwise superior genotypes will be able to survive and prosper with them. A female would be wise to choose such a male, because tail length has already done the work of weeding out weaker genotypes for her, but since the long tail of the male is not genetically determined, the handicap would not be passed on to her male offspring. Given this situation, females that choose longer-tailed members of the male population would be strongly selected. Once that choice had become genetically established in the population, there would be strong selection of males having long tails arising from genetic factors, because their male offspring would then benefit from the previously established female choice. At this point, the process can take care of itself and runaway sexual selection is at hand.

Clearly, we have much to learn about sexual selection. Females may simply prefer odd males, as seems to be the case in *Drosophila* flies, perhaps because they are most likely to be genetically distinct and not cause inbreeding depression in the progeny. Also, comparisons of related taxa suggest that female preference changes over time and diverges between isolated populations. The many species of highly distinctive birds-of-paradise must have arisen from a common ancestor with highly sexually selected plumage and behaviour. Yet, in each isolated line of bird-of-paradise, female preference has wandered to focus upon different parts of the plumage, colours, feather shapes, and displays, as if saying to the male, "Do whatever you wish, but it better be good."

14

Molecular Ecosystem

As the diverse information gathered by naturalists during the last century coalesced into a unified picture of nature two concepts emerged, pushing the study of ecology in new directions. The first was the realization that species of plants and animals formed natural associations, each with distinctive members. Just as morphological data had allowed systematists to assign species to a hierarchy of taxonomic groups, detailed studies of the ecological distributions of plants led to the classification of biological communities.

A second new concept was the realization that organisms are linked both directly and indirectly by means of their feeding relationships. Humans have appreciated since their beginnings that one organism both preys upon and is preyed upon by others, but that these feeding relations linked species into a functional unit was a novel idea at the turn of the present century.

Analogy

From its inception, thinking about systems of interacting populations has been divided over a major issue: whether the whole exceeds the sum of its parts. On one hand, we may believe that the system as a whole has attributes that cannot be understood in terms of the workings of its parts. On the other, the system may be viewed merely as a collection of independently functioning populations.

Many ecologists believe that there is determinism (and perhaps purpose or causality) at the level of the multispecies system analogous to the determinism imparted at the organism level by natural selection. No matter how much one knows about the anatomy of elephants, one cannot explain the special relationship between the whole of an elephant and its surroundings. Is it also true, then, that all possible knowledge of grass and crickets and mice and lizards cannot unlock the secrets of the prairie?

The frequency with which ecologists have compared associations of species living together (biological communities) to organisms attests to a strong undercurrent of holistic sentiment. Functional similarities between associations and organisms—primary production and feeding, predation and metabolism, species and organs, the regular succession of stages from fallow field to mature forest and the development of the individual—are obvious. The influential American plant ecologist F. E. Clements extrapolated these similarities into a concept of biological communities as discrete vegetation types (climaxes), each occurring in a particular region defined by climate and soil and having a characteristic sequence of developmental stages (the sere) leading from bare or cleared ground to the mature state. Clements viewed plant communities as superorganisms. He had little need to question the analogy between organism and superorganism because his work was primarily descriptive and the organism concept adequately embraced what he saw in nature.

Clements's alter ego was also an American plant ecologist, H. A. Gleason (1926, 1939), for whom the local community, far from being a distinct unit like an organism, was merely the fortuitous association of species whose adaptations enabled them to live in a particular place. While recognizing that species do interact (all animals must eat!), Gleason argued that the presence or absence of any one species is independent of all others. In his view, we may define an association for convenience, but it does not represent a natural unit and it has no functional significance beyond the roles played by each of its members.

Feeding Relationships

Following the tradition of floristic analysis in plant community ecology, Clements and Gleason were concerned with the species composition of communities. By the mid-1920s, however, other ecologists had begun to consider functional patterns within communities. Foremost among the proponents of this viewpoint was the English ecologist Charles Elton. During his student days at Oxford, Elton accompanied an ecological expedition to Spitsbergen Island in the North Atlantic Ocean, where, in collaboration with the botanist V. S. Summerhayes, he worked out the feeding relationships among inhabitants of a simple tundra community.

By the time Elton was twenty-six, he had developed a new concept of communities organized by the feeding relationships within them. In his book *Animal Ecology* (1927), which was to become a landmark for modem ecology, he wrote:

> Food is the burning question in animal society, and the whole structure and activities of the community are dependent upon questions of food-supply . . . animals have to depend ultimately upon plants for their supplies of energy, since plants alone are able to turn raw sunlight and chemicals into a form edible to animals. Consequently herbivores are the basic class in animal society. . . . The herbivores are usually preyed upon by carnivores, which get the energy of the sunlight at third-hand, and these again may be preyed upon by other carnivores, and so on, until we reach an animal which has no enemies, and which forms, as it were, a terminus on this food-cycle. There are, in fact, chains of animals linked together by food, and all dependent in the long run upon plants. We refer to these as "food-chains," and to all the food-chains in a community as the "food-cycle."

Food-cycle" became "food web" over the years, but Elton's basic concept has otherwise survived unchanged.

Elton prefaced ("The Animal Community") with three Chinese proverbs:

"The large fish eat the small fish; the small fish eat the water insects; the water insects eat plants and mud."

"Large fowl cannot eat small grain."

"One hill cannot shelter two tigers."

The first proverb is Elton's "food-chain." The second and third are cornerstones of another of Elton's general principles, the pyramid of numbers. As one goes up the food chain, one also ascends a more or less regular procession of body sizes because most predators consume prey somewhat smaller than themselves. Progressively larger animals require progressively more space to find food; hence their numbers are lower. Elton noted that an oak wood harbors "vast numbers of small herbivorous insects like aphids, a large number of spiders and carnivorous ground beetles, a fair number of small warblers (insectivorous birds), and only one or two hawks. Similarly in a small pond, the numbers of protozoa may run into millions, those of Daphnia and Cyclops into hundreds of thousands, while there will be far fewer beetle larvae, and only a very few small fish."

Elton explained this pyramid of numbers, large at the base of the food chain and small at the top, by using arguments from population biology and the scaling of biological functions to body size.

> The small herbivorous animals which form the key-industries in the community are able to increase at a very high rate (chiefly by virtue of their small size), and are therefore able to provide a large margin of numbers over and above that which would be necessary to maintain their population in the absence of enemies. This margin supports a set of carnivores, which are larger in size and fewer in numbers. These carnivores in turn can only provide a still smaller margin, owing to their large size which makes them increase more slowly, and to their smaller numbers.

And so on to the tiger on the hill.

During the 1930s, the idea of the community as an association of interacting species more and more became the focus of ecological thinking, but was far from universally accepted.

Meaning of Ecosystem

By the mid-1930s, the superorganism concept of the community was still very much alive. Clements remained influential. In a three-part article, "Succession, Development, the Climax and the Complex Organism: An Analysis of Concepts," English ecologist John Phillips (1934, 1935) elaborated the Clementsian view that the community has unique properties of function and organization, a "newness springing from the interaction, interrelation, integration and organization of qualities . . . (which) could not be predicted from the sum of the particular qualities or kinds of qualities concerned: integration of the qualities thus results in the development of a whole different from, unpredictable from, their mere summation." But Phillips also took a strong philosophical stance on causation in ecology by attributing "operative causes" and "inherent, dynamic characteristics" to communities; he stressed the "fundamental nature of the factor of holism innate in the very being of community, a factor of cause." He further explained:

> At different levels the whole reacts upon habitat, changing (ameliorating) this for higher level wholes: the reaction of a whole, taken into account with its particular habitat and with the interrelations existing among its constituent organisms, shows as emergent changes in the habitat that are different from the sum of the changes that the constituent organisms would undergo were these not in communal association.

The English plant ecologist A. G. Tansley (1936) rejected the superorganism notion of Clements and Phillips, preferring to regard the animals and plants in associations, together with the physical factors of their surroundings, simply as systems. He wrote that

> The more fundamental conception is, as it seems to me, the whole *system* (in the sense of physics), including not only the organism-complex, but also the whole complex of physical factors forming what we call the environment of the biome—the habitat factors in the widest sense. Though the organisms may claim our primary interest, when we are trying to think fundamentally we cannot separate them from their special environment, with which they form one physical system.

This Tansley called the *ecosystem*.

Flow of Energy

The controversy over the superorganism analogy went unnoticed by A. J, Lotka, a chemist by training, whose different vision of biological systems still influences us through the legacy of his book *The Elements of Physical Biology*, published in 1925. Lotka was the first to treat populations and communities as thermodynamic systems. In principle, he said, each system can be represented by a set of equations that govern transformations of mass among its components. Such transformations include the assimilation of carbon dioxide into organic carbon compounds by green plants and the consumption of plants by herbivores and animals by carnivores.

Lotka believed that the size of a system and the rate of transformations within it were determined according to certain thermodynamic principles. In the same sense that heavy machines and fast machines require more fuel to operate than their lighter and slower counterparts, and efficient machines require less fuel than inefficient ones, the energy transformations of ecosystems grow in direct relation to their size (roughly the total masses of their constituent organisms), productivity (rate of transformations), and inefficiency. Lotka regarded the ecosystem as a part of the world machine responsible for the transformation of the energy of sunlight reaching the surface of the earth. Not all the energy enters biological pathways of transformations; in fact, most of it drives the circulation of winds and ocean currents and the evaporation of water. But the portion that plants assimilate by photosynthesis drives a part of the world machine, what Lotka referred to, by way of analogy, as the mill-wheel of life.

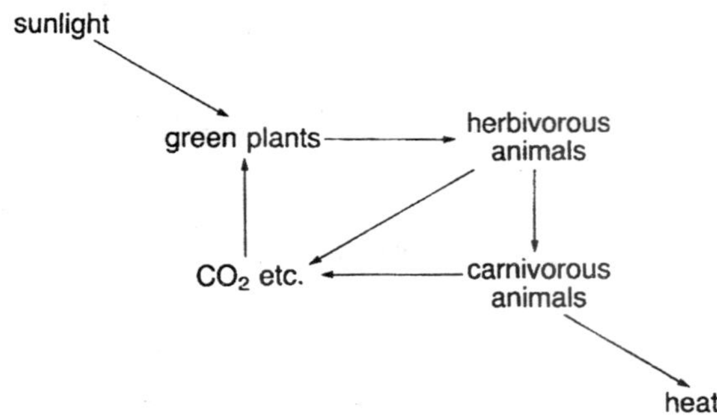

Fig. 14.1. Lotka's simple diagram of the mill-wheel of life.

When Lotka published his book, most ecologists were preoccupied with the problem of species associations and missed the implications of a thermodynamic characterization of the natural world. Tansley never referred to *The Elements of Physical Biology*, even though Lotka had provided the mechanical analogy—thus a tangible model—for his concept of the ecosystem.

Trophic Relationship

The idea of the ecosystem as an energy-transforming system was brought to the attention of many ecologists for the first time in a paper published in 1942 by Raymond Lindeman, a young aquatic ecologist from the University of Minnesota. The full story behind this historic paper has been recounted by Robert Cook (1977): the journal *Ecology* first rejected the manuscript on the advice of reviewers who felt the treatment was too theoretical; but strong advocacy by Yale ecologist G. E. Hutchinson finally led to its publication.

Lindeman's framework for understanding ecological succession based on sound thermodynamic principles made a deep impression. He adopted Tansley's notion of the ecosystem as the fundamental unit in ecology and Elton's concept of the food web, including inorganic nutrients at the base, as the most useful expression of ecosystem structure.

The food chain consisted of steps—primary producer, herbivore, carnivore—that Lindeman referred to as trophic levels. But rather than seek regularity in a trophic pyramid of numbers, as Elton had, he visualized a pyramid of energy transformation. He argued that less energy is available to each higher trophic level owing to the work performed and to the inefficiency of biological energy transformations on the next lower trophic level. Thus, of the light impinging on a lake (Λ_0), plants assimilate only a fraction (Λ_1), the primary production of

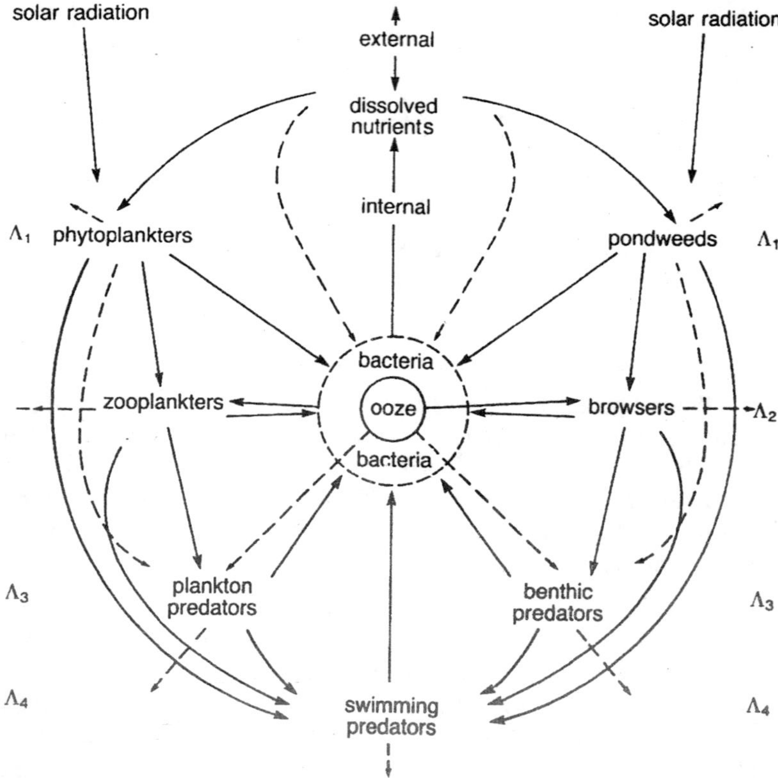

Fig. 14.2. Lindeman's diagram of the generalized "food-cycle" relationships within a temperate-zone lake.

the system. Herbivores assimilate less energy (Λ_2) than do plants because plants utilize some of their own production to maintain themselves before herbivores consume them. The ratio of production on one trophic level to that on the level below it (for example, Λ_2/Λ_1) is the biological efficiency of that link in the food chain.

By the 1950s, the ecosystem concept had fully pervaded ecological thinking and provided the beginnings for a new branch of ecology. University of Michigan ecologist Francis C. Evans (1956) summarized the essential points in a brief essay:

> In its fundamental aspects, an ecosystem involves the circulation, transformation, and accumulation of energy and matter through the medium of living things and their activities. Photosynthesis, decomposition, herbivory, predation, parasitism, and other symbiotic activities are among the principal biological processes responsible

> for the transport and storage of materials and energy, and the interactions of the organisms engaged in these activities provide the pathways of distribution. The food-chain is an example of such a pathway. . . . The ecologist, then, is primarily concerned with the quantities of matter and energy that pass through a given ecosystem and with the rates at which they do so.

Thus, the cycling of matter and the associated flux of energy through the ecosystem provided a basis for characterizing its structure and function. Currencies of energy and the masses of such elements as carbon allowed direct comparison of plants, animals, microbes, and abiotic sources of energy and elements in the ecosystem. Taxonomic lists of species and the numbers of individuals in populations gave way to measurements of energy assimilation and energetic efficiencies in this new thermodynamic concept of the ecosystem.

Ecological Relationship

Although Evans did not explicitly address the issue of the ecosystem as a superorganism, his writing expressed a mechanical notion of the function of the ecosystem and its regulation. Each population was an individual machine and the work it performed added to that performed by all the others to yield the work performed by the ecosystem as a whole. In Evans's view, probably the prevalent one in the 1950s, energy transformation by each population depended on factors in its environment that directly affected the activities of individuals in the population.

> Ecosystems are further characterized by a multiplicity of regulatory mechanisms, which, in limiting the numbers of organisms present and in influencing their physiology and behaviour, control the quantities and rates of movement of both matter and energy. Processes of growth and reproduction, agencies of mortality (physical as well as biological), patterns of immigration and emigration, and habits of adaptive significance are among the more important groups of regulatory mechanisms.

Lotka (1925) had formalized such relationships in a system of equations, one for each population. Each equation described the rate of change in the numbers of individuals (one could substitute biomass or its energy equivalent) in terms of physical and biological factors in the environment, including the species it fed upon and which fed upon it. Formally, Lotka considered a system with components X_1 through X_n: "In general the rate of growth dX/dt of any one of these components will depend upon, will be a function (F) of the topography, climate, etc. If these latter features are defined in terms of a set of parameters $P_1\ P_2\ .\ .\ .\ P_j$, we may write, . . .

$$\frac{dX_1}{dt} = F_1(X_1, X_2, \ldots X_n; P_1, P_2, \ldots P_j)$$

. . . In general there will be *n* such equations, one for each of the *n* components." When *Xs* are expressed in terms of energy, the sum of the $X_i s$ is the total energy content of the system at a given time and the sum of the dX_i/dts is the overall rate of change in the structure of the system. Energy transformations are dictated by individual terms of the functions *F*, which include all the increments of gain and loss to *X*.

In Raymond Lindeman's symbolism, Λ_n is the total energy content of trophic level *n* and the rate of change in Λ_n ($d\Lambda_n/dt$) is the sum of the gains (λ_n) and losses (λ'_n) of energy from the trophic level; hence,

$$\frac{d\Lambda_n}{dt} = \lambda_n + \lambda'_n$$

"where λ_n is by definition positive and represents the rate of contribution of energy from Λ_{n-1} (the previous level) to Λ_n, while λ'_n is negative and represents the sum of the rate of energy dissipated from Λ_n and the rate of energy content handed on to the following level Λ_{n-1}. The more interesting quantity is λ_n which is defined as the true productivity of level Λ_n".

According to Lotka's equations, the dynamics of a single link in the food web of the ecosystem—for example, between components (species, physical sources) *i* and *j*, are defined by terms in F_i and F_j. "The dynamical properties of the entire system are governed by all the terms of the individual functions F_i. The whole is equal to, and can be understood in terms of, the sum of its parts.

Bioenergetics

With a clear conceptual framework for the ecosystem and a currency of energy to describe its structure, ecologists began to measure energy flow and the cycling of nutrients in the ecosystem. One of the strongest proponents of this approach has been Eugene P. Odum of the University of Georgia, whose text *Fundamentals of Ecology*, first published in 1953, influenced a generation of ecologists.

Odum depicted ecosystems as simple energy flow diagrams. For any one trophic level, such a diagram consisted of a box representing the biomass (or its energy equivalent) at any given time and pathways

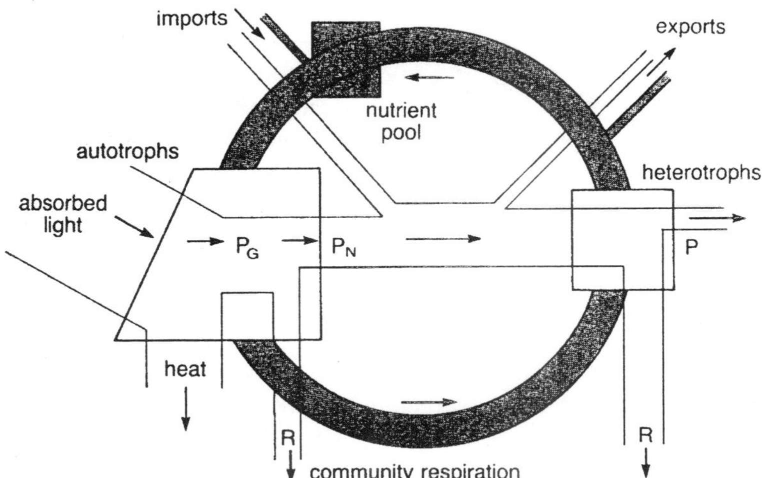

Fig. 14.3. E.P. Odum's flow diagram of an ecosystem showing the one-way flow of energy and the recycling of materials.

through the box representing the flow of energy. Feeding relationships linked energy flow diagrams into a food web.

Energy flow diagrams were elaborated to include the cycling of elements. Unlike energy, which ultimately comes from sunlight and leaves the ecosystem as heat, nutrients are regenerated and retained within the system. In the development of ecosystem studies, the cycling of elements has assumed nearly equal standing with the flow of energy. One reason for this prominence is that the amounts of elements and their movement between components can provide a convenient index to the flow of energy, which is difficult to measure directly. Carbon, in particular, bears a close relationship to energy content because of its intimate association with the assimilation of energy by photosynthesis.

A second reason for the prominence of nutrient cycling is the fact that levels of certain nutrients regulate primary production. In deserts, plant growth reflects amount of water rather than sunlight or minerals in the soil in most areas. By contrast, the open oceans are deserts by virtue of their scarce nutrients. Understanding how elements cycle between components of the ecosystem seems crucial to understanding the regulation of ecosystem structure and function.

Systematic Approach

General acceptance of the ecosystem viewpoint provided yet another point of departure for interpreting biological systems. On one hand, many ecologists feel that understanding an ecosystem requires analysis of the interactions of most of its individual components. On the other, a quite distinct group of ecologists feels that the properties of ecosystems transcend the behaviour of their individual components and can be understood only at the system level. Each component has its particular behaviour, of course, just as each organ of the body has its particular function, but in a holistic concept these behaviours are subjugated to the behaviour of the system as a whole.

The individualistic viewpoint, which has been referred to as the ecosystem viewpoint or *systems ecology*, relies heavily on mathematical analysis to handle the complexity of natural systems. George Van Dyne (1966) summarized the approach and its difficulties.

> The tools and processes required for systems ecology are different from those for conventional phases of ecology because of the complexity of the total ecosystem as compared with a segment of it. When we consider the totality of interactions of populations with one another and with their physical environs—i.e., ecosystem ecology—we face a new degree of complexity. . . . One of the major problems in systems ecology is that of analyzing and understanding interactions. Events in nature are seldom, if ever, caused by a single factor. They are due to multiple factors which are integrated by the organism or the ecosystem to produce an effect which we observe. To further complicate the matter, various combinations of factors and their interactions may be interpreted and integrated by the ecosystem to produce the same result.

According to Van Dyne, the goal of the ecosystem approach is to provide a model of the ecosystem capable of mimicking its dynamical properties. Disturbances, such as pollution, grazing, fire, introduced pests, hunting, and severe climate conditions, alter Lotka's system of equations dX_i/dt for certain species by changing the conditions of the environment (P) or the abundances of interacting species (X_j). By their effect on X_i, such disturbances indirectly affect other species or components (x_k) to which i is linked in the system. Only a detailed model that faithfully depicts the interactions among all components can assess the impact of a stress or other disturbance on the structure and function of the ecosystem as a whole.

Organismic Feature

Systems ecologists understand ecosystems by taking them apart and seeing how each piece functions in the whole. Their philosophy asserts that just as one cannot understand a house without detailed study of the blueprints and direct nail-by-nail observation of its construction, one cannot understand an organism without a map of its genes and step-by-step knowledge of its development and functioning.

But other ecologists view large systems as having a purposeful structure that cannot be comprehended by detailed examination of the parts. An understanding of how an organism functions in its particular environment differs from an understanding of the biochemistry of cell processes or the control of gene expression during development. Darwin did not have a detailed, or even correct, knowledge of genetics and yet he was able to appreciate the mechanism of evolution and the purpose in adaptation. An architect designs the structure of a house with purpose in mind. The detailed arrangement of the bricks and plumbing are merely the means to achieve this design; they do not offer insight into the purpose of the house as a complete, integrated structure that serves the purposes of human habitation.

It is only natural that organismic thinking should influence our concepts of ecosystems, as earlier in this century it influenced our concepts concerning associations of species. Odum (1959) expressed the beginnings of such a viewpoint:

> Some attributes, obviously, become more complex and variable as we proceed from (the cell to the ecosystem level of organization), but *it is an often overlooked fact that other attributes become less complex and less variable as we go from the small to the large unit*. Because homeostatic mechanisms, that is, checks and balances, forces and counter forces, operate all along the line, a certain amount of integration occurs as smaller units function within larger units. . . . *When we consider the unique characteristics which develop at each level*, there is no reason to suppose that any level is any more difficult or any easier to study quantitatively. . . . Furthermore, the findings at any one level *aid in the study of another level, but never completely explain the phenomena occurring at that level*.

In a paper entitled "The Strategy of Ecosystem Development," Odum (1969) developed this theme more explicitly, suggesting that evolution of components (species) with respect to each other has resulted in (perhaps, even, is directed toward) increased integrity and stability of the system:

> (Ecological succession) culminates in a stabilized ecosystem in which maximum biomass (or high information content) and symbiotic function between organisms are maintained per unit of available energy flow. In a word, the "strategy" of succession as a short-term process is basically the same as the "strategy" of

> long-term evolutionary development of the biosphere—namely increased control of, or homeostasis with, the physical environment in the sense of achieving maximum protection from its perturbations.

Information content in the ecosystem has been compared to the genetic information that directs the development of the organism and specifies the physiological apparatus that maintains its integrity. In ecosystems as in organisms, information content, specificity of ecosystem responses, and stability in the face of perturbation are interrelated.

Odum went on to describe how the adaptations of species to others in a system could contribute to overall stability:

> The time involved in an uninterrupted succession allows for increasingly intimate associations and reciprocal adaptations between plants and animals, which lead to the development of many mechanisms that reduce grazing—such as the development of indigestible supporting tissues (cellulose, lignin, and so on), feedback control between plants and herbivores, and increasing predatory pressure on herbivores. Such mechanisms enable the biological community to maintain the large and complex organic structure that mitigates perturbations of the physical environment. . . . there can be little doubt that the net result of community actions is symbiosis, nutrient conservation, stability, a decrease in entropy, and an increase in information. . . . The overall strategy is, as I stated at the beginning of this article, directed toward achieving as large and diverse an organic structure as is possible within the limits set by the available energy input and the prevailing physical conditions of existence (soil, water, climate, and so on).

If there was still any doubt about Odum's attraction to the organism analogy, he then posed the "intriguing question": "Do mature ecosystems age, as organisms do? In other words, after a long period of relative stability of 'adulthood,' do ecosystems again develop unbalanced metabolism and become more vulnerable to diseases and other perturbations?"

The resolution of the systems-versus-organism dichotomy poses a formidable challenge for several reasons. First, it is difficult to phrase an organismic theory of the ecosystems so as to make predictions that can be falsified by observation or experiment. The behaviour of an organism certainly suits the purpose of the organism; but we may also describe that behaviour mechanically in terms of the actions and reactions of the organism's component parts. At a higher level of organization, observed efficiencies of energy transfer between trophic levels may maximize the power output of the ecosystem, as H. T.

Odum and R. C. Pinkerton (1955) have suggested, but they may also follow naturally from adaptations of organisms to maximize their power output individually, provided that some connection links power output and evolutionary fitness. At what level of system—organism or ecosystem—is purpose being served? How can we know?

G. M. Weinberg (1975) suggested that ecosystems pose conceptual difficulties because they fall in the range of "middle-number systems" that cannot be treated appropriately either by differential equations, as Lotka proposed, or by statistical approaches whereby we may characterize the properties of a collection of components by calculating means and variances. Simple physical systems with few components, such as the gravitational attraction among the nine planets of our solar system and contrived laboratory microcosms with few species, are "small-number simple systems." "Large-number simple systems" consist of immense numbers of similar items—a quantity of gas consisting of billions and billions of atoms, for example. The temperature and pressure of the gas depend' on the motion and collisions of individual atoms, but averages of these properties adequately describe the whole system. Ecosystems are too complex to model as small-number systems of differential equations. Also, their parts differ so in function that simple averages, even within trophic levels, would obscure essential features of ecosystem structure. Thus, the ecosystem is what Weinberg calls a middle-number system—its parts are too numerous to describe fully, and too few and diversified to average meaningfully.

New Concepts

A few biologists have recently attempted to find new ways of conceptualizing ecosystems. We cannot at present judge the success of these endeavors; certainly their proponents have yet to attract large followings. These approaches suffer from borrowing analogies from inappropriate branches of study—physics, psychology, sociology, linguistics, and communication—and they may provide little more than a new set of jargon in the place of old conceptual frustrations.

Proponents of these new concepts attribute independent levels of causation to each level in the hierarchy of organization of systems, from organism to ecosystem. T. F. H. Alien and Thomas B. Starr (1982), of the University of Wisconsin, elaborated the notion of hierarchical structure in ecological systems, borrowing heavily from Arthur Koestler's 1967 book *The Ghost in the Machine*:

> The salient feature of an entity is that it is an integration of all its parts. Koestler uses the image of a doorway between the parts

> of the structure and the rest of the universe. The entity has a duality in that it looks inward at the parts and outward at an integration of its environment; it is at once a whole and a part. At every level in a hierarchy there are these entities, and they have this dual structure. As in taxonomy, where each level in the hierarchical classification is called a "taxon," Koestler call his two-faced entities "holons." . . . (Quoting Koestler:) "Every holon has dual tendency to preserve and assert its individuality as a quasi autonomous whole; and to function as an integrated part of (an existing or evolving) larger whole. . . . The self-assertive tendencies are the dynamic expression of holon wholeness, the integrative tendencies of its partness."

Thus, ecosystems can be viewed as parts within parts, each level of the hierarchy having its own scale of time and space. The more disparate the scales of any two levels, the less likely they are to be dynamically coupled. According to this insight, if the behaviour of large systems with dynamics of long period is purposeful, it can be understood only in terms of response to physical processes of similar scale. To attribute structural and functional "adaptations" of ecosystems to processes that vary over periods of days or even seasons makes little sense; the dynamics of individuals and populations are far closer in scale.

This is new ground for ecology. Perhaps it will be infertile; at worst, some "concepts" may turn out to be means of legitimizing a mysticism that has no place in science. Caution against the quick conclusion was raised, perhaps in self-defense, by Bernard Patton (1982), one of the few strong advocates of these new approaches:

> In constructing environ theory to its present form, the various plateaus were marked as they were reached with new words, to hold the place, and make it possible to explore further avenues and then return. The new terms tend to be resisted, which impedes acceptance of the theory. However, they are not ready to be discarded because they do express key ideas. The reader is asked to be tolerant; ignore the terms if necessary but hold the thoughts. The words in question are holon, creaon, genon, taxon, and environ.

Then, seemingly having given himself license, Patton one-ups the organismic concept with a more extreme analogy to the cell:

> In addition, one further lexical liberty is taken for didactic reasons. The environ center will be referred to as its nucleus and the remainder as enviroplasm.

15

FLOW OF ENERGY

Organisms dissipate energy in two ways. First, they perform work on the system in which they live. Thus, for example, as an animal runs it transfers kinetic energy to the atmosphere and to the ground. Second, the biochemical transformations required for movement, biosynthesis, secretion, and cell maintenance are inefficient. Organisms lose energy in the form of heat at each biochemical step.

In spite of these losses, the energy budget of the ecosystem is balanced because plants assimilate energy from the sun. Plants then use this energy, stored in the chemical bonds of carbohydrates, for their metabolic needs. But herbivores also take a portion, and thus start some of the energy assimilated by plants on its way up the food chain. In this chapter, we shall see how the flux of energy through the ecosystem depends on rates at which plants assimilate energy, rates of consumption at each trophic level, and energetic efficiencies of transforming food into biomass. A logical place to start is at the beginning of the food chain with plant production.

ENERGY TRAPPING BY PLANTS

Photosynthesis is the process by which plants capture light energy and transform it into the energy of chemical bonds in carbohydrates. Glucose and other organic compounds (starch and oils, for example) may be stored conveniently, and their energy later released in respiratory metabolic pathways. Photosynthesis chemically unites two common inorganic compounds, carbon dioxide (CO_2) and water (H_2O), to form glucose ($C_6H_{12}O_6$), with the release of oxygen (O_2). The overall stoichiometry (chemical balance) of photosynthesis is

$$6CO_2 + 12H_2O \rightarrow C_6H_{12}O_6 + 6O_2 + 6H_2O$$

Photosynthesis transforms carbon from an oxidized state in CO_2 to a reduced state in carbohydrate. Because work is performed on carbon atoms, photosynthesis requires energy, which is provided by visible light. For each gram of carbon assimilated, the plant gains 39 kilojoules (kJ) of energy. But because of inefficiencies in the various biochemical steps of photosynthesis, no more than 34 per cent, and usually much less, of the light energy absorbed by photosynthetic pigments eventually appears in carbohydrate molecules.

Photosynthesis supplies the carbohydrate building blocks and energy the plant needs to synthesize tissues and grow. Rearranged and joined together, glucose becomes fats, oils, and cellulose. Combined with nitrogen, phosphorus, sulfur, and magnesium, simple carbohydrates derived ultimately from glucose produce an array of proteins, nucleic acids, and pigments. Plants cannot grow unless they have all these basic building materials. Chlorophyll, for example, contains an atom of magnesium; even though all other necessary elements might be present in abundance, a plant lacking magnesium cannot produce chlorophyll and thus cannot grow.

Plants build and maintain tissues by complex, energy-requiring biochemical transformations. Because plants utilize much of the energy assimilated by photosynthesis to supply these needs, they always contain in their tissues substantially less energy than the total assimilated. Therefore, ecologists must distinguish two measures of assimilated energy: gross production, the total energy assimilated by photosynthesis, and net production, the accumulation of energy in plant biomass; that is to say, plant growth and reproduction. Because plants are the first link in the food chain, ecologists refer to these measures as gross or

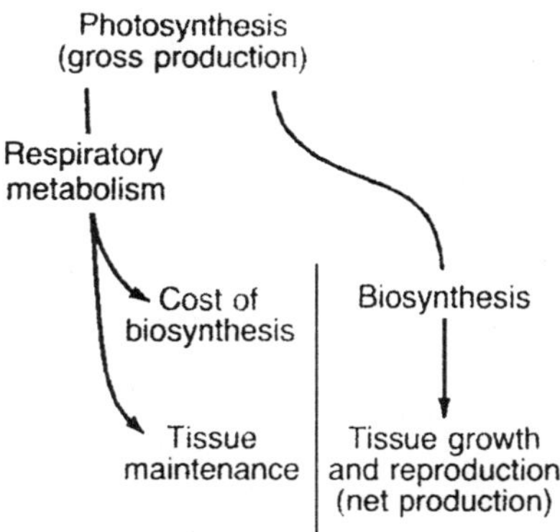

Fig. 15.1. A diagram of the partitioning of energy by plants.

net primary production. The difference between them is the energy of respiration, that utilized for maintenance and biosynthesis.

Measuring Plant Production

Because primary production involves fluxes of carbon dioxide, oxygen, minerals, and water, on one hand, and the accumulation of plant biomass, on the other, the rates of any of these should provide an index to the overall rate of plant production. In practice, the appropriate measure depends on habitat and growth form. Furthermore, one uses different measures to estimate gross and net production, or the production of an entire system and that of a small part of a single plant.

Net production can be expressed conveniently as grams of carbon assimilated, dry weight of plant tissues, or their energy equivalents. Ecologists use such indices interchangeably because they have found a high degree of correlation among them. The energy content of an organic compound depends primarily on its carbon content (assimilated at a cost of 39 kJ per gram) and the energy added or subtracted during various transformations. For example, glucose contains 40 per cent carbon by mass, and should therefore contain an equivalent of $0.40 \times 39 = 15.6$ kJ g^{-1}. The observed value is about 17.6 kJ g^{-1}, as determined by burning samples in devices called bomb calorimeters.

A calorimeter has a small chamber where the sample is burned, in a fast chemical version of cellular respiration, to oxidize it completely to carbon dioxide and water. Oxygen is forced under high pressure into the chamber to ensure complete combustion. A water jacket surrounds the chamber and absorbs the heat produced. The increase in temperature of a known amount of water in the jacket provides a direct estimate (4.2 kJ $°C^{-1}$ kg^{-1}) of the heat energy released by combustion. Measured in this way, amounts of energy released in complete oxidation average approximately 17.6 kJ g^{-1} of carbohydrate, 23.8 kJ g^{-1} of protein, and 39.7 kJ g^{-1} of fat. Thus, by measuring the composition of plant tissues, one can estimate their energy content.

In terrestrial ecosystems, ecologists usually estimate plant production by the annual increase in plant biomass, although measurements of CO_2 flux also are practical. In areas of seasonal production, annual growth is determined by cutting, drying, and weighing the plants at the end of the growing season. Root growth usually is ignored because roots are difficult to remove from most soils; thus, harvesting measures the annual aboveground net productivity (AANP), the most common basis for comparing terrestrial communities.

Because the atmosphere contains little carbon dioxide (0.03 per cent), uptake by plants can measurably reduce its concentration in enclosed chambers within short periods. Ecologists have used this principle to measure production in terrestrial habitats, beginning with the classic study of H. N. Transeau (1926) on field corn. The most convenient application of the method is to enclose samples of vegetation (whole herbaceous plants or branches of trees) in clear chambers (light must penetrate for photosynthesis) and to measure the change in concentration of CO_2 in air passed through the chamber. Nowadays ecologists measure the absorbance of infrared light by the airstream inasmuch as CO_2 strongly absorbs light in that part of the spectrum. CO_2 uptake per gram of dry weight or square centimeter of surface area of leaves in the chamber can then be extrapolated to the entire tree or forest.

Carbon dioxide flux during the light period of the day includes both assimilation (uptake) and respiration (output) and thus measures net production. Respiration can be estimated separately by carbon dioxide production during the night, when photosynthesis shuts down. Daniel Botkin et al. (1970) measured CO_2 uptake (grams per square meter of ground area per day) of the three most important species of trees (two oaks and a pine) in the Brookhaven National Laboratory forest over the course of a single growing season. During daylight hours plants assimilated 4336 g m^{-2} d^{-1}. Over 24 hours the value was reduced by nighttime respiration to 3702 g m^{-2} d^{-1}, a difference of 635 g m^{-2} d^{-1}. Because the average chemical composition of plant (tree) carbohydrate is $C_6H_{10}O_5$ one gram of CO_2 assimilated is equivalent to 0.614 g of carbohydrate produced. Using this ratio, Botkin and his colleagues estimated the net dry matter production of the forest to be $3702 \times 0.614 = 2273$ g m^{-2} d^{-1}, which agreed well with the harvest technique applied to the same forest.

The radioactive isotope carbon-14 (^{14}C) provides a useful variation on the gas exchange method of measuring productivity. When one adds a known amount of ^{14}C-carbon dioxide to an airtight enclosure, plants assimilate the radioactive carbon atoms in the same proportion as they occur in the air inside the chamber. Thus, one may calculate the rate of carbon fixation by dividing the amount of ^{14}C in the plant by the proportion of ^{14}C in the chamber at the beginning of the experiment. For example, if a plant assimilated 10 milligrams of ^{14}C in an hour, and the proportion in the chamber was 0.05, we could calculate that the plant assimilated carbon at a rate of 200 mg h^{-1} (10 divided by 0.05).

Aquatic Primary Production

Although production of large aquatic plants, such as kelps, can be estimated by harvesting, small size and rapid turnover of phytoplankton preclude the method as a general approach to aquatic production. But whereas high levels of atmospheric oxygen preclude using the production of oxygen by photosynthesis as a measure in terrestrial habitats, the low natural concentration of oxygen dissolved in water makes the measurement of small changes in oxygen concentration practical in most aquatic systems.

To measure production, samples of water containing phytoplankton are suspended in pairs of sealed bottles at desired depths beneath the surface of a lake or the ocean; one of the pair (a "light bottle") is clear and allows sunlight to enter; the other (a "dark bottle") is opaque. In the light bottles, photosynthesis and respiration occur together, and part of the oxygen produced by the first process is consumed by the second. In the dark bottles, respiration consumes oxygen without its being replenished by photosynthesis. Thus, to estimate gross production one adds the change in oxygen concentration in the light bottle (photosynthesis—respiration) to the amount consumed in the dark bottle.

In unproductive waters, such as those of deep lakes and the open ocean, changes in oxygen concentration due to photosynthesis and respiration are small compared to the amount dissolved, and light and

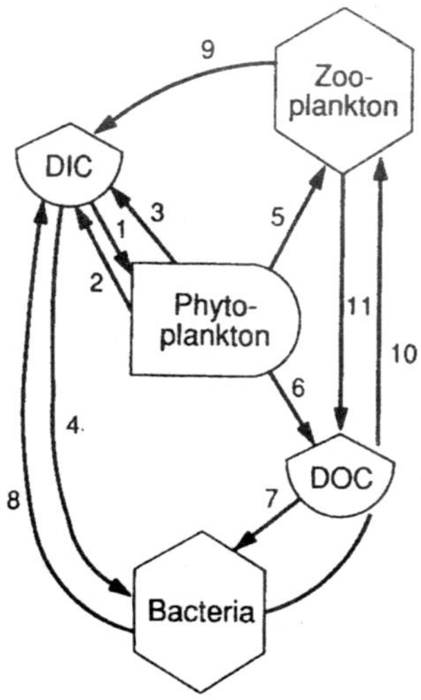

Fig. 15.2. Carbon pools and pathways influencing ^{14}C-CO_2 uptake experiments with natural plankton communities.

dark bottle measurements are not practical. Steeman Nielsen (1951) pioneered the measurement of ^{14}C uptake to estimate production under such conditions. The principle of the ^{14}C method is the same in aquatic systems as in terrestrial ones, except that the isotope is usually provided in the form of hydrogen carbonate ion (HCO_3^-). In practice, however, the interpretation of ^{14}C uptake values is plagued by problems, including photorespiration at high light intensities and the uptake and release of carbon in both organic and inorganic forms by bacteria and zooplankton.

A final method for estimating plant production in aquatic habitats is based on the idea that the concentration of chlorophyll sets an upper limit to the rate of photosynthesis at high light intensities. J. H. Ryther and C. S. Yentsch (1957), who advocated application of the method, determined that marine algae assimilate a maximum of 3.7 grams of carbon per gram of chlorophyll per hour (with variation between 2.1 and 5.7 over several studies). With experimentally

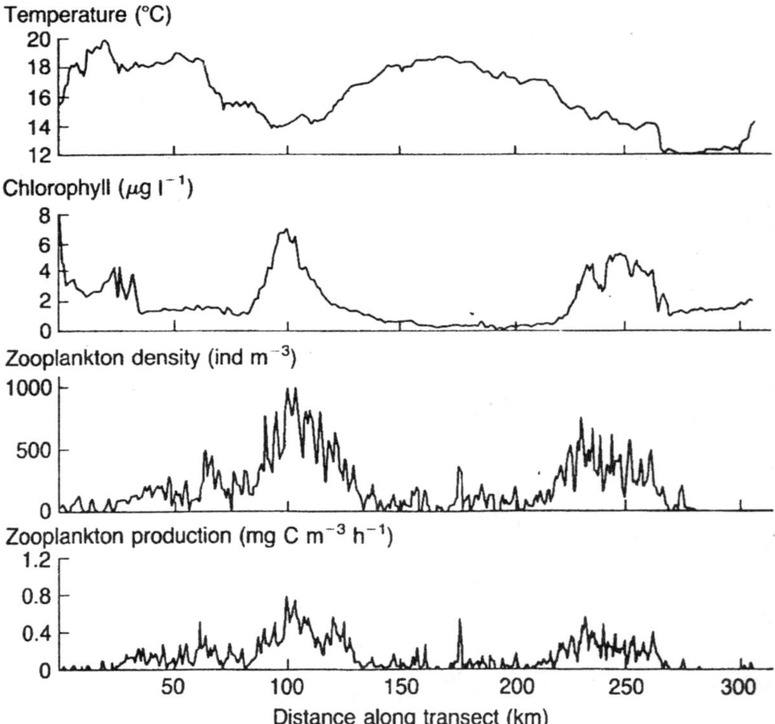

Fig. 15.3. Temperature, chlorophyll concentration, zooplankton density, and estimated zooplankton production at a depth of 4 meters.

determined relationships between light intensity and rate of photosynthesis by known concentrations of phytoplankton, and measurements of chlorophyll concentrations and light penetration down through the water column, one can estimate total production per unit of surface area. What the chlorophyll method lacks in precision it makes up in speed and simplicity. It has been used with transect surveys and satellite observations of surface water to map productivity and follow its temporal change over vast areas.

Effect of Light and Temperature

The rate of photosynthesis varies in direct proportion to light at low intensity, usually less than one-quarter that of full sunlight. Brighter light saturates the photosynthetic pigments, and as intensity increases rate of photosynthesis increases more slowly or levels off. In many plants, extremely bright light impairs photosynthesis owing to deactivation of photosynthetic reactions and enhanced photorespiration.

The response of photosynthesis to light intensity has two reference points. The first, called the compensation point, is the level of light intensity at which photosynthetic assimilation of energy just balances respiration. Above the compensation point, the energy balance of the plant is positive; below it, its energy balance is negative. The second reference point is the saturation point, above which the rate of photosynthesis no longer responds to increasing light intensity. Among terrestrial plants, the compensation points of species that normally grow in full sunlight (approximately 500 watts per square meter) occur between 1 and 2 W m^{-2}. The saturation points of such species usually are reached between 30 and 40 W m^{-2}, less than a tenth of the energy level of bright, direct sunlight.

Like most other physiological processes, photosynthesis proceeds most rapidly within a narrow range of temperature; the optimum varies with environment, from about 16°C in many temperate species to as high as 38°C in tropical species. Optimum temperature also varies with light intensity in some species, such as the alpine heath *Loiseleuria* of Austria. Net production depends on the rate of respiration as well as that of photosynthesis, and respiration generally increases with increasing leaf temperature.

Photosynthetic efficiency, which is the per cent of the energy in incident radiation converted to net primary production during the growing season, provides a useful index to rates of primary production under natural conditions. Where water and nutrients do not severely limit

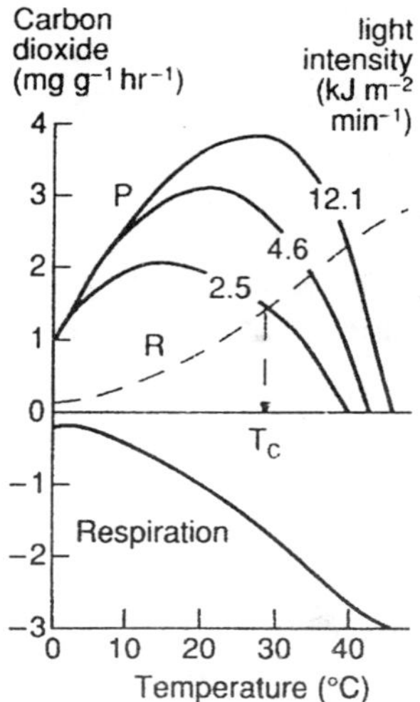

Fig. 15.4. Net photosynthestic rate as a function of leaf temperature and light intensity in the heath Loiseleuria.

plant production, photosynthetic efficiency varies between 1 and 2 per cent.

What happens to the remaining 98 to 99 per cent of the energy? Leaves and other surfaces reflect anywhere from one-quarter to three-quarters. Molecules other than photosynthetically active pigments absorb most of the remainder, which is converted to heat and either radiated or conducted across the leaf surface or dissipated by the evaporation of water from the leaf (transpiration). For example, in a study of an oak forest, annual incident photosynthetically active radiation was 1.9×10^6 kJ m^{-2} (an average of 60 W m^{-2})· Of this, 56 per cent was absorbed by leaves. Annual transpiration was approximately 0.5×10^6 g m^{-2}. At 2.24 kJ g^{-1} of water evaporated (the heat of vapourization), transpiration accounted for 1.1×10^6 kJ m^{-2}, which was approximately the total light energy absorbed.

Transpiration

The tiny openings (stomates) through which leaves exchange carbon dioxide and oxygen with the atmosphere also allow passage of water vapour (transpiration). As the moisture content of soil decreases, plants obtain water with increasing difficulty. As soil moisture approaches the wilting point, leaves close their stomates to reduce water loss; this prevents uptake of CO_2 and photosynthesis slows to a standstill. Consequently, the rate of photosynthesis depends on a plant's ability to

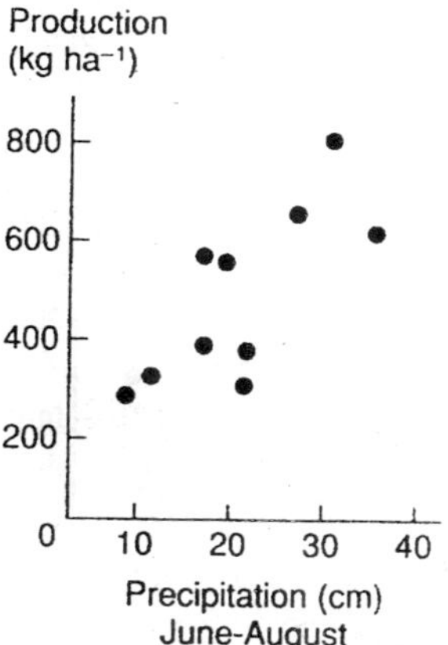

Fig. 15.5. Relationship between production and summer precipitation on perennial grassland.

tolerate water loss, the availability of soil moisture, and the influence of air temperature and solar radiation on the rate of transpiration.

Agronomists have devised the concept of transpiration efficiency, which is the ratio of net production to transpiration, as an index to the drought resistance of plants. Transpiration efficiencies are less than 2 grams of production per kilogram of water transpired in most plants, but they may be as high as 4 g kg^{-1} in drought-tolerant crops. Because the transpiration efficiency varies within narrow limits across a wide variety of plants, production is directly related to water availability in the environment. For example, annual production of perennial grasses in southern Arizona varies in direct relation to summer rainfall during the year of production.

Limitation

Fertilizers stimulate plant growth in most habitats. For example, when G. S. McMaster et al. (1982) applied nitrogen and phosphorus fertilizers singly and in combination to chaparral habitat in southern California, most species responded by increased leaf and stem production to the application of nitrogen, but not of phosphorus, indicating that the availability of nitrogen limited production. In contrast, production of the California lilac (*Ceanothus greggii*), which harbors nitrogen-fixing bacteria in root nodules, responded to the application of phosphorus, but not of nitrogen. The productivity of annual plants (forbs and grasses) in the same habitat increased when nitrogen was applied (biomass production of 66 g m^{-2} versus 30 g m^{-2} for the control), but was depressed somewhat by the application of phosphorus

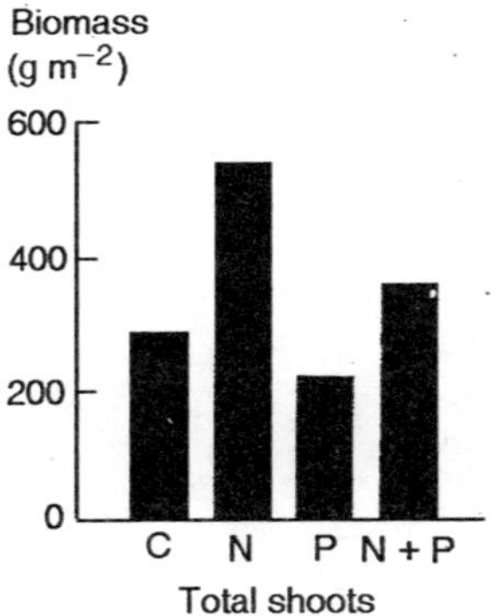

Fig. 15.6. Response of the chaparral shrub Adenostema to nitrogen (N) and phosphorus (P) fertilization (C=control).

alone (to 16 g m^{-2}). When the same amounts of nitrogen and phosphorus were added to a single plot, however, production soared to 252 g m^{-2} indicating that the plants could take advantage of increased phosphorus only in the presence of high levels of nitrogen.

Nutrient limitation generally is felt most strongly in aquatic habitats, particularly the open ocean where scarcity of dissolved minerals reduces production far below terrestrial levels. Even in shallow coastal waters, where vertical mixing, upwelling currents, and runoff from the land maintain nutrients at high concentrations, the addition of fertilizer (often inadvertently through water pollution) may greatly enhance aquatic production. In a study along the southern coast of Long Island, New York, Ryther and Dunstan (1971) found a close correlation between phytoplankton abundance and the level of inorganic phosphorus, the latter being a general indicator of pollution. But the addition of nutrients to standard cultures of the alga *Nannochloris* in water samples taken from different areas demonstrated that nitrogen rather than phosphorus limited primary production in both polluted and relatively clean waters.

Terrestrial Productivity

The favourable combination of intense sunlight, warm temperature, and abundant rainfall in the humid tropics results in the highest productivity on the earth. Low winter temperatures and long winter nights curtail production in temperate and arctic ecosystems. Within a given latitude belt, where light and temperature do not vary appreciably from one locality to the next, net production is directly related to annual precipitation. W. Webb et al. (1978) have shown that above a

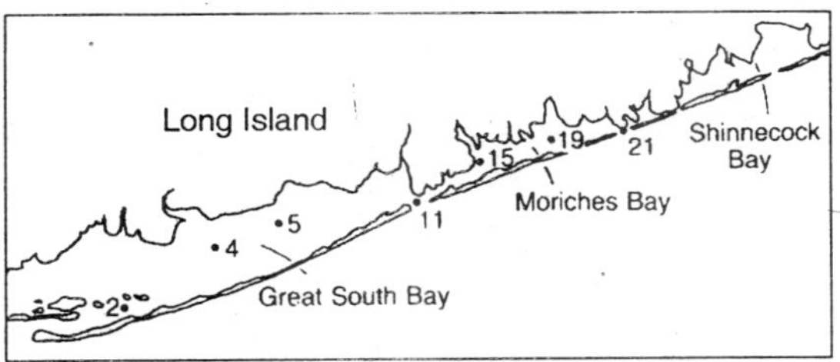

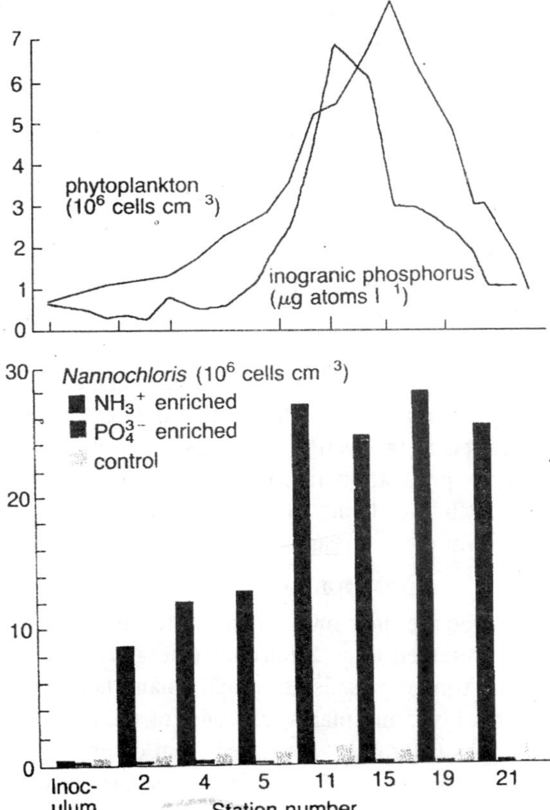

Fig. 15.7. Center: phytoplankton and inorganic phosphorus concentrations in water. Bottom: growth of standard cultures of the alga Nannochloris in water.

certain threshhold of water availability, production increases by 0.4 grams per kilogram of water in hot deserts and 1.1 g kg^{-1} in shortgrass prairies and cold (Great Basin) deserts. Forest ecosystems, whose water use efficiencies are on the order of 0.9 to 1.8 g kg^{-1}, are relatively insensitive to variation in water availability, but do not develop unless precipitation exceeds at least 50 cm annually in the cooler regions of the United States and perhaps 100 cm in the warmer eastern and southeastern areas.

Ecologists Robert Whittaker and Gene Likens (1973) estimated the net primary production for representative terrestrial and aquatic ecosystems. Their values were based on the results of many studies employing a wide variety of techniques, but what these lack in strict comparability does not override the general patterns they reveal. Whittaker and Likens's summary shows that production of terrestrial vegetation is greatest in the wet tropics and least in tundra and desert habitats. Swamp and marsh ecosystems, which occupy the interface between terrestrial and aquatic habitats, can be as productive as tropical forests.

Aquatic Productivity

The open ocean is a virtual desert, where scarcity of mineral nutrients limits productivity to a tenth or less that of temperate forests. Upwelling zones (where nutrients are brought to the surface from deeper waters) and continental-shelf areas (where exchange between shallow, nutrient-rich bottom sediments and surface waters is well developed) support greater production. In shallow estuaries, coral reefs, and coastal algal beds, production approaches that of adjacent terrestrial habitats. Primary production in freshwater habitats is comparable to that of marine habitats, being greatest in rivers, shallow lakes, and ponds, and least in clear streams and deep lakes.

Ecological Efficiencies

Plants manufacture their own "food" from raw inorganic materials. Hence they are referred to as autotrophs (literally, "self-nourishers"). Animals and most microorganisms, which obtain their energy and most of their nutrients by eating plants, animals, or their dead remains, are called heterotrophs (literally, "nourished from others"). The dual roles of living forms as food producers and food consumers give the ecosystem a trophic structure, determined by feeding relationships, through which energy flows and nutrients cycle. The food chain from grass to caterpillar to sparrow to snake to hawk delineates one particular path of energy through the trophic structure. But with each link in the food chain, much energy is dissipated before it can be consumed by organisms feeding on the next higher trophic level. All the grass in Africa piled together would dwarf a mound of all the grasshoppers, gazelles, zebras, wildebeests, and other animals that eat grass. But that mound of herbivores would be overwhelming beside the pitful heap of all the lions, hyenas, and other carnivores that feed on them.

As Lindeman (1942) pointed out, the amount of energy reaching each trophic level is determined by the net primary production and the efficiencies with which food energy is converted to biomass energy within each trophic step. Of the light energy assimilated by plants,. 15 to 70 per cent is used for maintenance and therefore is unavailable to consumers. Most herbivores and carnivores are more active than plants and expend correspondingly more of their assimilated energy on maintenance. As a result, the productivity of each trophic level is usually no more than 5 to 20 per cent that of the level below it. The percentage transfer of energy from one trophic level to the next is called both the ecological efficiency and the food chain efficiency.

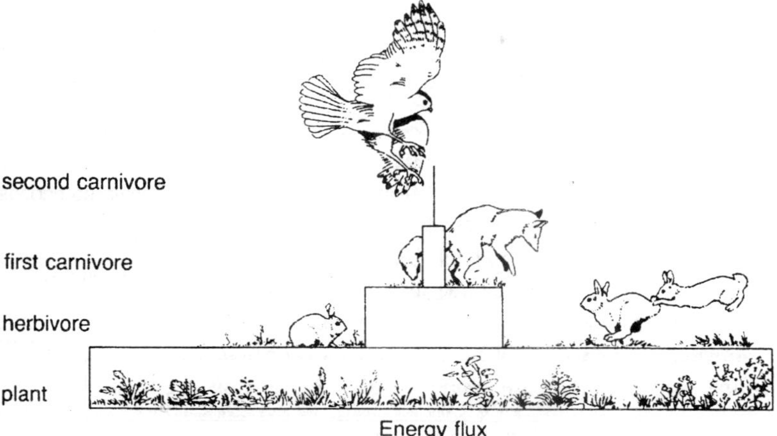

Fig. 15.8. An ecological pyramid in which the breadth of each bar represents the net productivity of each trophic level in the ecosystem.

Unit of Trophic Structure

Once food is eaten, its energy follows a variety of paths through the organism. Regardless of an organism's source of food, what it digests and absorbs is referred to as assimilated energy, which supports maintenance, builds tissues, or is excreted in unusable metabolic by-products. The energy used to fulfill metabolic needs, most of which is lost as heat, is known as respired energy. A smaller fraction of assimilated energy is excreted in the form of organic, nitrogen-containing wastes (primarily ammonia, urea, or uric acid) produced when the

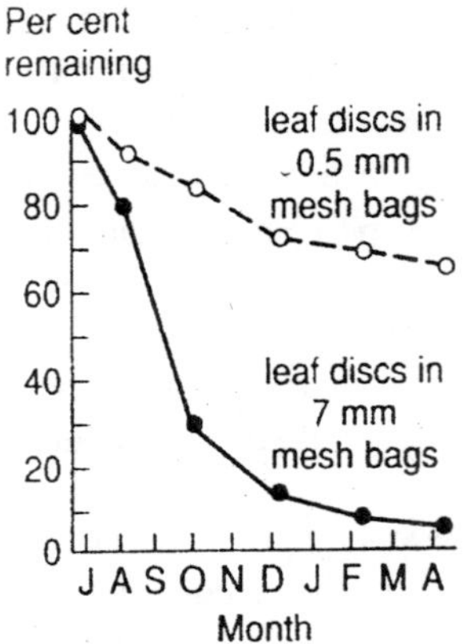

Fig. 15.9. The influence of mechanical breakdown of plant litter by large detritus feeders on its overall consumption.

diet contains an excess of nitrogen. Assimilated energy retained by the individual organism is available for the synthesis of new biomass (production) through growth and reproduction, which may then be consumed by herbivores, carnivores, and detritivores.

Not all food can be fully assimilated; hair, feathers, insect exoskeletons, cartilage and bone in animal foods, and cellulose and lignin in plant foods resist digestion by most animals. These materials are egested either by defecation or by regurgitation of pellets of undigested remains. Some egested wastes are substances that have been relatively unaltered chemically during their passage through an organism, but nearly all have been mechanically broken up into fragments by chewing and by contractions of the stomach and intestines, which makes them more readily usable by detritus feeders.

Assimilation and Productivity

Ecological efficiency is the product of efficiencies with which organisms exploit their food resources and convert them into biomass available to the next higher trophic level. Because most biological production is consumed, exploitation efficiency is 100 per cent overall, and ecological efficiency depends on two factors: the proportion of consumed energy assimilated, and the proportion of assimilated energy incorporated in growth, storage, and reproduction. The first proportion is called the *assimilation efficiency* and the second, the *net production efficiency*. The product of the assimilation and net production efficiencies is the gross production efficiency: the proportion of food energy that is converted to consumer biomass energy.

For plants, net production efficiency is defined as the ratio of net to gross production. This index has been found to vary between 30 and 85 per cent, depending on habitat and growth form. Rapidly growing plants in temperate zones, whether trees, old-field herbs, crop species, or aquatic plants, have uniformly high net production efficiencies (75 to 85 per cent). Similar vegetation types in the tropics exhibit lower net production efficiencies, perhaps 40 to 60 per cent. Thus, respiration increases relative to photosynthesis at low latitudes.

Animal food is more easily digested than plant food; assimilation efficiencies of predatory species vary from 60 to 90 per cent. Vertebrate prey are digested more efficiently than insect prey because the indigestible exoskeletons of insects constitute a larger proportion of the body than the hair, feathers, and scales of vertebrates. Assimilation efficiencies of insectivores vary between 70 and 80 per cent, whereas those of most carnivores are about 90 per cent.

The nutritional value of plant foods depends upon the amount of cellulose, lignin, and other indigestible materials present. Herbivores assimilate as much as 80 per cent of the energy in seeds, and 60 to 70 per cent of that in young vegetation. Most grazers and browsers (elephants, cattle, grasshoppers) assimilate 30 to 40 per cent of the energy in their food. Millipedes, which eat decaying wood composed mostly of cellulose and lignin (and the microorganisms that occur in decaying wood), assimilate only 15 per cent.

The effect of food quality on assimilation can be seen in the efficiencies with which a single animal retains energy from different portions of its diet. In studies with mountain hares (*Lepus timidus*), Ake Pehrson (1983) determined that the efficiency of energy assimilation on a diet of small willow twigs was 39 per cent, of which 5 per cent was lost through urinary nitrogen excretion. Assimilation efficiencies were lower on larger twigs (31 per cent), presumably because of their thick, less digestible bark, and on birch twigs (23 to 35 per cent), on which hares could not maintain a constant weight. By measuring the content of fiber (cellulose and lignin) in the food and feces, Pehrson found that the digestibility of fiber was between only 15 and 25 per cent. Assimilation of nutritional components of the diet, estimated by similar input-output measurements (for example, on willow twigs 3 to 5 mm in diameter), varied between 9 per cent for phosphorus and 81 per cent for magnesium.

Maintenance, movement, and, in warm-blooded animals, heat production require energy that otherwise could be utilized for growth

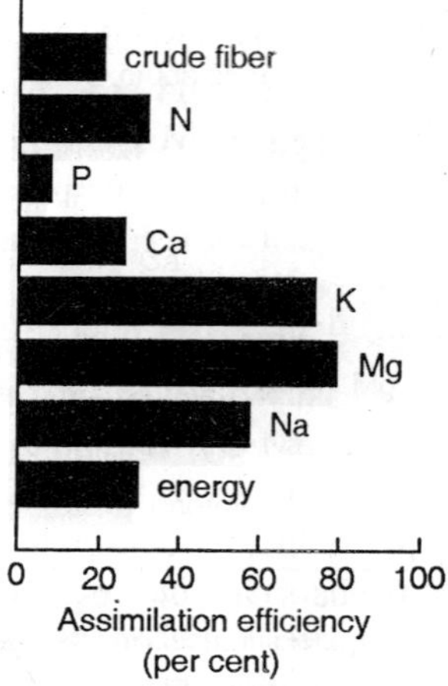

Fig. 15.10. Proportion of dietary components assimilated by caged mountain hares (Lepus timidus) during the winter.

and reproduction. Active, warm-blooded animals exhibit low net production efficiencies: birds less than 1 per cent, small mammals with high reproductive rates up to 6 per cent. More sedentary, cold-blooded animals, particularly aquatic species, channel as much as 75 per cent of their assimilated energy into growth and reproduction. The extreme high value approaches the biochemical efficiency of egg production and tissue growth, between 70 and 80 per cent in domesticated animals.

The efficiency of biomass production within a trophic level (gross production efficiency) is the product of assimilation efficiency and net production efficiency. Gross production efficiencies of warm-blooded, terrestrial animals rarely exceed 5 per cent, and those of some birds and large mammals fall below 1 per cent. Gross production efficiencies of insects lie within the range of 5 to 15 per cent, and those of some aquatic animals exceed 30 per cent.

Terrestrial plants, especially woody species, allocate much of their production to structures that are difficult to ingest, let alone digest. As a result, most terrestrial plant production is consumed as detritus by organisms specialized to attack wood and leaf litter. This partitioning between herbivory and detritus feeding establishes two food chains in terrestrial communities. The first originates as relatively large animals feed on leafy vegetation, fruits, and seeds; the second originates with relatively small animals and microorganisms consuming detritus in the litter and soil layer. These separate food chains sometimes mingle

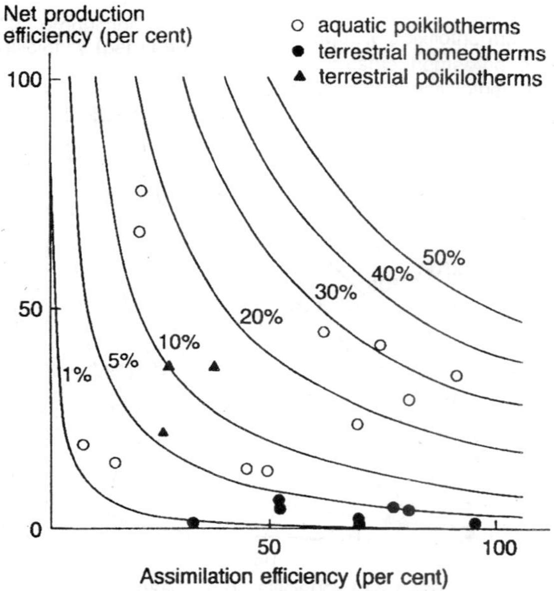

Fig. 15.11. Relationships between assimilation efficiency and net production efficiency for a variety of animals.

considerably at higher trophic levels, but the energy of detritus tends to move into the food chain much more slowly than the energy consumed by herbivores.

The relative importance of predatory-based and detritus-based food chains also varies greatly among communities. Predators predominate in plankton communities, detritus feeders in terrestrial communities. The proportion of net production that enters herbivore-predator strands of the food web depends on the relative allocation of plant tissue between structural and supportive functions, on one hand, and growth and photosynthetic functions, on the other. Herbivores consume 1.5 to 2.5 per cent of the net production of temperate deciduous forests, 12 per cent of that in old-field habitats, and 60 to 99 per cent in plankton communities.

Energy Flow in Community

Food chain efficiencies indicate the amount of energy that eventually reaches each trophic level of the community. A second component of the energy dynamics of the ecosystem is the rate of transfer of energy or, inversely, its residence time in each trophic level. For a given rate of production, the residence (or transit) time of energy in the community and the storage of energy in living biomass and detritus are directly related: the longer the residence time, the greater the accumulation.

The average residence time in a particular link in the food chain is equal to the energy stored divided by the rate of flux, or

$$\text{residence time (yr)} = \frac{\text{biomass (kJ m}^{-2})}{\text{net productivity (kJ m}^{-2}\ \text{yr}^{-1})}.$$

(the residence time defined by this equation is sometimes calculated in terms of mass rather than energy and is then referred to as the biomass accumulation ratio). According to Whittaker and Likens (1973), wet tropical forests produce an average of 2000 grams of dry matter per square meter per year and have an average living biomass of 45,000 grams per square meter. Inserting these values into the equation for average residence time, we obtain 22.5 years (45,000/2000). Average residence times of energy in plants in representative ecosystems vary from more than 20 years in forested terrestrial environments to less than 20 days in aquatic plankton-based communities. These figures underestimate the total residence time of energy in primary production, however, because they do not include the accumulation of dead organic matter in the litter.

The residence time of energy in accumulated litter can be estimated by an equation analogous to that for the biomass accumulation ratio

$$\text{residence time (yr)} = \frac{\text{litter accumulation (g m}^{-2})}{\text{rate of litter fall (g m}^{-1})}$$

In forests, the average residence time for energy varies from 3 months in the wet tropics to 1 to 2 years in dry and montane tropical habitats, 4 to 16 years in the southeastern United States, and more than 100 years in temperate mountains and boreal regions. Warm temperature and abundance of moisture in lowland tropical regions create optimum conditions for rapid decomposition.

Radioactive Tracers

Organic compounds containing energy can be labeled with a radioactive element, and their movement followed. In radioactive tracer studies, investigators label plants (or water in an aquatic system) with a radioactive isotope, usually phosphorus-32 (^{32}P) applied in a phosphate solution. They then collect consumer species at intervals after the initial labeling and examine them with a radiation counter.

Eugene P. Odum and his co-workers at the University of Georgia used this technique to follow the movement of ^{32}P through components of an old-field community. They labeled a common plant—telegraph weed (*Heterotheca*)—with drops of ^{32}P-phosphate solution placed directly on leaves, and then collected insects, snails, and spiders at intervals of several days for about 5 weeks. Certain herbivores, notably crickets and ants, began to accumulate ^{32}P within a few days of its initial application and attained peak amounts of the tracer at 2 to 3 weeks. Ground-living, detritus-feeding insects (carabid and tenebrionid beetles, and gryllid crickets), and predatory spiders did not accumulate peak amounts of tracer until 3 weeks after the start of the experiment. Thus, the ^{32}P label appeared first in herbivores and later in detritivores and predators. Moreover, even though the average residence time of primary production in temperate grasslands is on the order of 3 years, these experiments show that a portion begins to move up the food chain very quickly.

A similar study on a small trout stream in Michigan gave results comparable to those of Odum's: ^{32}P-phosphate was added to the stream at one point and its accumulation in plants and animals downstream from the release site was monitored for 2 months. The median time for each population to accumulate its maximum concentration of ^{32}P varied from a few days for aquatic plants to 1 to 2 weeks for filter feeders and other herbivores, 3 to 4 weeks for omnivores, 4 to 5 weeks for detritus feeders, and 4 weeks to more than 2 months for most predators. The results suggest that most of the energy assimilated in aquatic ecosystems is dissipated within a few weeks, although a small portion may linger for months in the predator food chain, and perhaps for years in organic sediments on the bottoms of streams and lakes.

Function of Ecosystems

Flux of energy and its efficiency of transfer summarize certain aspects of the structure of an ecosystem: the number of trophic levels, the relative importance of detritus and predatory feeding, the steady-state values for biomass and accumulated detritus, and the turnover rates of organic matter in the community. The importance of these measures was argued by Lindeman (1942), who constructed the first energy budget for an entire biological community — that of Cedar Bog Lake in Minnesota. The proliferation of energy flow studies during the 1950s and 1960s more clearly reflected energy's value as a universal currency, a common denominator to which all populations and their acts of consumption could be reduced.

The overall energy budget of the ecosystem reflects the balance between income and expenditure, just as it does in a bank account. The ecosystem gains energy through photosynthetic assimilation of light by green plants and the transport of organic matter into the system from outside. Sources of the latter are referred to as allochthonous inputs (from the Greek *chthonos*, of the earth, and *allos*, other); local production is referred to as autochthonous.

In Root Spring, near Concord, Massachusetts, the energy flux of herbivores is 0.31 W m^{-2}, but the net productivity of aquatic plants is only 0.09 W m^{-2}, the balance being transported into the spring by leaf fall from nearby trees. G. W. Minshall (1978) surveyed the relative amounts of autochthonous and allochthonous inputs in aquatic systems: large rivers, lakes, and most marine ecosystems are dominated by autochthonous production; allochthonous imports were most important in small streams and springs under the closed canopies of forests; life in caves and abyssal depths of the oceans, to which no light penetrates, subsists entirely on energy transported in from outside.

Lindeman constructed the Cedar Bog Lake energy budget from measurements of the harvestable net production on each of three trophic levels (all carnivores were lumped together) and from laboratory determinations of respiration and assimilation efficiencies. The animals and plants collected at the end of the growing season constituted the net production of the trophic levels to which each species was assigned.

	Harvestable production	
Trophic level	(kJ m^{-2} y^{-1})	(W m^{-2})
Primary producers (green plants)	2944	0.0934
Primary consumers (herbivores)	293	0.0093
Secondary consumers (carnivores)	54	0.0017

Lindeman estimated the energy dissipated by respiration from ratios of respiratory metabolism to production measured in the laboratory: 0.33 for aquatic plants, 0.63 for herbivores, and 1.4 for the more active carnivores in the lake. He calculated the gross production of carnivores as the sum of their harvestable production (54 kJ m^{-2} y^{-1}) and respiration ($54 \times 1.4 = 76$ kJ m^{-2} y^{-1}), a total of 130 kJ m^{-2} y^{-1}. He then estimated the gross production of primary consumers as the sum of their harvestable production, respiration (production $\times$ 0.63), and the consumption of primary consumers by secondary consumers. In making the last calculation, Lindeman assumed that the assimilation efficiencies of secondary consumers were 90 per cent. By backtracking another step in this manner, he calculated the production of primary producers as well.

Lindeman's energy budget is somewhat startling in that organisms at the next higher trophic level failed to consume 83 per cent of the net primary production and 70 per cent of the net secondary production. This surplus production is transported out of the system by sedimentation; the lake is filling with layers of organic detritus.

Even with sedimentation, the overall ecological efficiency of energy transfer between trophic levels in Cedar Bog Lake was about 12 per cent. After comparing similar analyses for five aquatic communities, D. G. Kozlovski (1968) concluded that (1) assimilation efficiency increases at higher trophic levels; (2) net production efficiency decreases at higher levels; (3) gross production efficiency also decreases; and (4) ecological efficiency (assimilation or gross production at level n divided by that at level $n - 1$) remains constant between trophic levels at about 10 per cent.

Length of Food Chains

Kozlovski's 10 per cent generalization is not a fixed law of ecological thermodynamics. In Silver Springs, Florida, Odum (1957) measured ecological efficiencies of 17 per cent between the producer and herbivore level but only 5 per cent between herbivores and secondary consumers. Ecological efficiencies are usually lower in terrestrial habitats, and a useful rule of thumb states that the top carnivores in terrestrial communities can feed no higher than the third trophic level on average, whereas aquatic carnivores may feed as high as the fourth or fifth levels. This is not to say that there can be no more than three links in a terrestrial food chain; a tiny fraction of the total energy may travel through a dozen links before it is dissipated by respiration. Such high trophic levels do not, however, contain enough energy to fully support a single predator population.

We can estimate the average length of food chains in a community from net primary production, average ecological efficiency, and average energy flux of a top predator population. The energy available [$E(n)$] to a predator at a given trophic level [n (plants being level 1)] is equal to the product of the net primary production (*NNP*) and the intervening ecological efficiencies (*Eff*). Thus

$$E(n) = NNP\ Eff^{n-1}$$

where *Eff* is the geometric mean of the efficiencies of transfer between each level. This expression can be rearranged to give

$$n = 1 + \frac{\log[E(n)] - \log(NPP)}{\log(Eff)}$$

Using this equation and some rough estimates for the values on its right-hand side, we can calculate the average number of trophic levels to be about 7 for marine plankton-based systems, 5 for inshore aquatic communities, 4 for grasslands, and 3 for wet tropical forests. These estimates should be taken with a grain of salt, to be sure, but they do indicate the general size of the pyramid of energy built upon a base of primary production within an ecosystem.

16

Elemental Cycles

Nutrients, unlike energy, are retained within the ecosystem, where they are continually recycled between organisms and the physical environment. Most of these nutrients originate in rocks of the earth's crust. But because they are made available very slowly, high rates of organic production require that materials assimilated by one organism can be reused by others.

All the energy assimilated by green plants is "new" energy received from outside the ecosystem. In contrast, most of the nutrients assimilated by plants have been used before, many of them quite recently. Nitrates absorbed from the soil by roots might have been released from decaying leaves on the forest floor during the same day. The carbon dioxide assimilated by every terrestrial green plant was produced by animal, plant, or microbial respiration. On average, a molecule of carbon dioxide resides in the atmosphere for about 8 years before being reassimilated, although carbon dioxide can be recycled almost instantaneously (for example, within the closed canopy of a forest) before it escapes to the general atmospheric circulation.

Each element follows a unique route, determined by its particular biochemical transformations, in its cycle through the ecosystem. Living systems transform elements and their compounds in order to provide nutrients for building structure and to mobilize the energy required by all life processes. Over long periods, processes that transform elements from one form to a second form must be balanced by processes that restore the first.

Elemental cycles sometimes do become unbalanced, and elements accumulate in or are removed from the system. For example, during periods of coal and peat formation, dead organic materials accumulate in the sediments of lakes, marshes, and shallow seas where anaerobic conditions slow their decomposition by microorganisms. In other instances, under intensive cultivation or following removal of natural vegetation, erosion can wash away nutrient-laden layers of soil that took centuries to develop. By and large, however, most systems exist in a steady state in which export of elements is approximately balanced by import. Furthermore, gains and losses usually are small compared to the rate at which nutrients are cycled within the system.

This chapter is devoted to describing the pathways of elements in relation to the chemical and biochemical transformations responsible for their cycling and the conditions of the environment that influence these transformations.

Movement of Elements

Transformations that result in the production of organic forms of a particular element are referred to as assimilatory processes. Photosynthesis, by which "inorganic" (oxidized) carbon (carbon dioxide) is reduced to the "organic" carbon of carbohydrates, is the most obvious assimilatory transformation of an element, although others involving nitrogen and sulfur, for example, are equally important. In the overall cycling of carbon, photosynthesis is balanced by respiration, a complementary dissimilatory process that involves the oxidation of organic carbon with the accompanying release of energy and the return of carbon to its available inorganic form.

Not all transformations of elements in the ecosystem are biological, nor do all involve the net assimilation or release of useful quantities of energy. Many chemical reactions take place in the air, soil, and water. Some of these, like the weathering of bedrock, release certain elements (potassium, phosphorus, and silicon, for example) to the ecosystem. Other physical and chemical process, such as the sedimentation of calcium carbonate in the oceans, remove elements from ecosystem circulation to rocks in the earth's crust, where they can remain locked up for eons.

Most energy transformations are associated with biochemical oxidation and reduction of carbon, oxygen, nitrogen, phosphorus, and sulfur. In each case, an energy-releasing transformation is coupled with an energy-requiring transformation, so that energy is transferred from the reactants in the first to the products in the second. When more energy is released by the first than required by the second, the

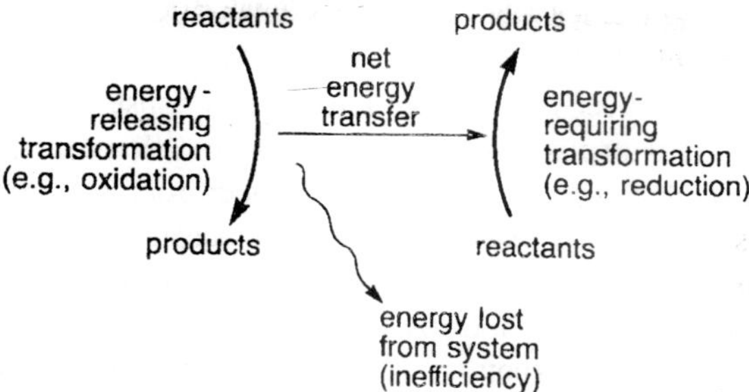

Fig. 16.1. The coupling of energy-releasing and energy-requiring transformations is the basis of energy flow in the ecosystem.

balance is lost as heat—hence the thermodynamic inefficiency of life processes. A typical coupling between transformations might involve oxidation of carbon in carbohydrate (perhaps glucose), with the release of energy, and reduction of nitrate-nitrogen to amino-nitrogen, which requires energy. These particular transformations are not directly linked; many intermediate steps required for the transfer of energy between them.

The initial input of energy into the ecosystem is accomplished by an assimilatory transformation—the reduction of carbon—in which the source of energy is light rather than a coupled dissimilatory process. A portion of that energy is lost with each subsequent transformation. Some of these involve assimilation of other elements required for growth and reproduction; most involve biochemical transformations required for maintaining the cell environment and for movement. The cycling of elements between living and physical parts of the ecosystem is related to energy flow by the coupling of the dissimilatory part of one cycle to the assimilatory part of another.

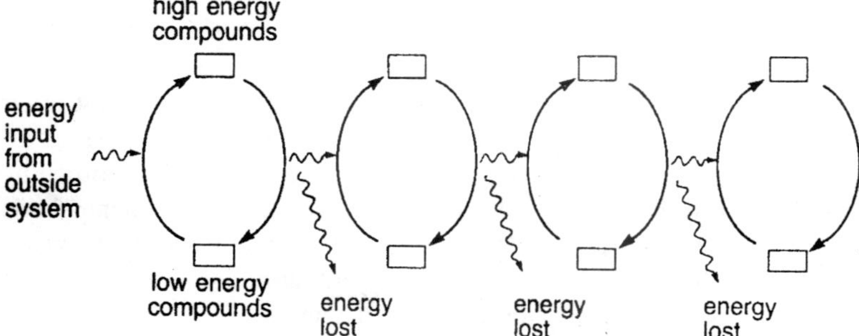

Fig. 16.2. As energy flows through the ecosystem, elements alternate assimilatory and dissimilatory transformations, thus going through cycles.

Cycling of Elements in the Ecosystem

Each form of an element can be thought of as occupying a separate compartment in the ecosystem, like a room in a house. In this analogy, biochemical transformations are fluxes of the elements among the compartments; that is, their movement between rooms. Carbon occurs in the forms of carbon dioxide both in the atmosphere and dissolved in water, carbonate and bicarbonate ions dissolved in water, calcium carbonate (lime-stone) sediments, and all organic molecules. Photosynthesis moves carbon from the carbon dioxide compartment to that containing organic forms of carbon (assimilation); respiration brings it back (dissimilation). The compartment of organic carbon has many subcompartments: animals, plants, microorganisms, and detritus. Herbivory, predation, and detritus feeding move carbon between subcompartments.

Some transformations involve changes in energy state. Photosynthesis adds energy to carbon, which may be thought of as lifting the element to the second floor of a house. In descending the respiration "staircase," carbon releases this stored chemical energy, which can then be used for other purposes.

Both organic and inorganic forms of elements may be removed from rapid circulation within the ecosystem to compartments that are not readily accessible to transforming agents. For example, coal, oil, and peat contain vast quantities of carbon removed from the ecosystem. Sedimentation removes inorganic (oxidized) carbon from ecosystem circulation primarily by sedimentation as calcium carbonate, which forms thick layers of limestone over large areas presently or formerly covered by seas. These forms of carbon are returned to the rapid cycling compartments only by the slow geological processes of uplift and erosion. Of course, man has greatly accelerated one component of this flux by burning fossil fuels.

Water Cycle

Although water is chemically involved in photosynthesis, most water flux through the ecosystem occurs by the physical processes of evaporation, transpiration, and precipitation. Light energy absorbed by water performs the work of evaporation. The condensation of atmospheric water vapour to form clouds releases the potential energy in water vapour as heat. Thus, evaporation and condensation resemble photosynthesis and respiration thermodynamically.

More than 90 per cent of all water is locked up in rocks in the earth's core and in sedimentary deposits near the surface. Water from

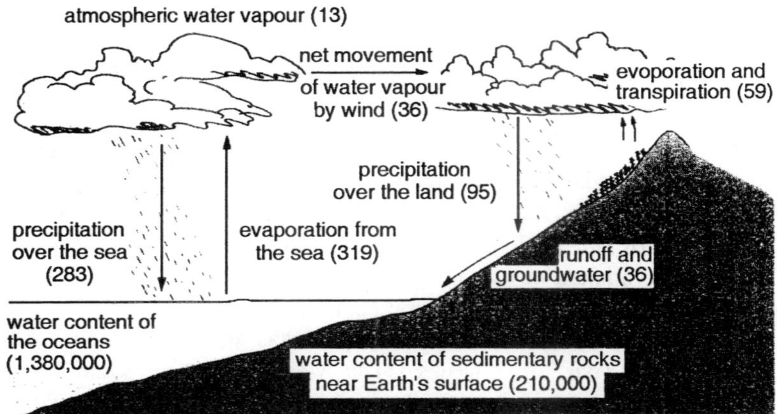

Fig. 16.3. The water cycle, with its major components expressed on a global scale.

this compartment enters the hydrological cycles of ecosystems through such geological processes as volcanic outpourings of steam; most of the water at the earth's surface originated in this way, although the great reservoirs in the interior contribute little to contemporary water flux in the ecosystem.

Over land surfaces, precipitation (23 per cent of the global total) exceeds evaporation and transpiration (16 per cent). Evaporation of water from the oceans exceeds replenishment from precipitation. Much of the water vapour that winds transport from the oceans to the land condenses and falls as precipitation over mountainous regions and where rapid heating of the land surface creates vertical air currents. The net flow of atmospheric water vapour from ocean to land is balanced by runoff from the land into ocean basins.

The energy that drives the hydrological cycle approximately equals the global evaporation of water (378×10^{18} g yr^{-1}) times the energy required to evaporate 1 gram of water (2.24 kJ). The product, 8.5×10^{20} kJ yr^{-1}, represents about one-fifth of the total energy income from the sun's radiation striking the earth. The remaining energy is either reflected, or absorbed and reradiated as heat. Of the total radiation striking temperate forests, for example, plants absorb about 40 per cent, of which they dissipate two-thirds through evapotranspiration. Only a few per cent, at most, are assimilated by photosynthesis.

Evaporation, not precipitation, determines the flux of water through the ecosystem. The absorption of radiant energy by liquid water couples an energy source to the hydrological cycle. Evaporation and precipitation are closely linked because the atmosphere has a limited capacity to hold water vapour; any increase in evaporation of water into the atmosphere creates an excess of vapour and results in an equal increase in precipitation.

Water vapour in the atmosphere at any one time (the compartment size) corresponds to an average of 2.5 centimeters (1 inch) of water spread evenly over the surface of the earth. An average of 65 cm (26 in) of rain or snow falls each year (the water flux), which is 26 times the average amount of water vapour. The steady-state content of water in the atmosphere—the atmospheric pool—is therefore replaced 26 times each year on average. (Conversely, water has an average transit time of 1/26 of a year, or 2 weeks.) The water content of soils, rivers, lakes, and oceans exceeds 100,000 times that of the atmosphere. Fluxes through both pools are the same, however, because evaporation balances precipitation. Thus, the average transit time of water in its liquid form at the earth's surface (about 3650 years) is 100,000 times longer than in the atmosphere.

Redox Potential

Water gains energy as it evaporates; hence, the energy content of water vapour and its capacity to release energy exceed those of liquid water. The chemical energy of a compound is indicated by its ability to reduce other compounds and, conversely, by its ability to become oxidized. In chemical reactions, an atom is oxidized when it donates an electron to another atom, which becomes reduced by accepting the electron. An oxidizing agent is therefore one that accepts electrons. The more readily a substance accepts electrons, the greater its oxidizing potential. A reducing agent donates electrons, and its strength depends on how readily it does so. Giving and taking electrons are opposite sides of the same coin; each substance may serve as an oxidizing or a reducing agent. But because strong oxidizers hold onto electrons tightly, they are always weak reducers, and vice versa. Put another way, a strong oxidizer can always oxidize (be reduced by) a weaker oxidizer.

Every oxidation-reduction (redox) reaction can be characterized by a pair of half-reactions, one for the reduction (electron-accepting) step and the other for the oxidation (electron-donating) step. For example, when molecular oxygen acts as an oxidizing agent, each atom accepts 2 electrons; i.e., $O_2 + 4e^- = 2O_2$. In this reduced form, oxygen readily combines with any of a variety of positive ions, such as H^+, giving

$$O_2 + 4e^- + 4H^+ = 2H_2O$$

or C^{4+}, giving

$$O_2 + 4e^- + C^{4+} = CO_2$$

These half-reactions include only the reduction step. The electrons must come from an oxidation half-reaction; for example,

$$CH_2O = C^{4+} + H_2O + 4e^-$$

In which case, the carbon atom donates electrons (it is oxidized) and the electrons are available to reduce an oxidizer. The overall reaction combining the last two half-reactions has the form

$$CH_2O + O_2 = CO_2 + H_2O$$

which is the familiar equation for respiration.

The relative strengths of oxidizers and reducers are quantified by their electrical potentials (Eh) measured in volts (V). (An electrical current represents the movement of electrons; a chemical battery generates an electrical potential by a pair of redox reactions, one at each electrode.) Because oxygen has a high redox potential (0.81 volts at pH 7 and 25 °C), it is an excellent oxidizer. So are nitrate (NO_3^-), ferric iron (Fe^{3+}), and, to a lesser extent, sulfate (SO_4^{2-}). Near the bottom of the redox scale are carbon dioxide (CO_2) and molecular nitrogen (N_2). The electrical potentials of the half-reactions reducing CO_2-carbon (C^{4+}) to organic carbon (C^0, –0.42 V) and N_2 to organic (ammonia)-nitrogen (N^{3-}; –0.25 V) are unfavourable thermodynamically.

Any oxidization reaction having a high redox potential coupled with a reduction reaction having a low redox potential can proceed with the release of energy. Thus the oxidation of carbon by oxygen (O_2) releases energy because oxygen is a stronger oxidizing agent than carbon dioxide. For the reaction to proceed in the opposite direction (for example, fixation of carbon by photosynthesis), energy must be added.

Oxygen and carbon are not the only electron acceptors and donors of importance to biological reactions. Nitrate is a powerful oxidizer; hydrogen sulfide (H_2S) and ammonium (NH_3) are good reducing agents. Indeed, the synthesis and metabolism of many organic compounds, including amino acids, involve assimilatory and dissimilatory redox reactions of nitrogen and sulfur. In general, oxygen is the oxidizer of choice owing to its ubiquity and the fact that its common reduced form (H_2O) is innocuous. In contrast, the reduction of nitrate produces nitrite (NO_2^-), which most organisms can tolerate only in minute quantities. Oxidizers such as nitrate and ferric iron can nonetheless become important in soils and aquatic sediments when oxygen has been depleted (anaerobic conditions), as we shall see below.

When all other oxidizers have been exhausted (extremely reduced conditions), hydrogen ion can be used as an electron acceptor: $2H^+ + 2e^- = H_2$. This reaction is, however, thermodynamically very unfavourable (redox potential = –0.41 V) and therefore requires considerable energy. Organic carbon also can be an electron acceptor when other oxidizing agents are absent:

$$CH_2O + 2H^+ + 4e^- = CH_4 + O^{2-}$$

This reaction occurs in anaerobic fermentations in some oxygen-depleted sediments and, notably, in the rumens of ungulates, which lack inorganic electron acceptors. Methane, or natural gas (CH_4), is one of the end-products of this set of reactions.

Hydrogen can be a powerful electron donor, by the half-reaction $H_2 = 2H^+ + 2e^-$, depending on the hydrogen ion concentration of the environment, or pH. At lower pH (hydrogen ions more abundant, more acid conditions), hydrogen gives up electrons less readily and the electrical potential of the redox reaction increases (that is, hydrogen is a poorer reducing agent). As the pH increases, the supply of hydrogen ions (H^+) decreases, the equilibrium of the reaction $H_2 = 2H^+ + 2e^-$ shifts to the right, and electrode potentials decrease.

With respect to oxidation-reduction reactions, an environment may be characterized with respect to pH and electrical potential (Eh). The first measures the availability of hydrogen ions, the second the availability of electrons. Because hydrogen and oxygen are so abundant, their reactions limit naturally occurring values of pH and Eh. For a

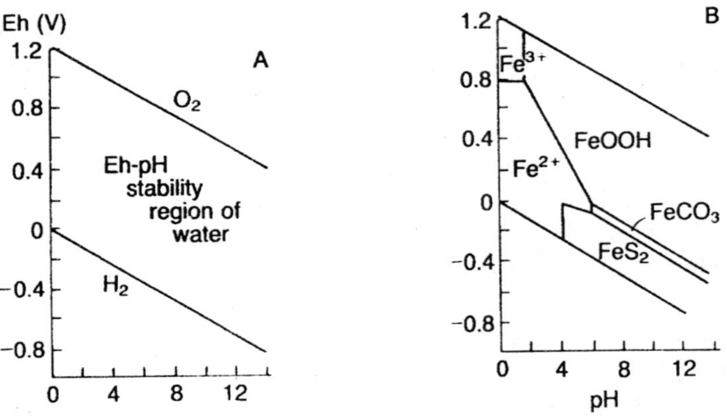

Fig. 16.4. Distribution of pH and electrical potentials of soils and the ion compounds associated with these condition. A—The stability region for water sets outer limits to soil conditions. B—Stability regions for various iron ions and compounds.

given pH, electrical potential has a practical upper bound set by the reaction $O_2 + 2e^- = 2O^{2-}$ (Eh = 0.81 V at pH 7) and a lower bound set by the reaction $2H^+ + 2e^- = H_2$ (Eh = –0.41 V). When a powerful oxidizer raises the Eh above 0.81 V, it oxidizes the O_2 in water, liberating oxygen (O^{2-}), until the oxidizing agent is fully consumed (reduced) and Eh drops to that of O_2. When a powerful reducing agent decreases the Eh below –0.41 V, it reduces hydrogen ions to molecular hydrogen (H_2) until the reducing agent is fully oxidized. In general, high values of Eh occur in well-oxygenated soils; lower values occur in waterlogged soils and the deep waters and sediments of lakes and estuaries, into which oxygen diffuses from the atmosphere much more slowly than it is consumed by the respiration of organisms.

Each element that undergoes redox reactions assumes a predominant form that may vary depending upon the Eh and pH of the environment. For example, iron occurs in its ferric form (Fe^{3+}) under oxidizing conditions but in its ferrous form (Fe^{2+}) under reducing conditions. Under acidic conditions, both forms tend to occur as dissociated ions, but as the hydrogen ion concentration decreases, they more readily form insoluble compounds (hydroxides, carbonates, and sulfides). As we shall see below, conditions of pH and Eh have important consequences for the cycling of many elements.

Carbon Cycle

Three major classes of process are involved in the cycling of carbon in aquatic and terrestrial systems. The first includes the assimilatory and dissimilatory redox reactions of carbon in photosynthesis and respiration. These are the major energy-transforming reactions of life. Approximately 10^{17} grams (10^{11} metric tons) of carbon enter into such reactions worldwide each year.

The second class of process is the physical exchange of carbon dioxide between the atmosphere and oceans, lakes, and streams. Carbon dioxide dissolves readily in water; indeed, the oceans contain about 50 times as much CO_2 as the atmosphere. Exchange across the air-water boundary links the carbon cycles of terrestrial and aquatic systems, but the two may be considered virtually independent because production balances consumption of CO_2 in both, and dissolved and atmospheric CO_2 achieve an approximate equilibrium globally.

The third class of process is the solution and precipitation (deposition) of carbonate compounds as sediments, particularly limestone and dolomite. On a global scale, these processes are approximately balanced, although certain conditions favouring precipitation have led to the deposition of extensive layers of calcium carbonate-rich sediments in the past. In aquatic systems, dissolution and deposition occur about two orders of magnitude more slowly than assimilation and dissimilation. Thus, exchange between sediments and the water column is relatively unimportant to the short-term cycling of carbon in the ecosystem. Locally, and over long periods, it can assume much greater importance; in fact, most of the ecosystem's carbon resides in sedimentary rocks.

When carbon dioxide dissolves in water, it forms carbonic acid

$$CO_2 + H_2O \rightleftharpoons H_2CO_3$$

which readily dissociates into bicarbonate and carbonate ions,

$$H_2CO_3 \rightleftharpoons H^+ + HCO_3^-$$
$$HCO_3^- \rightleftharpoons H^+ + CO_3^{2-}$$

At low pH, abundant hydrogen ions in the environment drive these reactions to the left. When calcium is present, it also equilibrates with the carbonate and bicarbonate ions

$$CaCO_3 \rightleftharpoons Ca^{2+} + CO_3^{2-}$$

Calcium carbonate ($CaCO_3$) has low solubility under most conditions and readily precipitates out of the water column.

Under acid conditions, carbonate ion is removed from the system as the equilibrium $H^+ + CO_3^{2-} \rightleftharpoons HCO_3^-$ shifts to the right and the dissociation of $CaCO_3$ increases. This reduces the amount of dissolved calcium carbonate in the water, which, in turn, enhances the dissolution of calcium carbonate rocks and sediments that the water passes over. By this process, slightly acid streams and groundwater dissolve limestone sediments (and acid rain erodes marble statuary). As these calcium-laden fresh waters enter the oceans and mix with water of nearly neutral pH, the equilibria shift back toward undissociated

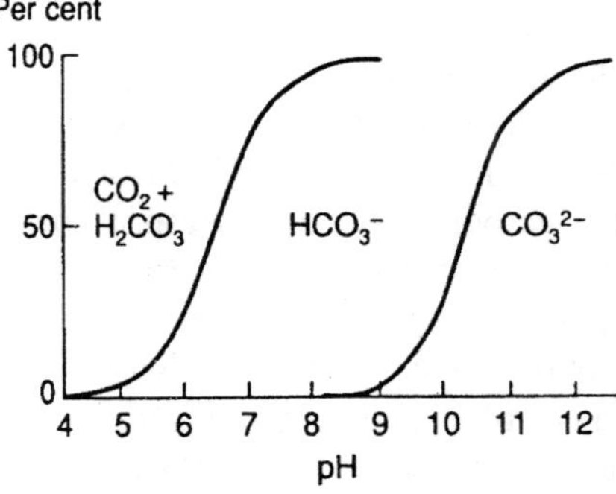

Fig. 16.5. Proportion of carbon as carbonic acid, bicarbonate, and carbonate in solution as a function of pH.

forms of carbonate, and calcium carbonate precipitates out of the water column to form sediments.

Dissolution and dissociation may be affected locally by the activities of organisms. In the marine system, under approximately neutral conditions, the carbonate system has the overall equilibrium

$$CaCO_3 \text{ (insoluble)} + H_2O + CO_2 \rightleftharpoons Ca(HCO_3)_2 \text{ (soluble)}$$

Removal of CO_2 by photosynthesis shifts the equilibrium to the left, resulting in the formation and precipitation of calcium carbonate. Many algae excrete this calcium carbonate to surrounding water, but reef-building algae and coralline algae incorporate the substance as hard structure. In the system as a whole, when photosynthesis exceeds respiration, as it does during algal blooms, calcium tends to be precipitated out of the system.

Nitrogen Cycle

The ultimate source of nitrogen to the ecosystem is molecular nitrogen in the atmosphere. This dissolves to some extent in water but none is found in native rock. So, were it not for biological processes, under the present oxidizing conditions at the earth's surface virtually all nitrogen would occur in its molecular form.

Molecular nitrogen enters biological pathways of the nitrogen cycle owing to assimilation (nitrogen fixation) by certain microorganisms.

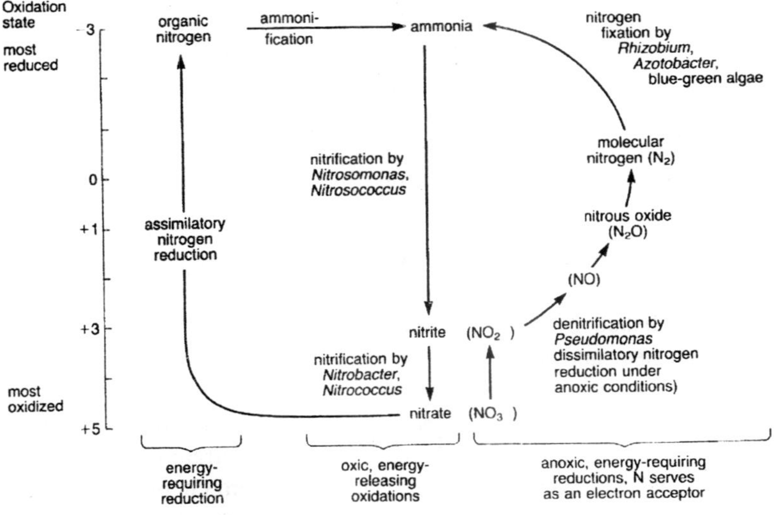

Fig. 16.6. Schematic diagram of transformations and oxidation states of compounds in the nitrogen cycle.

Although this pathway ($N_2 \rightarrow NH_3$) constitutes a small fraction of the earth's annual nitrogen flux, all biologically cycled nitrogen can be traced back to nitrogen fixation. Once in the biological realm, the cycling of nitrogen is much more complicated than that of carbon owing to its numerous oxidation states. Beginning with reduced (organic) nitrogen, the first step in the nitrogen cycle is ammonification: the hydrolysis of protein and oxidation of amino acids, which is accomplished by all organisms. During the initial breakdown of amino acids, some carbon is oxidized, releasing energy, but the oxidation state of nitrogen does not change.

Nitrification involves the oxidation of nitrogen, first from ammonia to nitrite, then from nitrite to nitrate. Both steps are carried out only by specialized bacteria: $NH_3 \rightarrow NO_2^-$ ($N^{3-} \rightarrow N^{3+}$) by *Nitrosomonas* in the soil and by *Nitrosococcus* in marine systems; $NO_2^- \rightarrow NO_3^-$ (N^{5+}) by *Nitrobacter* in the soil and *Ntrocccusm* the seas.

Since nitrification steps are oxidations, they require oxic conditions under which oxygen can act as an electron acceptor. Under the anoxic conditions of waterlogged soils and sediments, nitrate and nitrite can act as electron acceptors (oxidizers) and the nitrification reactions reverse. This denitrification occurs in soil when the redox potential is less than 0.2 V, and is accomplished by such bacteria as *Pseudomonas denitrificans*. Additional physical reactions can result in the production of molecular nitrogen; that is,

$$NO_3^- \rightarrow NO_2^- \rightarrow NO \rightarrow N_2O \rightarrow N_2$$

with the consequent loss of nitrogen from biological circulation. In soils, for example, nitrate produced by nitrification in oxygenated surface layers may be washed to deeper, anaerobic layers where denitrification may take place primarily by physical reactions.

Nitrogen Fixation

Denitrification is balanced in terrestrial and aquatic systems by nitrogen fixation. This assimilatory reduction of nitrogen ($N^0 \rightarrow N^{3-}$) is accomplished by bacteria, such as *Azotobacter*, which is a free-living species, and *Rhizobium*, which occurs in symbiotic association with the roots of some legumes (pea family), and by blue-green algae.

Nitrogen fixation is energetically expensive although no more so than the conversion of an equivalent amount of nitrate to ammonia by plants. The reduction of an atom of nitrogen from chemical valence 0 (N_2) to –3 (NH_3) requires approximately the amount of energy released by the oxidation of an atom of carbon from valence 0 (glucose, $C_6H_{12}O_6$) to +4 (CO_2).

Nitrogen-fixing microorganisms obtain the energy and reducing power (organic carbon) they need to reduce N_2 to NH_3 by oxidizing sugars or other organic compounds. Free-living bacteria must obtain these resources by metabolizing organic detritus in the soil, sediments, or water column. More abundant supplies of energy are available to bacteria that enter into symbiotic relationships with plants, which provide them with photosynthate. The best known of these associations is the infection of root nodules of leguminous plants (peas, alfalfa, and their relatives) by the nitrogen-fixing bacterium *Rhizobium*. The nodules are specialized structures of the root cortex, whose development is stimulated by the *Rhizobium*. The nodules provide an optimum environment of abundant photosynthate and low oxygen for nitrogen fixation. Nitrogen-fixing symbioses are by no means limited to legumes, and such associations of bacteria or blue-green algae have been identified in other terrestrial plants (alders, *Casaurina*), some marine algae, lichens, and shipworms.

On a global scale, nitrogen fixation approximately balances the production of N_2 by denitrification. The fluxes are about 2 per cent of the total cycling of nitrogen through the ecosystem. On a local scale, nitrogen fixation can assume much greater importance, especially in nitrogen-poor habitats. When land is exposed to colonization by plants (for example, areas left bare by receding glaciers) such species with nitrogen-fixing capabilities as the alder *Alnus* dominate newly established vegetation.

Phosphorus Cycle

Ecologists have studied the role of phosphorus in the ecosystem intensively because organisms require phosphorus at a high level (about one-tenth that of nitrogen) as a major constituent of nucleic acids, cell membranes, energy-transfer systems, bones, and teeth. Phosphorus is thought to limit plant productivity in many aquatic habitats, and the influx of phosphorus to rivers and lakes, in the form of sewage and runoff from fertilized agricultural lands, artificially stimulates production in aquatic habitats with undesirable consequences.

The phosphorus cycle has fewer steps than the nitrogen cycle: plants assimilate phosphorus as phosphate (PO_4^{3-}) directly from the soil or water and incorporate it into various organic compounds as phosphate esters. Animals eliminate excess organic phosphorus in their diets by excreting phosphorus salts in urine; phosphatizing bacteria also convert organic phosphorus in detritus to phosphate ion. Phosphorus does not enter the atmosphere in any form other than dust; thus, the phosphorus cycle involves only the soil and aquatic compartments of the ecosystem.

Phosphorus occurs in the environment as orthophosphate (PO_4^{3-}, valence +5) and it does not undergo oxidation-reduction reactions in its cycling through the ecosystem except in very limited microbial transformations. The availability, and consequently the uptake and assimilation, of phosphorus is greatly affected by the pH of the environment. In general, at low pH (acid conditions), phosphorus binds tightly to clay particles in soil and forms relatively insoluble compounds with ferric iron [strengite, $Fe(OH)_2H_2PO_4$] and aluminum [variscite, $Al(OH)_2H_2PO_4$]. At high pH, other insoluble compounds are formed; for example, with calcium [hydroxyapatite, $Ca_{10}(PO_4)_6(OH)_2$]. These solubility relationships can be portrayed in a solubility diagram, in which the concentration of dissolved $H_2PO_4^-$ in equilibrium with insoluble compounds is shown as a function of the acidity of the soil. At low pH and in the presence of ferric iron or aluminum (which are present in virtually all soils, sediments, and waters), the equilibrium is driven to a very low level of dissolved $H_2PO_4^-$. Similarly, in alkaline soils in the presence of calcium, the equilibrium also is very low. When both ferric iron or aluminum and calcium are present, the highest concentration of dissolved phosphate—that is, the greatest availability of phosphorus—occurs at a pH of between 6 and 7. Under anoxic conditions in aquatic sediments, phosphorus is solubilized when iron is

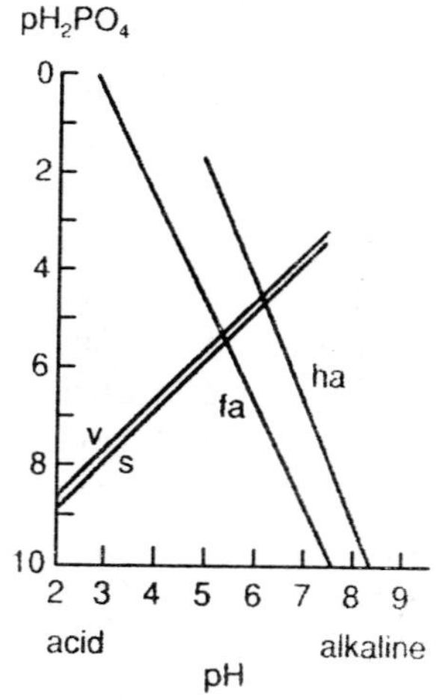

Fig. 16.7. The solubility of phosphate ion in equilibrium with several insoluble forms of phosphorus as a function of pH of the soil; v=variscite, s=stengite, fa=flouroapatite, ha=hydroxyapatite.

reduced from Fe^{3+} to Fe^{2+} because the iron then forms sulfides rather than phosphate compounds.

Sulfur Cycle

Sulfur is a component of the amino acids cysteine and methionine; hence, its requirement by plants and animals. But the importance of sulfur in the ecosystem goes far beyond its assimilation by plants. Like nitrogen, sulfur has many oxidation states. Like nitrogen, therefore, it follows complex chemical pathways and affects the cycling of other elements.

The most oxidized form of sulfur is sulfate (SO_4^{2-}, valence + 6); the most reduced are sulfide (S^{2-}) and the organic (thiol) form of sulfur (also S^{2-}). Under oxic conditions, assimilatory sulfur reduction ($S0_4^{2-}$ → organic S) is balanced by the oxidation of organic sulfur back to sulfate, either directly or with SO_3^{2-} (sulfite, S^{4+}) as an intermediate step. This oxidation is accomplished by most animals when they excrete excess dietary organic sulfur and by microorganisms when they decompose plant and animal detritus.

Under anoxic conditions (Eh < 0, dependent upon pH), sulfate may function as an oxidizer. The bacteria *Desulfovibrio* and *Desulfomonas* both couple dissimilatory sulfate reduction ($SO_4^{2-} \rightarrow S^{2-}$) to the oxidation of organic carbon in order to make energy available. The reduced S^{2-} may then be used as a reducing agent ($S^{2-} \rightarrow S^0$ +

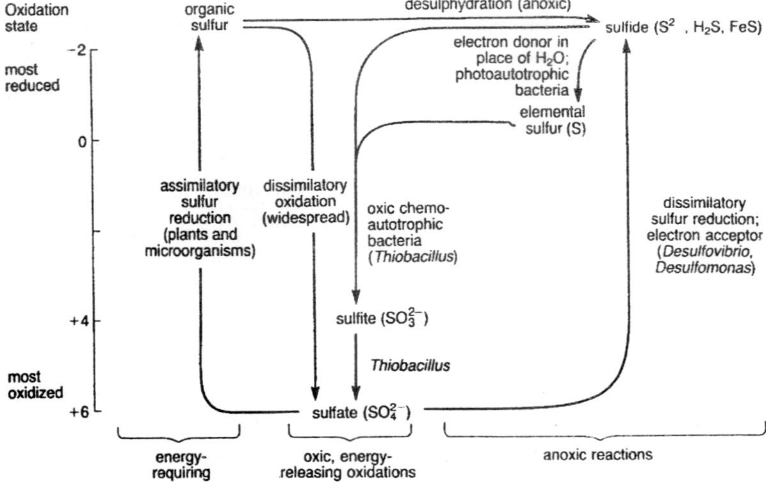

Fig. 16.8. Schematic diagram of transformations and oxidation states of compounds in the sulfur cycle.

$2e^-$) by photoautotrophic bacteria, which assimilate carbon by pathways analogous to photosynthesis in green plants. Sulfur takes the place of the oxygen atom in water as an electron donor. The elemental sulfur (S^0) accumulates unless the sediments are exposed to aeration or oxygenated water, at which point the sulfur may be further oxidized to SO_3^{2-} and SO_4^{2-}.

The fate of S^{2-} produced under oxic conditions depends on the availability of positive ions. Frequently hydrogen sulfide forms; it escapes from shallow sediments and mucky soils as a gas having the familiar smell of rotten eggs. Because oxic conditions generally favour the reduction of ferric iron (Fe^{3+}) to ferrous iron (Fe^{2+}), the presence of iron in sediments leads to the formation of iron sulfide (FeS). Sulfides are commonly associated with coal and oil. When exposed to the atmosphere in mine wastes or burned for energy, the reduced sulfur is oxidized, with the help of *Thiobacillus* bacteria in mine wastes, and the oxidized forms combine with water to form the sulfuric acid of acid mine drainage and acid rain.

Microorganism and Cycles

Many of the transformations discussed in this chapter are accomplished mainly or entirely by microorganisms. In fact, if it were not for the activities of bacteria and fungi, many element cycles would be drastically altered and the productivity of the ecosystem much reduced. For example, wood is broken down for the most part only by microorganisms, especially fungi. In their absence, organic carbon would accumulate as cellulose and lignin rather than passing through detritus-based food chains. Over long periods, primary production would drop as levels of carbon dioxide in the atmosphere decreased. For a second example, without the capacity of some microbes to utilize sulfur and iron as electron acceptors, little decomposition would occur in anoxic organic sediments, whose resulting accumulation would further reduce the amount of inorganic carbon in the ecosystem.

Nitrogen fixation, the sole source of biological nitrogen, is carried out only by microorganisms. Without them, life could not exist. Plants probably can assimilate ammonia-nitrogen as easily as, and at less energetic expense than, nitrate-nitrogen. Thus, nitrifying bacteria may actually reduce the primary productivity of terrestrial and aquatic plants by competing for nitrogen substrates in the soil and water.

Many of the transformations carried out by microorganisms, such as the metabolism of sugars and other organic molecules, are accomplished in similar ways by plants and animals. The bacteria and cyanobacteria (blue-green algae) are distinguished physiologically primarily by the ability of many species to metabolize substrates under anoxic conditions and to use substrates other than organic carbon as energy sources. The requirements of fungi for inorganic substrates as sources of nitrogen, phosphorus, potassium, sulfur, and other elements are similar to those of higher plants. What distinguishes the fungi from the bacteria is their greater ability to break down such complex polysaccharides as cellulose and lignin, which make up a large proportion of terrestrial plant litter.

Every organism needs above all a source of carbon for building organic structure and a source of energy to fuel the life processes. Organisms can be distinguished according to their source of carbon. Heterotrophs, or organotrophs, obtain carbon in reduced (organic) form by consuming other organisms or organic detritus. All animals and fungi, and many bacteria, are heterotrophs. Autotrophs assimilate carbon as carbon dioxide and expend energy to reduce it to an organic form. Photoautotrophs utilize sunlight as their source of energy for photosynthesis. These include all green plants and the cyanobacteria, which use H_2O as an electron donor ($O^{2-} \rightarrow O^0 + 2e^-$) and are aerobic, and the purple and green bacteria, which have different light-absorbing pigments from green plants, use H_2S or organic compounds as electron donors, and are anaerobic.

Organisms may obtain energy from three sources: sunlight (photoautotrophs), the oxidation of organic compounds (all heterotrophs), and the oxidation of inorganic compounds (chemoautotrophs). Oxidation of glucose, represented as $CH_2O + O_2 \rightarrow CO_2 + H_2O$, incorporates an oxidation reaction that releases energy, in this case $C^0 \rightarrow C^{4+} + 4e^-$, and an electron-accepting (reduction) reaction, $2O^0 + 4e^- \rightarrow 2O^{2-}$. Heterotrophs all use C^0 as reduced substrate for oxidation. Under oxic conditions, O_2 is used as an electron acceptor. But many bacteria (sulfate-reducing bacteria, methanogens) are either facultatively or obligately capable of utilizing SO_4^{2-}, Fe^{3+}, or CO_2 as electron acceptors under anoxic conditions.

Chemoautotrophs all use CO_2 as a carbon source, but obtain energy for its reduction by the oxic oxidation of inorganic substrates: methane (for example, *Methanosomonas*, *Methylomonas*), hydrogen (*Hydrogenomonas*, *Micrococcus*), ammonia (nitrifying bacteria *Nitrosomonas*, *Nitrosococcus*), nitrite (nitrifying bacteria *Nitrobacter*, *Nitrococcus*), hydrogen sulfide, sulfur, and sulfite (*Thiobacillus*), and ferrous iron salts (*Ferrobacillus*, *Gallionella*). The chemoautotrophs are almost exclusively bacteria, which apparently are the only type of organism that can become so specialized biochemically as to make efficient use of inorganic substrates and efficiently dispose of the products of chemoautotrophic metabolism.

In this chapter we have examined the cycling of several important elements from the standpoint of their chemical and biochemical reactions. Each type of habitat presents a different chemical environment, however, and it stands to reason that the patterns of element fluxes differ greatly among them.

17

Regenerative Processes

The paths of elements through the ecosystem are directed by their chemical properties, which in turn determine chemical and biochemical reactions in the ecosystem. We also have seen that utilizable forms of elements must be regenerated within the system to maintain high productivity. Elements enter the rapid-cycle compartments from external sources—for example, by nitrogen fixation and weathering of bedrock—too slowly to support the energy fluxes characteristic of most habitats. Thus the regeneration of assimilable forms of nutrients is a key to understanding the regulation of ecosystem function.

Processes of regeneration differ between terrestrial and aquatic systems. The chemical and biochemical transformations involved are basically the same; oxidation of carbohydrates, nitrification, chemoautotrophic oxidation of sulfur, among many others, result from similar biochemical mechanisms in both habitats. It is the material basis for nutrient regeneration that distinguishes terrestrial and aquatic habitats. In terrestrial systems, most nutrients cycle through detritus at the soil surface. In most aquatic systems, lakes and oceans in particular, sediments are the ultimate source of regenerated nutrients. Small streams have functional affinities with terrestrial habitats to the extent that production is based on the decomposition of terrestrial detritus, often under well-aerated conditions. To be sure, both aquatic and terrestrial systems exhibit great variation in element cycles.

In Terrestrial Ecosystem

While soil is difficult to define, it can be described as whatever overlies unaltered rock at the surface of the earth. It includes minerals derived from the parent rock, altered minerals formed anew within the zone of weathering, organic material contributed by plants, air and water within the pores of the soil, living roots of plants, microorganisms, and the larger worms and arthropods that make the soil their home. Five factors largely determine the characteristics of soils: climate, parent material (underlying rock), vegetation, local topography, and, to some extent, age.

Soils are in a dynamic state, changing as they develop on newly exposed rock material. But even after soils achieve stable properties, they remain in a constant state of flux. Groundwater removes some material; other material enters the soil from vegetation, in precipitation, and as dust from above and by the weathering of rock from below. Where little rain falls, the parent material weathers slowly and plant production adds little organic detritus to the soil. Thus arid regions typically have shallow soils, with bedrock lying close to the surface. Soils may not form at all where weathered material and detritus erode as rapidly as they form. Soil development also stops short on alluvial deposits, where fresh layers of silt deposited each year by flood waters bury weathered material. At the other extreme, weathering proceeds rapidly in parts of the humid tropics, where chemical alteration of parent material may extend to depths of 100 meters. Most soils of temperate zones are intermediate in depth, extending to about I meter, as a rough average.

Soil Horizons

Where a recent roadcut or excavation exposes soil in cross-section, one often notices distinct layers, called *horizons*. These horizons have been described with complex and sometimes conflicting terminology by soil classifiers. A generalized, and somewhat simplified, soil profile has four major divisions: O, A, B, and C horizons, with two subdivisions of the A horizon. Arrayed in descending order from the surface of the soil, the horizons and their predominant characteristics

O Primarily dead organic litter. Most soil organisms are found in this layer.

A_1 A layer rich in humus, consisting of partly decomposed organic material mixed with mineral soil.

A_2 A region of extensive leaching (or eluviation) of minerals from the soil. Because minerals are dissolved by water (mobilized) in this layer, plant roots are concentrated here.

B A region of little organic material whose chemical composition resembles that of the underlying rock. Clay minerals and oxides of aluminum and iron leached out of the overlying A_2 horizon are sometimes deposited here (illuviation).

C Primarily weakly weathered material, similar to the parent rock. Calcium and magnesium carbonates accumulate in this layer, especially in dry regions, sometimes forming hard, impenetrable layers.

Soil horizons demonstrate the decreasing influence of climate and biotic factors with increasing depth. Critical to soil formation is the movement of mineral elements upward and downward through the soil profile. But before considering these processes in detail, we shall examine the initial weathering of the bedrock and how it influences soil characteristics.

Weathering of Soil

Weathering of rock is fostered by the action of both physical and chemical agents. It occurs only where surface water penetrates. Repeated freezing and thawing of water in crevices breaks up rock into smaller pieces and exposes greater surface area to chemical action. Initial chemical alteration of the rock occurs when water dissolves some of the more soluble minerals, especially sodium chloride (NaCl) and calcium sulfate ($CaSO_4$). Other materials, particularly the oxides of titanium, aluminum, iron, and silicon, dissolve less readily.

The weathering of granite exemplifies some basic processes of soil formation. Granite, an igneous rock, forms when less dense molten material deep within the earth rises to the surface, cools, and crystallizes. The minerals making up the grainy texture of granite—feldspar, mica, and quartz—consist of various combinations of oxides of aluminum, ircn, silicon, magnesium, calcium, and potassium, along with other, less abundant compounds. The key to weathering is the displacement of certain elements in these minerals—notably calcium, magnesium, sodium, and potassium—by hydrogen ions, followed by the reorganization of the remaining oxides of aluminum, iron, and silicon into new minerals. For example, in the weathering of bedrock in the Hubbard Brook Experimental Forest of New Hampshire, virtually all the calcium is displaced and leached from the developing soil, as are 16 to 24 per cent of the sodium, magnesium, and potassium; in contrast, only 2 per cent of the aluminum and silicon are lost.

Feldspar, which consists of aluminosilicates of potassium, weathers rapidly owing to the displacement of potassium (K) by hydrogen ions according to

$$KAlSi_3O_8 + 4H_2O + 4H^+ \rightarrow K^+ + Al^{3+} + 3Si(OH)_4$$

The silicon hydroxide is soluble and subject to leaching under these conditions, but new insoluble materials, particularly clay particles, normally form and most of the aluminum and silicon stay in the soil:

$$2Al^{3+} + 2Si(OH)_4 + H_2O \rightarrow 5H^+ + Al_2Si_2O_5(OH)_4$$

This new mineral is called kaolinite; along with other clay minerals, it helps retain teachable ions (for example, Ca^{2+}, K^+) in the soil, as we shall see below.

Mica grains consist of aluminosilicates of potassium, magnesium, and iron. As in feldspar, the potassium and magnesium are displaced readily during weathering, and the remaining iron, aluminum, and silicon form various kinds of clay particles. Quarts, a form of silica (SiO_2), is relatively insoluble and therefore remains more or less unaltered in the soil as grains of sand. Changes in chemical composition as granite weathers from rock to soil in different climates show that weathering is most severe under tropical conditions of high temperature and rainfall.

An important factor in the initial weathering of parent material, regardless of the chemical nature of the rock, is the presence of hydrogen ions (H^+) in the water that percolates to the bedrock. These are derived from two sources. All precipitation contains dissolved carbon dioxide, which, as we have seen, forms carbonic acid, some of which dissociates to H^+ and HCO_3^-. In regions not affected by pollution-caused acidification, concentrations of hydrogen ions in rainwater are between 1 and 2×10^{-5} moles per liter, equivalent to a pH of about 5.

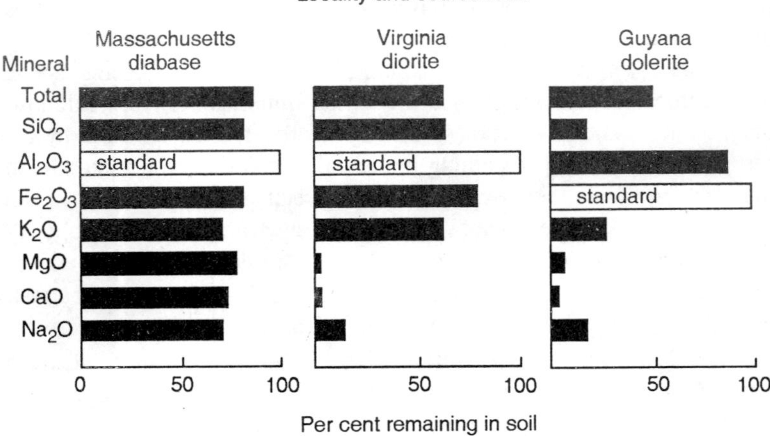

Fig. 17.1. Differential removal of minerals from granitic rocks.

The hydrogen ions in rainwater are supplemented by others generated by the oxidation of organic material in the soil. For example, nitrification of ammonia acidifies the soil by the reaction

$$NH_4^+ + 2O_2 \rightarrow 2H^+ + H_2O + NO_3^-$$

Similarly, the oxidation of carbohydrate results in the production of CO_2 and the generation of additional H^+ by the dissociation of carbonic acid. In the Hubbard Brook Forest, internal processes account for about 30 per cent of the hydrogen ions needed for weathering of bedrock, but internal sources may be much more important in the tropics, leading to more rapid weathering.

Clay and Humus

Plants obtain mineral nutrients from the soil in the form of dissolved ions, which are electrically charged atoms or compounds, whose solubility is determined by their electrostatic attraction to water molecules. Positively charged ions, such as Na^+, are called cations; negatively charged ones (Cl^-), anions. Because ions are dissolved in water, those not immediately taken up by plants and fungi may wash out of the soil profile if they are not strongly attracted to stable soil particles. Clay and humus particles, separately or associated in complexes, are large enough to form a stable component of the soil. These particles and complexes, known as *micelles*, have negative electrical charges at their surfaces that are critical to maintaining the smaller, more mobile ions in the soil.

The negative charges of micelles arise by two processes. First, during the formation of clay particles by crystallization, an atom of magnesium might substitute in the crystal lattice for one of aluminum, or one of aluminum for one of silicon. In each of these substitutions, a less positively charged atom replaces a more positively charged

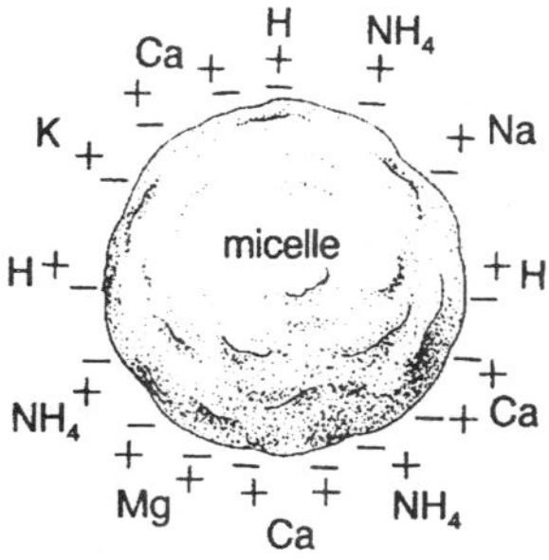

Fig. 17.2. Schematic representation of a clay or humus particle (micelle) with hydrogen ions and mineral ions attracted by negative charges at its surface.

atom (that is, Mg^{2+} for Al^{3+}, or Al^{3+} for Si^{4+}). The result is a net negative charge on the clay particle. The second process is the ionization of several types of functional groups; for example, hydroxide (—OH) in clay and the carboxyl (—COOH), amine (—NH_2), or phenolic (—C_6H_4OH) groups in organic compounds. The dissociation of such groups (for example, —COOH $\rightleftharpoons$ —COO^- + H^+) exposes negative charges that can then hold other ions in the soil. Charges resulting from the substitution of metal atoms in clay particles are called permanent charges because they are insensitive to the concentration of hydrogen ions (pH) in the soil. In contrast, the dissociation of functional groups varies inversely with hydrogen ion concentration and therefore is pH dependent.

The bonds between soil particles and such ions as K^+ and Ca^{2+} are relatively weak, so they constantly break and reform. When a potassium ion (K^+) dissociates from a micelle, its place may be taken by any other positive ion (cation) close by. Some ions cling more strongly to micelles than others: in order of decreasing tenacity, hydrogen (H^+), calcium (Ca^{2+}, magnesium (Mg^{2+}), potassium (K^+), and sodium (Na^+). Hydrogen ions thus tend to displace calcium and all other cations in the soil. If cations were not added to or removed from the soil, the relative proportions of the various cations associated with clay-humus particles would assume a steady state. But carbonic acid in rainwater and organic acids produced by the decomposition of organic detritus continually add hydrogen ions to the upper layers of the soil; these readily displace other cations, which are then washed out of the soil and into the groundwater. The influx of hydrogen ions in water percolating through the soil largely causes the mobility of ions in the soil and the differentiation of layers in the soil profile.

Negative ions important to plant nutrition, such as nitrate, phosphate, and sulfate, can be adsorbed onto clay particles by means of ion "bridges." These bridges form under more acid conditions by the association of an additional hydrogen ion with a functional group; for example, —OH + H^+ → —OH_2^+, which makes possible —$OH_2^+ \cdots NO_3^-$.

It should be clear from the cation-binding properties of clay and humus particles that the potential long-term fertility of soil—its capacity for storing nutrients—depends in large part on its clay content. Furthermore, the realized nutrient storage and the immediate availability of ions to plants depends to a large degree on the acidity of the soil.

Podsolization

Under mild, temperate conditions of temperature and rainfall, sand grains and clay particles resist weathering and form stable components of the soil skeleton. In acid soils, however, clay particles break down in the A horizon of the soil profile, and their soluble ions are transported downward and deposited in lower horizons. This process, known as podsolization, reduces the ion-exchange capacity and, therefore, the fertility of the upper layers of the soil.

Acid soils occur primarily in cold regions where coniferous trees dominate the forests. The slow decomposition of plant litter produces organic acids. In addition, in regions of podsolization rainfall usually exceeds evaporation. Under these moist conditions, water continually moves downward through the soil profile, so little clay-forming material is transported upward from the weathered bedrock below.

In North America, podsolization advances farthest under spruce and fir forests in New England and the Great Lakes region and across a wide belt of southern and western Canada. A typical profile of a podsolized soil reveals striking bands corresponding to regions of leaching (eluviated horizons) and redeposition (illuviated horizons). The topmost layers of the profile (O and A_1) are dark and rich in organic matter. These are underlain by a light-coloured horizon (A_2), which has been leached of most of its clay content. As a result, A_2 consists mainly of sandy skeletal material that holds neither water nor nutrients well. One usually finds a dark band of deposition immediately under the eluviated A_2 horizon. This is the uppermost layer of the B horizon, where iron and aluminum oxides are redeposited. Other, more mobile minerals may accumulate to some extent in lower parts of the B horizon, which then grades almost imperceptibly into a C horizon and the parent material.

Because of the warm, wet climate, tropical soils tend to be deeply weathered, and the deeper the ultimate source of nutrients in the unaltered bedrock the poorer the surface layers tend to be. Rich soils do develop in many tropical regions, particularly in mountainous areas where erosion continuously removes nutrient-depleted surface layers of the soil. But the soils of other areas, especially those in low-lying regions (the Amazon Basin, for example) and those that develop on parent material deficient in quartz (SiO_2) but rich in iron and magnesium (basalt, for example), contain little clay and therefore do not hold nutrients well. Clay fails to form in any abundance because of a lack of silicon. Instead, iron and aluminum oxides predominate in the soil horizon. This type of weathering is known as laterization; oxides give lateritic soils (latisols) their typical red colour.

Development of Soil

Natural weathering of bedrock occurs under deep layers of soil where it is inaccessible to direct investigation. The process may be studied indirectly, however, by measuring the net efflux of certain elements from the system. When a soil is in a steady state, as it is in undisturbed areas, the efflux of an element equals the weathering of that element from the parent material plus any influx from other sources, such as precipitation. The basic cations—calcium, potassium, sodium, and magnesium—are good candidates for studying such balances because they are readily leached from the soil profile and leave the system in stream water as dissolved ions, which are easily measured. Elements that form less soluble compounds—silicon, iron, and aluminum—leave the system as solid particles ranging in size from suspended silt to large boulders carried down stream beds during exceptionally heavy runoff. These are more difficult to monitor.

Cation budgets of large areas are most easily studied in a small watershed whose entire drainage can be sampled at a single point along the stream that drains it. Ideally, the watershed should be uniform in bedrock, soil, and vegetation so that measurements reflect processes in only a single type of habitat. In addition, the watershed must lie upon impervious bedrock so that groundwater cannot move into and out of the area except by streams.

Many such watersheds have been studied in detail, particularly in temperate regions. The best-known is the Hubbard Brook Forest, where G. E. Likens, F. H. Bormann, and their colleagues have investigated patterns of water and nutrient cycles since the early 1960s. Detailed cation budgets were obtained for several small watersheds in the area by measuring the inputs in rainwater collected at various locations in the watershed and outputs in water leaving the watershed by way of the stream that drains it.

At Hubbard Brook, during the period 1963-1974, annual input of calcium in precipitation averaged 2.2 kg ha^{-1} while loss of dissolved Ca^{2+} in stream flow was 13.7 kg ha^{-1}; therefore, the net loss to the system was 11.5 kg ha^{-1}. The investigators also determined that living and dead biomass increased in the watershed during this period, owing to the fact that the forest was recovering from earlier clearing; net assimilation of calcium in vegetation and detritus brought its overall removal from the mineral soil to 21.1 kg ha^{-1} yr^{-1}. Because calcium constitutes about 1.4 per cent of the weight of the bedrock in the area, its annual loss equaled the weathering of 1500 kg (21.1/0.014) of bedrock per hectare, or approximately 1 millimeter of depth per year. The same calculations based on elements other than calcium gave quite different (typically lower) estimates of weathering. Nevertheless, such studies agree in concluding that weathering in cool, temperate regions proceeds very slowly compared to the annual uptake of soil cations by plants and their release by leaching and decomposition of detritus within the system.

Role of Detritus

Plants assimilate elements from the soil far more rapidly than they are generated by weathering of the parent material. Most basic cations (Ca^{2+}, Mg^{2+}, K^{+}, Na^{+}) do not figure prominently in biochemical transformations and, for the most part, are merely taken up with the water that plants need in such quantity. Excess uptake of these elements is either sequestered or excreted, and their depletion from the soil probably has little direct effect on plant production. Not so with supplies of such important nutrients as nitrogen, phosphorus, and sulfur, which are poorly represented in parent material. Igneous rocks, such as granite and basalt, contain no nitrogen, only 0.3 per cent of phosphate (P_2O_5), and 0.1 per cent of sulfate (SO_3) by mass. Hence, weathering adds little of these nutrients to the soil; inputs from precipitation and nitrogen fixation also are small. Plant production therefore depends on the rapid regeneration of nutrients from detritus.

Organic detritus occurs everywhere, most conspicuously in terrestrial systems where the resistance of woody plant parts to herbivory results in the accumulation of abundant plant remains. Regardless of the habitat, however, this reservoir of nutrients is regenerated by the activities of a wide variety of worms, snails, insects, mites, bacteria, and fungi that consume detritus for food—their primary source of carbon and energy.

Of all the detritus-based communities, the organisms that consume the litter of leaves and branches on the forest floor are probably best known. The breakdown of leaf litter occurs in three ways: (1) leaching of soluble minerals and small organic compounds by water; (2) consumption by large detritus-feeding organisms (millipedes, earthworms, woodlice, and other invertebrates); and (3) further attack by fungi and eventual mineralization of phosphorus, nitrogen, and sulfur by bacteria. Between 10 and 30 per cent of the substances in newly fallen leaves dissolve in cold water; leaching rapidly removes most of these (salts, sugars, amino acids) from the litter, leaving behind complex carbohydrates and other organic compounds. Although large detritus feeders assimilate no more than 30 to 45 per cent of the energy available in leaf litter, and even less from wood, they nonetheless speed the decay of litter because they macerate leaves in their digestive tracts, and the finer particles in their egested wastes expose new surfaces to microbial feeding.

Leaves of different species of trees decompose at different rates, depending on their composition. For example, in eastern Tennessee, weight loss of leaves during the first year after leaf fall varies from 64 per cent for mulberry, to 39 per cent for oak, 32 per cent for sugar maple, and 21 per cent for beech. The needles of pines and other conifers also decompose slowly. Differences between species depend to a large extent upon the lignin content of the leaves. Lignins are a heterogeneous class of phenolic polymers. They lend wood many of its structural qualities, and are even more difficult to digest than cellulose. In fact, only the "white rot" fungi can break down lignin.

The resistance of some types of litter to degradation points up the unique role of fungi in the regeneration of nutrients. The familiar mushrooms and shelf fungi are merely fruiting structures produced by the mass of the fungal organism deep within the litter or wood. Most fungi consist of a network, or mycelium, of hyphae, threadlike elements that can penetrate the woody cells of plant litter that bacteria cannot reach. Like bacteria, fungi secrete enzymes into the substrate itself

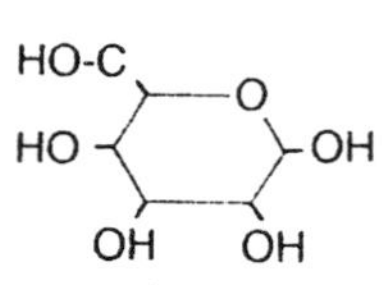

glucose

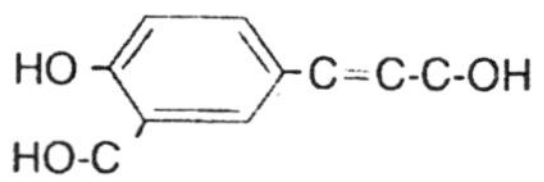

coniferyl alcohol

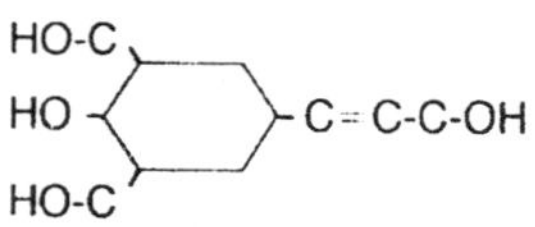

syringenin

Fig. 17.3. Chemical structure of the subunits of cellulose (top) and lignin (center and bottom).

and absorb the simple sugar and amino-acid breakdown products of this exocellular digestion. Fungi differ from bacteria in being able to digest cellulose (which a few bacteria, protozoa, and snails also can accomplish) and, especially, lignin. Cellulose digestion begins by hydrolyzing polysaccharides (long chains of sugar subunits) into simple sugars that can be absorbed into the hyphae. The breakdown of lignin by fungi apparently is initiated by an oxidation that cleaves the aromatic (phenolic) ring structure.

Symbiotic Association

In addition to their role in decomposing detritus, some kinds of fungi grow on the surface of or inside the roots of many types of plants, especially woody species. This association, called a mycorrhiza (literally, fungus root), enhances the plant's ability to extract mineral nutrients from the soil. Although many forms of the association are recognized, they are classified as endomycorrhizae when the fungus penetrates the root tissue and ectomychorrhizae when it forms a sheath over the root surface. Mycorrhizae, whose fundamental importance to production and element cycling is gaining increasing appreciation, have become the subject of an extensive literature. General accounts may be found in Rovira (1965), Wilde (1968), Marks and Koziowski (1973), Smith (1980), and Harley and Smith (1983).

Mycorrhizae occur everywhere, but their importance is best demonstrated experimentally by growing plants on sterile soil to which

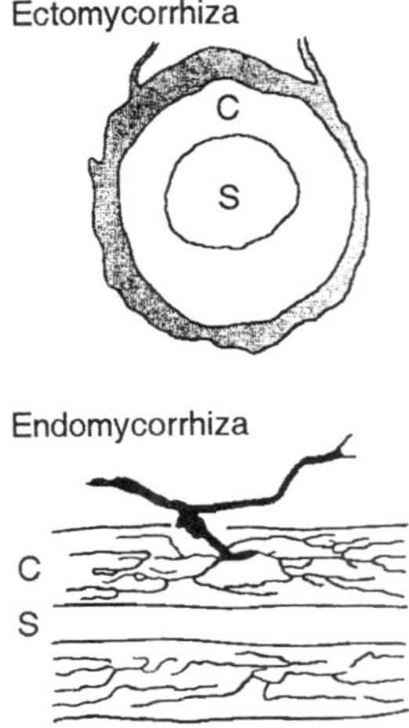

Fig. 17.4. Generalized structure of ectomycorrhiza and endomycorrhiza; C=root cortex, S=vascular tissue of stele.

spores of mycorrhizae-forming fungi are either added or not added. For example, in an experiment with *Pinus strobus* seedlings, ectomycorrhizae increased by two to three times the uptake of nitrogen, phosphorus, and potassium per unit of root mass and greatly improved the growth of the plant. The infections seem to be more successful in soils that are relatively depleted of nutrients.

Mycorrhizae increase a plant's uptake of minerals by penetrating a greater volume of soil and increasing the total surface area for nutrient assimilation. Mycorrhizae, especially external forms, may also protect the plant root from infection by pathogens by physically excluding them or producing antibiotics (antibacterial toxins). What do the fungi derive from the association? The main advantage appears to be a reliable source of carbon in the form of photosynthates transported to the roots.

The commonest type of endomycorrhiza is the vesicular-arbuscular type, so-called because of structures developed within the host tissue. These do not grow freely in the soil, but infect plant roots from spores left behind from dead, previously infected roots. (Presumably spores blow into virgin soils along with other dust.) Fungi that form vesicular-arbuscular mycorrhizae derive their carbon exclusively from the plant root. Mineral nutrients are derived from the soil, into which the fungi send long hyphae. They apparently can use sources of phosphorus not available to plants, such as the highly insoluble "rock phosphorus" $Ca_3(PO_4)_2$; plants can use only the more soluble forms $CaHPO_4$ and KH_2PO_4. Perhaps the hyphae excrete hydrogen ions or organic acids, or grow in association with phosphate-solubilizing bacteria.

Ectomycorrhizae are very widespread, especially among trees and shrubs. In addition to the hyphae, which extend out into the soil, the fungus forms a tough sheath around the root that may account for as much as 40 per cent of the weight of the mycorrhizal association. The sheath stores large quantities of soil-derived nutrients and carbon compounds, which may be one of the chief advantages of the mycorrhizae to the plant. Ectomycorrhizae are found predominantly in association with vegetation experiencing seasonal or intermittent growth, where they may serve to extend the productive period. Regardless of how they function, they do account for a major part of energy and nutrient budgets in some habitats. K. A. Vogt et al. (1982) estimated that the carbon assimilated by mycorrhizae in a fir (*Abies*) forest accounted for 15 per cent of the total net primary production. The mycorrhizal fungus is undoubtedly responsible for a much greater share of the mineral nutrient uptake.

Regeneration of Nutrients

Nutrient cycling in tropical and temperate ecosystems differs because of the effects of climate on weathering and regeneration of detritus. The deep weathering and poor cation exchange capacity of many tropical soils result in relatively low fertility. But rapid regeneration of nutrients from detritus under warm, humid conditions and efficient nutrient retention supports high productivity. Ecologists have concluded that in the "typical" tropical system, most of the nutrients occur in the living biomass and elements are regenerated and assimilated very rapidly. This has important implications for tropical agriculture and conservation.

Many rich and fertile tropical soils are well-suited to the removal-type agriculture practiced commonly in temperate regions. Even with clearing of the forests and frequent removal of crops, such soils maintain their fertility. But over extensive regions within the tropics, planting such crops as corn on clear-cut land has had disastrous consequences. The practice of cutting and burning vegetation releases many mineral nutrients to the soil. But while these may support a year or two of crop growth, they are readily leached without the natural tropical vegetation to assimilate them, and soil fertility declines rapidly. Furthermore, as the exposed soil dries, upward movement of water draws iron and aluminum oxides to the surface where they form a concrete-like substance called *laterite*; surface runoff over the impenetrable laterite accelerates erosion, further depleting nutrients and choking streams with sediment.

The ecological lesson taught by such experience is that vegetation is critical to the development and maintenance of soil fertility. Even in temperate zones, removal of vegetation points up its important role in retention of soil nutrients. Clear-cutting of small watersheds in the Hubbard Brook Forest increased stream flow several times owing to removal of transpiring leaf surfaces; losses of cations increased 3 to 20 times over comparable undisturbed systems. The nitrogen budgets of the cutover watersheds sustained the most striking change. Plants assimilate available soil nitrogen so rapidly that undisturbed forest gained nitrogen at the rate of 1 to 3 kg ha^{-1} yr^{-1}. In the clear-cut watershed net loss of nitrogen as nitrate soared to 54 kg ha^{-1} yr^{-1}, a value comparable to the annual assimilation of nitrogen by vegetation, and many times the precipitation input (7 kg ha^{-1} yr^{-1}). The loss of nitrate resulted from nitrification of organic nitrogen at the normal annual rate by soil microorganisms without simultaneous rapid uptake by plants.

Comparative studies of temperate and tropical forests shed further light on their differences. Of the total biomass of vegetation and detritus, litter on the forest floor comprises an average of about 20 per cent in temperate coniferous forests, 5 per cent in temperate hardwood forests, and only 1 to 2 per cent in tropical rain forests. The ratio of litter to the biomass of living leaves is between 5 and 10 in temperate forests, but less than I in tropical forests. Of the total organic carbon, more than 50 per cent occurs in the soil and litter in northern forests, but less than 25 per cent in tropical rain forests. Clearly, dead organic material decomposes rapidly in the tropics and does not form a substantial nutrient reservoir, as it does in temperate regions.

Fewer data are available on the relative proportions of nutrients in the soil compared to the living vegetation. Although these data may not be typical, two points stand out. First, the accumulation of nutrients in vegetation, on a weight for weight basis, is somewhat greater in the tropical forest. For example, the total dry weight of living vegetation in the Belgian ash-oak forest exceeds that of the tropical deciduous forest in Ghana by 14 per cent, but the accumulation of the three elements per gram of dried vegetation is 32 to 38 per cent lower in the temperate forest. Second, the ratio of each element in soil to its level in biomass is much lower in the tropics; more than 90 per cent of the phosphorus in the tropical system resides in the living biomass.

C. F. Jordan and R. Herrera (1981) recognized the general nutrient poverty of many tropical soils but also distinguished nutrient-rich and nutrient-poor soils within the tropics. The former, which they called eutrophic (literally "well-nourished"), develop in geologically active areas where natural erosion is high and soils are relatively young. With the bedrock closer to the surface, weathering adds nutrients more rapidly and soils retain nutrients more effectively. In the Western Hemisphere, such eutrophic soils occur widely in the Andes Mountains, in Central America, and in the Caribbean. By contrast, nutrient-poor (oligotrophic) soils develop in old, geologically stable areas, particularly on sandy alluvial deposits (as in much of the Amazon Basin), where intense weathering removes clay and reduces nutrient retention.

Comparison of an oligotrophic and a eutrophic forest illustrates that although production and nutrient flux are similar, the distributions of nutrients (calcium, in this case) in the oligotrophic forest shift from the soil to biomass relative to the eutrophic forest. Moreover, nutrients are held more tightly; that is, loss of calcium from the eutrophic forest in Puerto Rico is equivalent to about half the annual flux through vegetation, whereas the oligotrophic system appears to gain calcium through precipitation input.

Especially in nutrient-poor areas, nutrient retention by vegetation is the key to the productivity of tropical ecosystems. This is accomplished by a very dense mat of roots (and associated fungi) close to the surface and even extending up the trunks of trees to intercept nutrients washing down from the canopy. Data from Africa revealed between 68 and 85 per cent of the root biomass of forests concentrated within the top 25 to 30 centimeters of the soil. Application of isotopically labeled compounds showed that nutrients regenerated by the leaching and decomposition of detritus are intercepted by the root mat before they can penetrate into the mineral soil and be washed out of the system.

Regeneration of Photic Zone

Because most chemical and biochemical processes involved in the cycling of elements take place in an aqueous medium, the processes themselves do not differ markedly in terrestrial and aquatic systems. What is unique and distinctive about most rivers, lakes, and oceans is the sedimentation of nutrients into bottom deposits from which they are regenerated and returned to zones of productivity very slowly.

The sediments in aquatic systems are comparable to the detritus layer in terrestrial systems. But terrestrial detritus differs from sediments in two important ways. First, the regeneration of nutrients from terrestrial detritus takes place near plant roots. Aquatic plants assimilate nutrients directly from the water column in the uppermost sunlit (photic) zones, often far removed from sediments at the bottom. Second, decomposition of terrestrial detritus is accomplished, for the most part, aerobically, hence relatively rapidly. Aquatic sediments often become anoxic, greatly slowing most biochemical transformations.

The maintenance of high aquatic production depends in part on the proximity of bottom sediments to the photic zone at the surface, or the presence of upwelling currents bringing nutrients regenerated from sediments back to the surface. A map of productivity of the oceans shows that the rate of carbon fixation is greatest in shallow seas, in both the tropics and high latitudes, and in zones of upwelling. The latter occur along the western coasts of Africa and the Americas, where winds blow surface waters away from shore, thus establishing a vertical current to replace them.

Regeneration of nutrients in the water column occurs by excretion and microbial decomposition, just as it does in terrrstrial systems. For some elements, rates of regeneration are highly correlated with rates of production. For example, in a study of the nitrification of ammonia ($NH_4^+ \rightarrow NO_3^-$) off the western coast of North America, W. G. Harrison (1978) found that phytoplankton assimilated about half the ammonia directly and bacteria nitrified the other half, regardless of the overall flux. Studies in temperate freshwater systems indicate turnover rates of organic nitrogen on the order of 0.2 to 2 days; in temperate marine systems, turnover varies between 1 and 10 days, suggesting lower productivity overall.

Plants assimilate regenerated nitrogen very rapidly, especially in nutrient-depleted waters. J. J. McCarthy and J. C. Goidman (1979) demonstrated that phytoplankton can take up more nitrogen than they need for growth. This ability to store nitrogen enables algae to take advantage of local "pockets" of nutrients made available by discrete bursts of excretion of decomposition. While we think of water as homogeneous, the concentration of nutrients in organisms makes the distribution of elements such as nitrogen and phosphorus extremely heterogeneous. Excretion and decomposition produce transient local abundances of nutrients that may be the primary practical source of nutrients. As McCarthy and Goidman point out, once turbulent mixing disperses them, they may become too sparse to be assimilated.

Aquatic systems can be highly productive only where there is strong interchange between bottom layers and the surface, at least on occasion. Nitrogen budgets measured by C. F. Liao and D. R. S. Lean (1978) in the Bay of Quinte, Lake Huron, Ontario, illustrated the relative magnitudes of assimilation and regeneration. The studies were conducted within columns of water in "limnocorrals," which are triangular or circular in cross-section, and enclosed by sheets of plastic suspended by floats at the surface and entrenched in sediment at the bottom, in this case 4 meters beneath the surface. Such enclosures allow studies of the fluxes of elements by addition of isotopically labeled compounds, but they may disrupt normal patterns of mixing of water layers; thus the vertical profiles of production and nutrient cycling within enclosures may differ from those in adjacent open water.

Ammonia accounted for about 65 per cent of the regenerated nitrogen available to plants and nitrate for the other 35 per cent, both early in the growing season (5 June) and late (4 September). But although levels of nitrogen were similar, gross production in September was nearly 7 times the level in June, probably due to a combination of differences in water temperature and some other, limiting nutrient. Short-term uptake of nitrogen by plants was about one-tenth the level of carbon fixation. Sedimentation of nitrogen in sinking particulate matter amounted to 14 per cent as much as uptake in June and 28 per cent as much as uptake in September. Accordingly, during the September sample period when the total nitrogen in the system (NH_4^+, NO_3^-, and particulate) was 586 g l^{-1} and sedimentation was 36 g l^{-1} d^{-1}, physical removal of nitrogen from the system could deplete the resource quickly. But sedimentation of particulate nitrogen was approximately balanced throughout the season by return of ammonia, because total nitrogen in the water column varied little relative to internal cycling.

Thermal Stratification

The vertical mixing of water requires an input of energy to accelerate water masses and keep them moving. Winds supply a part of this energy, causing turbulent mixing of shallow water. Mixing can be prevented in such systems when freshwater floats over more dense saltwater, as in the bay of a large river, or when the sun heats surface water to establish a warmer layer of less density overlying a cooler layer of greater density. Other processes promote vertical mixing. In marine systems, when evaporation exceeds freshwater input, surface layers become more saline, hence denser, and literally fall through the lighter water below. Similarly, when surface layers of water cool during the autumn, the now denser surface water tends to fall through the warmer, hence less dense, layers below.

Vertical mixing of water affects production in two opposing ways. On one hand, mixing can bring nutrient-rich water from the depths to the sunlit surface and thereby promote production. On the other hand, mixing can carry phytoplankton below the zone of photosynthesis and thereby reduce production. Indeed, if vertical mixing extends far below the photic zone, the algae of the phytoplankton may not be able to maintain themselves, much less reproduce. Under such conditions, primary production may shut down altogether, resulting in the seeming contradiction of nutrient-rich water without primary production.

The more typical situation in many temperate-zone freshwater systems is one in which thermal stratification during the summer prevents vertical mixing; then, as sedimentation removes nutrients from the surface layers, production decreases. We may think of the cycle as beginning during the winter, when the temperature of the entire water column of a temperate lake is uniformly cold. There is little or no production owing to low light, particularly when the lake is frozen over. In early spring, the ice melts and the sun rises higher in the sky each day. But strong winds keep the water column thoroughly mixed and production, at least in deep lakes, remains low.

As the surface layers of the lake begin to warm, however, thermal stratification ensues, and a layer of nutrient-rich water becomes "trapped" at the surface, thus creating optimal conditions for high production. The resulting spurt of algal growth is often referred to as the spring bloom of phytoplankton production. When stratification continues into the summer, nutrients are depleted in the surface layers and production decreases as phytoplankton are consumed by larger zooplankton and the excreta and dead remains of these zooplankton sediment out of the water. Nutrients may be regenerated in the deeper layers of the lake, but these cannot reach the surface.

Thermal stratification develops only weakly, if at all, at both high and low latitudes. In arctic and subarctic regions, the heat input into lakes is not sufficient to counter turbulent mixing and the water column heats up uniformly to the extent that water temperature rises at all. In the tropics, the lack of a pronounced seasonal temperature cycle reduces the sharpness of thermal stratification, although surface layers of water may be warmed by the sun during the day.

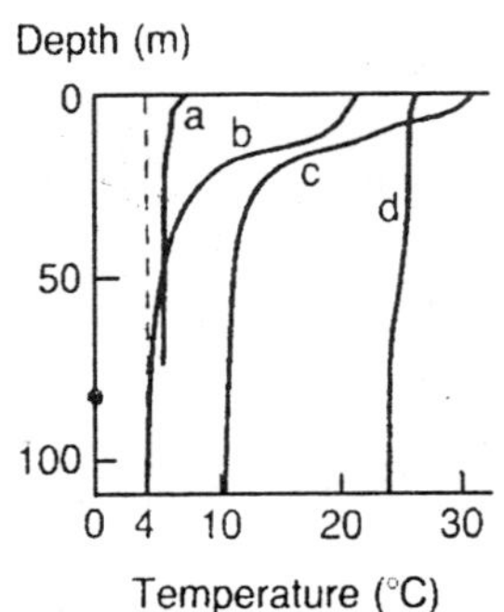

Fig. 17.5. Temperature profiles of lakes at the height of summer stratification.

In temperate zones, as temperatures begin to decrease in the fall, the surface layers of water become more dense as they cool and therefore begin to fall through the water column, commencing a period of vertical mixing that is accelerated by wind-driven turbulence. At

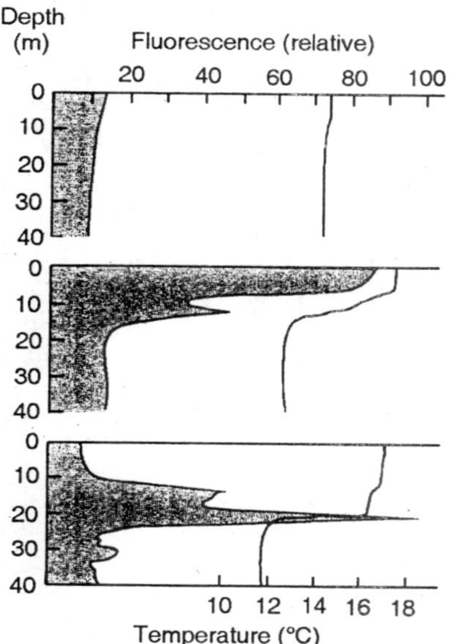

Fig. 17.6. Vertical profiles of chlorophyll concentration (flourescence) and temperature.

this time, bottom water with regenerated nutrients rises to the surface, where it may support a new rush of phytoplankton growth, referred to as the fall bloom. Further cooling of the lake and the darkening days of the oncoming winter complete the annual cycle.

In marine systems, two very different water masses may meet at a "front" where intermixing creates special conditions for high production. Sometimes at the boundary of a shallow-water system and a deep-water system, mixed (deep) and stratified (shallow) water masses are brought together felicitously. On the mixed side, nutrients may be abundant but phytoplankton do not remain within the photic zone. On the stratified side, nutrients may have been depleted from the surface waters. Where the two meet, some of the nutrient-laden mixed water enters the stratified layer, creating ideal conditions for photosynthesis and nutrient assimilation.

Regeneration in Hypolimnion

Lake ecologists call the region above the sharp change (*thermocline*) in a stratified temperature profile the epilimnion; the zone below the thermocline is the hypolimnion. During periods of stratification, bacterial respiration in the hypolimnion depletes the oxygen supply and the water may become anoxic if stratification is prolonged and there is abundant organic matter to oxidize.

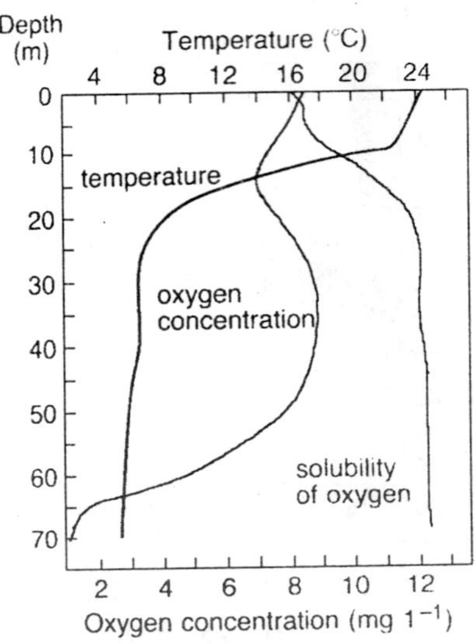

Fig. 17.7. Profiles of temperature and oxygen concentration in Green Lake, illustrating oxygen depletion in the hypolimnion during summer stratification.

In the low redox environment in bottom sediments and waters immediately over them, the shift of such redox elements as iron, manganese, and nitrogen from oxidized to reduced forms greatly affects their solubility. In particular, as ferric iron (Fe^{3+}) is reduced to ferrous iron (Fe^{2+}), insoluble iron-phosphate complexes become solubilized and both elements may move into the water column. This was first demonstrated in C. H. Mortimer's classic (1941, 1942) studies on the exchange of dissolved substances between the mud and water of lakes, and reconfirmed in many more recent studies.

Changes in the water chemistry of the hypolimnion of an English lake, Esthwaite Water, during the course of a single season show the effects of anoxic conditions. After stratification, the level of oxygen at the deepest level of the lake decreased gradually while dissolved carbon dioxide increased. The water became anoxic by early July and remained so until the end of stratification and the onset of vertical mixing in late September. During the period of anoxia, levels of ferrous iron, phosphate, and ammonia increased dramatically as they were released by reduction processes in the sediments and at the sediment-water boundary. The return of oxidizing conditions in the fall completely altered the chemistry of the bottom water, initially because of the replacement of bottom with surface water but ultimately because oxidized forms of several redox elements formed insoluble compounds,

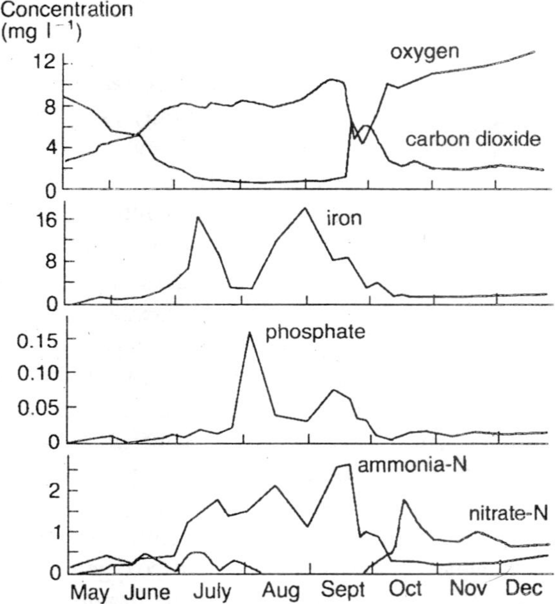

Fig. 17.8. Seasonal course of water chemistry of the hypolimnion, showing the solubilizing of reduced phosphorus and iron compounds during the period of summer anoxia.

which precipitated out of the water column. Nitrogen was a conspicuous exception; under oxic conditions nitrifying bacteria convert ammonia to nitrate, which generally remains in solution.

Phosphorus in Eutrophication

Phosphorus is often scarce in the well-oxygenated surface waters of lakes, and low levels of phosphorus limit the production of an aquatic system. In small lakes on the Canadian Shield, productivity increased dramatically in response to the addition of phosphorus, but not nitrogen or carbon (in the form of sucrose). Natural lakes exhibit a wide range of productivity, depending on inputs of nutrients from outside (external loading: rainfall, streams) and the regeneration of nutrients within the lake (internal loading). In shallow lakes lacking a hypolimnion internal loading occurs continuously through resuspension of bottom sediments. In somewhat deeper lakes, where the thermocline is weakly developed, vertical mixing may occur periodically as a result of occasional strong winds or cooling periods during the summer. Such mixing returns regenerated nutrients to the surface and stimulates production. In very deep lakes, bottom waters rarely mix with the surface and production depends almost entirely on external loading.

Aquatic ecologists classify lakes on a continuum ranging from poorly nourished (*oligotrophic*) to well-nourished (*eutrophic*) depending on their nutrient status and production. Naturally eutrophic lakes have characteristic temporal patterns of production and nutrient cycling that maintain the system in a steady state. Artificial addition of nutrients in sewage and drainage from fertilized agricultural lands can cause inappropriate nutrient loading and greatly alter natural cycles.

Increased production is not bad in and of itself; indeed, many lakes and ponds are artificially fertilized to increase commercial fish production. But overproduction can lead to imbalance when natural regeneration processes cannot handle the increased demands on cycling. Heavy organic pollution, such as that which results from dumping raw sewage into rivers and lakes, creates what is called *biological oxygen demand* (BOD) due to oxidative breakdown of the detritus by microorganisms. Inorganic nutrients stimulate the production of organic detritus, adding to the BOD. In its worst manifestations, this type of pollution can deplete the surface water of its oxygen, leading to the suffocation of fish and other obligately aerobic organisms.

In spite of what can be devastating effects of external nutrient loading, culturally eutrophied lakes can recover their original condition if inputs are shut off. Eventually, oxic conditions return, phosphorus precipitates out of the water column, and the normal cycles of assimilation and regeneration are restored. A spectacular and convincing demonstration of this occurred after diversion of sewage from Lake Washington, in Seattle, where an advanced and rather ugly case of eutrophication was quickly reversed.

Estuaries and Marshes

Shallow estuaries, which are semi-enclosed coastal regions subject to both freshwater inputs from rivers and tidal inputs from the sea, are among the most productive ecosystems on the earth. Salt marshes, intertidal areas with emergent vegetation, combine the most favourable attributes of aquatic and terrestrial systems, resulting in similarly high production. But in addition to these local attributes of estuaries and coastal marshes, their significance for marine systems extends seaward through net export of production. From a Georgia salt marsh, nearly 10 per cent of the gross primary production and almost half the net primary production is exported into surrounding marine systems in the form of organisms, particulate detritus, and dissolved organic material carried out with the tides. Because of their high productivity and, owing to their structural complexity, the hiding places offered prey

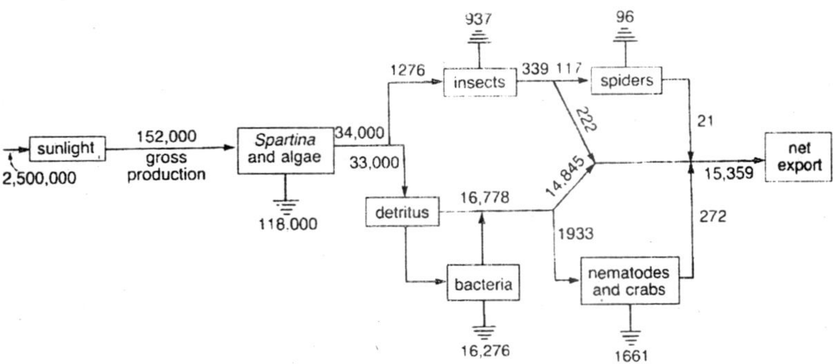

Fig. 17.9. Energy-flow diagram for a Georgia salt marsh; units are in kJ m^{-2} yr^{-1}.

organisms, coastal marshes and estuaries are important feeding areas for the larvae and immature stages of many fish and invertebrates that later complete their life cycles in the sea.

The high production of coastal systems must be supported by high nutrient levels as well. Since a large fraction of the production of the coastal habitat is carried out to sea, exported nutrients must be replaced by imports to maintain a balance. The nitrogen budget of Great Sippewissett Marsh on Cape Cod reveals inputs to the marsh through precipitation (minor), groundwater flow from surrounding terrestrial systems, and local fixation of atmospheric nitrogen, which are approximately balanced by losses through denitrification, local accumulation of sediments, and net export in tidal water. The high rate of denitrification, which occurred primarily in the creek bottom draining the salt marsh, underscores the role of anoxic processes in salt marsh metabolism. The rich organic sediments are mostly anoxic just below their surfaces owing to high rates of microbial decomposition of organic matter. As a result, oxidations based upon denitrification ($N^{5+} \rightarrow N^0$) and sulfate reduction become important for the regeneration of nutrients.

As we have seen, the basic chemical and biochemical transformations of the element cycles are uniquely molded by each type of terrestrial and aquatic system according to physical and chemical conditions created in these environments. Paths of elements discussed in this chapter describe patterns of nutrient cycling.

18

MOLECULAR REGULATION

Primary production indicates the magnitude of energy flux and cycling of elements within ecosystems. As we have seen, the most productive systems are tropical rain forests, coral reefs, and estuaries, where favourable combinations of high temperature, abundant water, and intense sunlight promote rapid photosynthesis and assimilation. But plant growth also depends on the regeneration of nutrients through biological processes, which themselves are sensitive to temperature, moisture, and other conditions. How, then, can we untangle the factors responsible for the regulation of ecosystem function? The correlation of production with physical factors does not provide a cause-effect explanation, because physical factors may exert their influence on production through a variety of pathways.

Discerning how ecosystem function is regulated would be relatively simple if, for example, a single resource necessary for production were used up; whatever controlled the supply of that resource would control ecosystem function, the way an accelerator pedal controls the speed of a car. But under the most favourable circumstances plants assimilate only 1 to 2 per cent of the light energy striking the earth's surface. Not all the water entering the system leaves by transpiration from plant surfaces. Furthermore, plants do not exhaust nutrient elements from the soil.

It is not surprising that ecologists have yet to resolve the control of productivity in most ecosystems. Their data allow broad comparisons of function, but simple correlations between resources, physical conditions, and production show only that external factors are regulators, not how or where they act. A second approach to understanding ecosystem function is experimentation. But experiments are difficult to conduct on the scale of ecosystems. How, for example, could one eliminate nitrifying bacteria from soil without changing related conditions and resources? Where experiments have been performed (for example, by adding water or nutrients to a system), it is difficult to ascertain which step or steps in the cycling of critical elements are affected or whether observed responses mimic natural changes in systems.

A third approach, after comparison and experiment, is to model ecosystem function by studying the behaviour of mathematical or electrical analogies of systems in response to alteration of symbolic or electrical equivalents of resources and conditions. This approach requires detailed knowledge of all processes critical to regulating ecosystem function (which could be interpreted as everything that happens in the system!), including validation of each part of the model by field observation and experiment.

The systems modeling approach was a major focus of American activities in the International Biological Program (IBP) during the 1960s and early 1970s, particularly in the Tundra and Grassland Biome Projects. But the models became so complex as to diminish their general value. Even those developed to mirror specific ecosystems depended upon the sometimes erratic behaviour of particular processes. To make a model resemble the complexities of nature, it must to be adjusted with respect to a long list of unverifiable variables. The unfulfilled promise of the IBP systems models soured many ecologists on the value of modeling for general applications, although the approach maintains a strong following and in relation to certain types of problems.

In this chapter, we shall develop simple systems models with the limited aim of obtaining some insights into the control of ecosystem function. We shall then use these insights to interpret particular observations and experimental results.

Internal Feedback

Ecosystems include populations that interact through predation, pollination, seed dispersal, mutualism, and competition, and physical components that exchange material with other physical components through chemical transformations and with biological components through assimilation and regeneration. Furthermore, ecosystems function in the context of factors such as light, temperature, precipitation, bedrock, and atmosphere, which are external in the sense that they exist independently of the development of the ecosystem.

In systems terminology, understanding ecosystem function requires knowledge of how external forcing functions and internal control feedbacks are integrated. External forcing functions simply are material inputs from outside the system and physical conditions of the environment that influence the system's structure and function. Consider the following analogy. The rate at which a bucket fills with water depends on an external forcing function—the flow of water from the tap. If the bucket has a hole in its bottom, the hole, as a property of the system, could be considered an internal control feedback.

Applied to ecosystems, external forcing functions include light, temperature, weathering and precipitation inputs, salinity, and other kindred factors. Internal control feedbacks result from the chemical behaviour of elements in the physical part of the system and the responses of organisms to the physical environment and to each other. Ecosystem production depends on the regeneration of nutrients by microorganisms and others; their activities form an internal feedback control in the sense that they determine the availability of nutrients to plants. Because external forcing functions can act on both assimilatory and regenerative phases of ecosystem processes, one cannot understand the control of ecosystem function without a detailed appreciation of the influence of external forcing functions on all parts of the feedback control loops within the system.

There are two issues here. The first is the degree to which variation in ecosystem function is driven by variation in external forcing variables as opposed to unique internal properties of ecosystems—essentially the difference between variation in the flow of water coming from the tap as opposed to variation in the number and size of holes in the bucket. The second issue concerns the particular means by which external forcing variables exert their influence on internal feedback controls.

External Feedback

Ecologists are reasonably agreed that basic aspects of ecosystem structure and function are determined by external forcing variables. Maps of worldwide terrestrial production have been drafted by using predictive equations, based on a small number of studies, to translate temperature and rainfall into grams of dry matter. As early as 1862, Justus Liebig speculated that levels of production could be generalized from local studies to global patterns: "If we think of the surface of the earth as being entirely covered with a green meadow yielding annually 5000 kg/ha, the total CO_2 content of the atmosphere would be used up within 21-22 years if the CO_2 were not replaced." Remarkably,

Liebig's estimate of world production, 230 – 240 $\times$ 10^9 metric tons of CO_2 assimilated annually, is very close to present estimates and better than any other estimate in the intervening century and a quarter.

Liebig's was a lucky guess; continents are not covered uniformly with green meadows. But as more measurements of local production have become available, it has been possible to test the relationship between production and external forcing variables (particularly precipitation and temperature) on a global scale. H. Lieth (1973) summarized annual total dry matter production (g m^{-2}), average temperature, and annual precipitation for 53 localities distributed throughout the world. Estimates of production varied 50-fold, from 70 to over 3500 g m^{-2}, precipitation varied 48-fold, from 94 to 4500 mm annually; and temperature ranged from –14.2 to 27.1°C.

We can use Lieth's tabulation to develop a predictive equation relating production to temperature and precipitation. Because we are interested in factorial rather than absolute differences between sites, we first transform the values for productivity and precipitation to logarithms. We cannot treat temperatures in the same way because some of the values are negative (average temperatures below zero Celcius!), and one cannot take the logarithm of a negative number. So we must leave them untransformed. Now we fit a simple algebraic model to the data:

$$P = a + bR + cR^2 + dT + eT^2 + fRT$$

where P is the logarithm of productivity, R is the logarithm of precipitation, and T is the temperature. If the squared (quadratic) terms were significant, they would indicate nonlinearity in the relationship. A significant R-times-T interaction would indicate any synergism between temperature and precipitation. The results of the analysis provide two types of information: one concerning the contribution of variation in temperature and rainfall to variation in production; the other concerning the amount of variation not explained by the relationship.

Variation in terrestrial production clearly is strongly related to the physical environment. All the terms of the predictive model make significant, unique contributions to variation in production in spite of the fact that temperature and rainfall are, themselves, strongly correlated. The signs of the coefficients b and c show that productivity increases in direct relation to precipitation over the lower part of the range (b positive) and then levels off (c negative), presumably because some other factor becomes more limiting. When variation in precipitation is factored out statistically, we find that higher temperatures lead to decreased production (*d* and *e* negative), confirming the notion that water goes farther in cold climates than in warm climates. But there is also a positive synergism between temperature and rainfall (*f* positive) that partially balances this effect. That is, hot, humid climates have higher productivity than one would predict from the separate relationships of production to precipitation and temperature.

The analysis shows that external forcing variables are in some way responsible for much of the variation in terrestrial production. But how much? Statistically, their effect is estimated by the coefficient of determination (R^2), which is the proportion of the variation in production that is related to variation in temperature and precipitation. For Lieth's set of data, $R^2 = 0.73$; that is, 73 per cent of the variation in productivity is explained by the model, and 27 per cent is not. Another way to consider the unexplained variation is to quantify the deviations of individual data points from the predictive model. In the present example, these have a standard deviation of 0.20 $\log_{10}$ units, which means that two-thirds of the observations fall within 0.2 units of the line and 95 per cent fall within 0.4 units of the line. A range of 0.4 units (0.2 above and 0.2 below) corresponds to a factor of 2.5, a range of 0.8 units to a factor of 6.3.

Temperature and precipitation clearly "control" terrestrial production to a large extent. The unexplained variation could be attributable to measurement error, external factors not included in the model (seasonal variation in temperature and precipitation, solar radiation, and nutrient status of the soil, to name a few), and internal

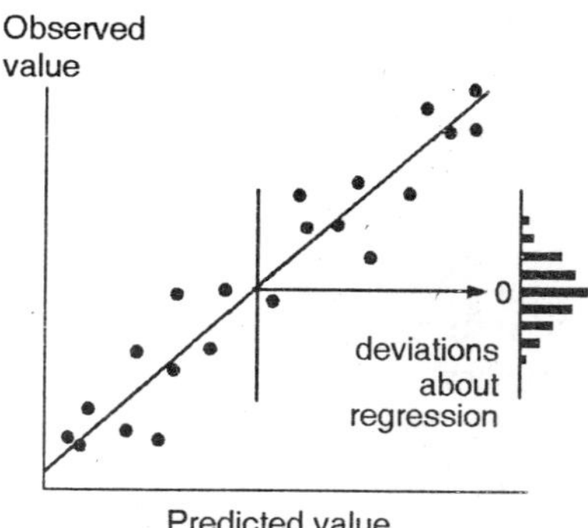

Fig. 18.1. An example of a regression of observed values upon values predicted by several independent variables.

control feedbacks unique to each locality. The latter may derive from particular species, perhaps herbivores, present in one locality but not others. My intuition tells me that additional external factors and inconsistent measurements account for most of the unexplained variation. It is evident, however, that one index to ecosystem function—net aboveground primary productivity—is responsive to external forcing functions.

ELEMENT FLUXES

Energy is the most generalizable currency of ecosystem function, but availability of energy—the external forcing function, light—may have little to do with differences between systems. In this chapter, our interest focuses upon factors that regulate ecosystem function and cause its diversification. But how does one identify the component or components of the system upon which regulators act?

Peter Vitousek, a plant ecologist at Stanford University, has attempted to find these components by comparing the flux of individual elements with the flux of energy through the system as a whole. One might suppose that the element whose flux shows the strongest correlation with primary production exercises predominant control. In his search, Vitousek (1982) brought together data from studies on forests throughout the world, from subarctic to tropical latitudes.

Because energy and nutrient fluxes are difficult to measure directly, Vitousek substituted indirect but nonetheless reasonable indices. When a forest achieves a steady state, net aboveground primary production is approximately equal to litter production. Similarly, the flux of each element is approximately equal to its amount in the litter. Vitousek's analyses compared the flux of total dry matter in the litter (proportional to carbon, hence to energy content) to the flux of several elements.

As you can see, production more closely parallels the flux of nitrogen than that of phosphorus (or similarly, calcium.) Vitousek concluded, quite reasonably it would seem, that factors regulating the cycling of nitrogen in forests predominantly controlled primary production.

Vitousek's study also showed that production is not directly proportional to nitrogen flux. Rather, forests with low fluxes have relatively greater production per unit of nitrogen cycled (nutrient use efficiency) than those with high fluxes. Two factors result in increased nutrient use efficiency: trees assimilate more energy per unit of nutrient assimilated; alternatively, trees retain nutrients for reuse by drawing them back into the stem before leaves are dropped. Nonetheless nutrient

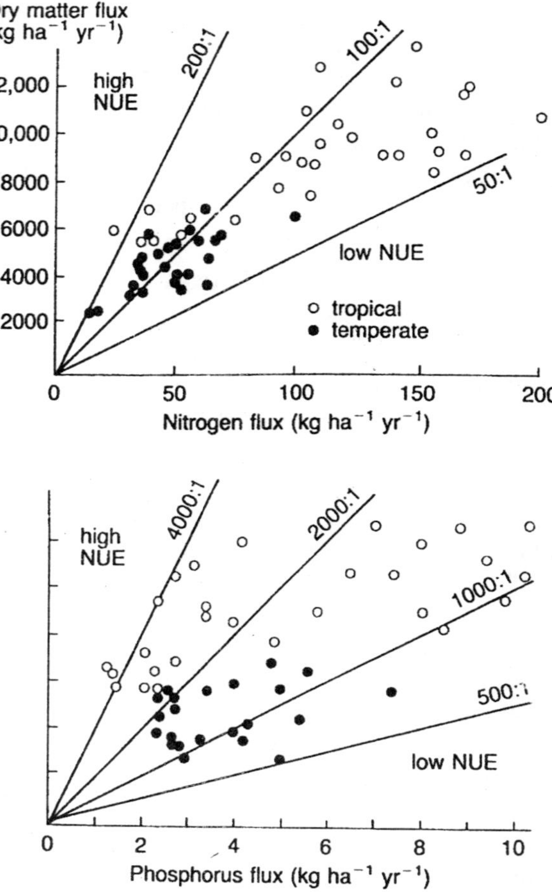

Fig. 18.2. The relationship of dry matter flux in litter fall to nitrogen and phosphorus fluxes in temperature and tropical forests; NUE=nutrient use efficiency.

use efficiency clearly is greater where nutrient flux is lower and, for phosphorus, it is greater in the tropics than in temperate latitudes.

Because nutrient use efficiency partly expresses certain adaptations of the plant, and partly may be a fortuitous consequence of nutrient availability, it is difficult to ascribe a regulatory role to an element based on the relationships. Just as nutrient use efficiency varied so much for nitrogen, whose flux supposedly limits production, the patterns for phosphorus could also reflect different adaptations of plants under different conditions of phosphorus limitation. Lacking a mechanistic model of ecosystem function, these patterns remain ambiguous.

This ambiguity is emphasized by another analysis based only on tropical forests. In this study, Vitousek first related litter production (dry weight) to temperature and rainfall, much as we did earlier with Lieth's data on primary production. As reported in other studies, such as those of Meentemeyer et al. (1982) and Spain (1984), these external forcing functions account for a large part of the variance in litter production of tropical forests (42 per cent). Vitousek then compared deviations from values of production predicted from rainfall and temperature to the nutrient use efficiencies of phosphorus and nitrogen; the residuals were significantly related to the nutrient use efficiency of phosphorus, but not to that of nitrogen. Does this suggest that phosphorus regulates production?

Vitousek's analyses are made more difficult to interpret by the effect of nutrient retention on nutrient use efficiency. The nitrogen and phosphorus contents of living vegetation are not so well known as the corresponding contents of litter. But several studies indicate little difference between tropical and temperate species or between forests within a latitude belt. The lowest nutrient use efficiencies calculated from litter correspond to the nitrogen and phosphorus contents of living vegetation, suggesting that higher nutrient use efficiencies result from greater retention of nutrients in the vegetation rather than lower nutrient requirements for production.

In order to sort out the factors responsible for the regulation of ecosystem function, we shall develop simple systems models that incorporate expressions for the effects of these factors. Then we shall ask whether variation in these factors produce diagnostic, measurable variation in ecosystem properties. First, however, some basics.

Ecosystem Function

Each form of a nutrient or energy within a system is a distinct compartment, which we shall designate X_i for the *i*th form (organic-N, ammonia, nitrate, for example), and each compartment has inflows and outflows, which we shall designate by *Js*. If the compartment is the water in our bucket, then J_0 is the flow from the tap and J_1 is what leaks out through the hole in the bottom. The laws of conservation of matter and energy dictate that the rate of change in the amount of water in the bucket (dX_1/dt) is equal to the difference between the input and the output—algebraically

$$\frac{dX_1}{dT} = J_0 - J_1$$

The amount of water in the bucket is in a steady state (that is, $dX_1/dt = 0$) when inflow equals outflow ($J_1 = J_0$).

Each flux (J) may be a constant or it may vary depending on the state of other factors, including the value of X and the values of external forcing functions. We may express such variable fluxes in symbolic notation as $J = f(X; S; P)$, where f denotes that J is a function of the values inside the parentheses; S represents the state of the system (number and size of holes in the bucket, leaf area available for photosynthesis, population density of nitrifying bacteria); and P stands for various parameters that are the external forcing functions.

In the bucket example, suppose that J_0 is a constant but that J_1 increases in direct proportion to the volume of water in the bucket (that is, $J_1 = k_1X_1$). (As the bucket fills, the water pressure at the bottom increases the flow of water through the hole. In practice, this will not be a linear function, but portraying it as such will serve our purpose.) The amount of water in the bucket (X_1) reaches a steady state (providing the bucket is large enough) when $J_0 = k_1X_1$. This may be rearranged to show that the steady state amount of water is equal to J_0/k_1. Hence when the value of the forcing function (J_0) increases, water reaches a higher level in the bucket. Reducing the size of the hole at the bottom (the state variable k_1) has the same effect.

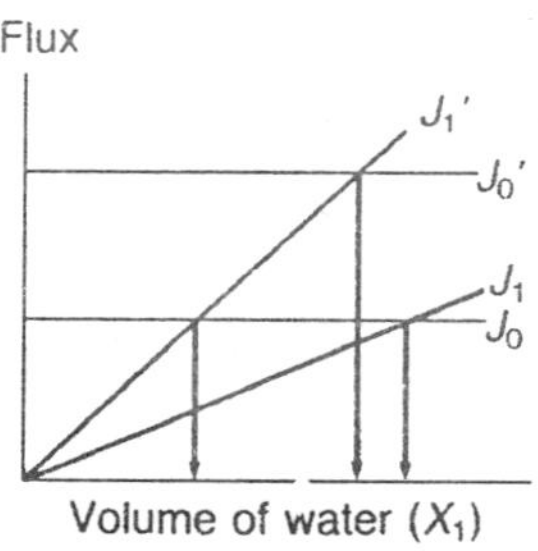

Fig. 18.3. The effect of changing input (J_0) and output (J_1) rates on the value of water (X_1) in a bucket.

Single compartment models may be used to describe organisms or populations, but at least two compartments are required to depict the internal cycling of elements within ecosystems, and realistic systems models can become much more complex. Consider the simplest case, two compartments that cycle an element between them. Compartment sizes are X_1 and X_2, and fluxes are J_1 and J_2. As we have seen above, the Js represent functions. These may be of zero order, in which case J is a constant ($J = c$); first order (linear), in which case J_1 is a

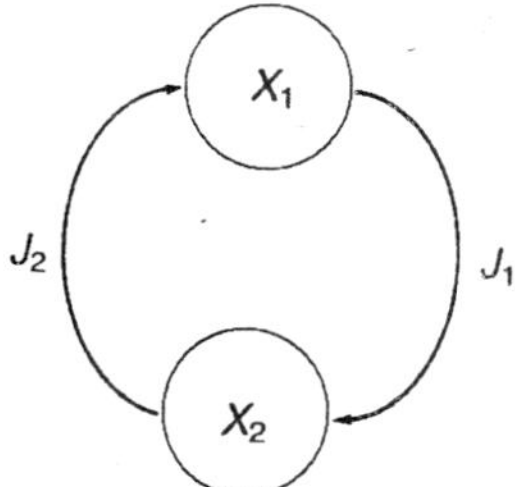

Fig. 18.4. Diagram of a closed, two-compartment system.

function of X_1; or second order (quadratic), J_1 being a function of both X_1 and X_2.

Little can be said of the functioning of a particular system without knowing the details of flux functions. Simple models do, however, lead to a general understanding of some broad features. The global water cycle, for example, can be thought of as two compartments, vapour (X_1) and liquid (X_2). Knowledge of the physics of state change of water allows us to describe functions for the fluxes in very general terms. Precipitation (J_1) is a first order equation depending only on X_1 and various external forcing functions such as air temperature. The amount of water at the earth's surface (X_2) probably has little direct effect on precipitation. Air has a limited capacity to hold water vapour at a given temperature, and so condensation increases disproportionately as water vapour increases toward this limit.

Evaporation (J_2) is a second order equation depending on the surface area of water (some function of X_2), vapour pressure of water in the atmosphere (proportional to X_1), and the external forcing functions, temperature, and insolation (intensity of sunlight). On a global scale, precipitation probably has a smaller effect on the surface area of water than evaporation has on atmospheric water vapour, so it might

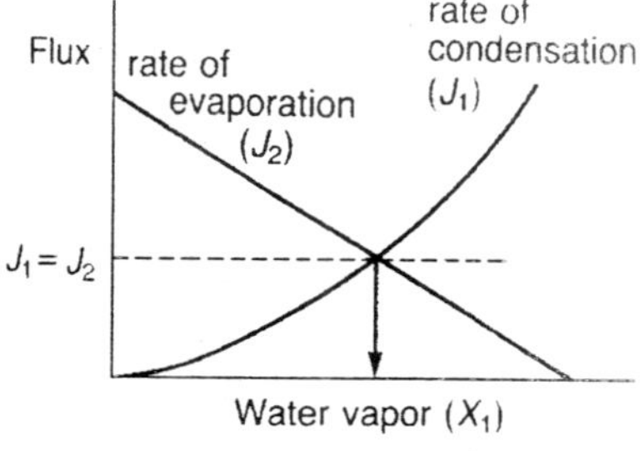

Fig. 18.5. Rate of evaporation decreases and rate of condensation increases with increasing water vapour in the atmosphere.

be possible to ignore X_2 in the function J_2. Thus we may portray the general features of the hydrological cycle on a graph relating J_1 and J_2 to X_1. The model shows that the system assumes a steady state with $J_1 = J_2$ at X_1 and that changes in the forcing functions affecting J_1 or J_2 (change in temperature, for example) adjust the equilibrium point for the entire system. The model can't be used to predict the local weather.

Model of Ecosystem Function

In his book *The Elements of Physical Biology* (1925), Lotka investigated the behaviour of biological systems using the insights of thermodynamics and the tools of mathematical modeling borrowed from the study of chemical equilibria and other physical phenomena. In the space of a few pages, he outlined the application of systems modeling to the interpretation of nutrient cycling in ecosystems. Lotka treated the specific case of three compartments X_1, X_2, X_3 through which some element or other material cycles with fluxes J_1, J_2, J_3.

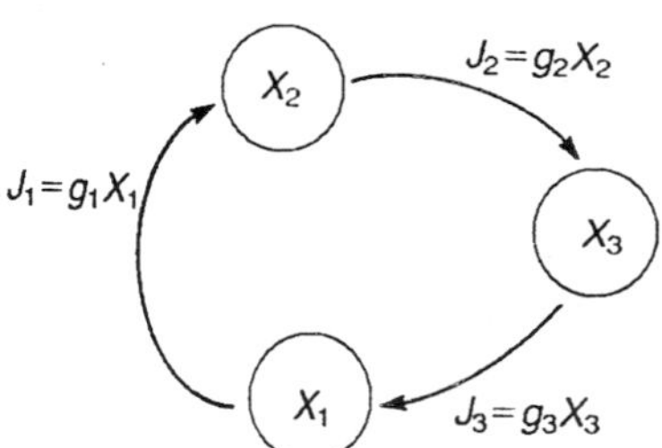

Fig. 18.6. A diagram of Lotka's system of three compartments in which the flux between one compartment and the next is a function only of the first compartment.

Change in any compartment—for example, dX_1/dt—equals the difference between the fluxes into and out of the compartment (J_3 and J_1 in the case of compartment 1). To illustrate the behaviour of such a system, Lotka described each flux as a first order term $f_i(X_i)$, arbitrarily having the form g_iX_i, g_i is the rate at which material in compartment i is transferred to the next compartment. Thus $dX_1/dt = g_3X_3 - g_1X_1$. When the system is in a steady state, all the fluxes are equal; hence, $J_1 = J_2 = J_3$, or $g_1X_1 = g_2X_2 = g_3X_3$. Because $X_i = J_i/g_i$ and all the Js are equal, the sizes of the compartments are in the relative proportions.

$$X_1 : X_2 : X_3 = \frac{1}{g_1} : \frac{1}{g_2} : \frac{1}{g_3}$$

Now, if all the compartments of the system contain a total M (= X_1 + X_2 + X_3) of the cycling material, the amount in any one compartment is the proportion in that compartment (P_i) times the total (that is, P_iM). The proportion depends strictly on the functions g_i because

$$P_i = \frac{X_i}{X_1 + X_2 + X_3} = \frac{1/g_i}{1/g_1 + 1/g_2 + 1/g_3}$$

A little algebraic rearrangement provides the following expressions for compartment size and flux:

$$X_1 = M\left(\frac{g_2 g_3}{g_1 + g_2 + g_3}\right)$$

and

$$J = g_1 X_1 = M\left(\frac{g_1 g_2 g_3}{g_1 + g_2 + g_3}\right)$$

These equations tell us that the structure (*Xs*) and function (*Js*) of the system are defined by the transfer functions (*gs*), which incorporate the external forcing variables and internal feedback controls.

The derivative of the equation for J with respect to g_i tells us how sensitive J is to changes in that expression. That derivative, which has the same form for all the *gs* in the model above, is

$$\frac{dJ}{dg_1} = M\left[\frac{g_2 g_3 (g_2 + g_3)}{(g_1 + g_2 + g_3)^2}\right]$$

Can we determine whether changes in rate (g) of one transformation influences overall cycling (J) more than change in another? Because g_1 appears only in the denominator of the expression dJ/dg_1, the larger the value of g_1 relative to g_2 and g_3, the smaller the sensitivity of J to g_1. Alternatively, as g_1 becomes very small, the sensitivity of J to changes in g_1 approaches a maximum of $Mg_2g_3/(g_2 + g_3)$. As Lotka pointed out, the relative sensitivities of J to g_1 and g_2, for example, are determined by the ratio

$$\frac{dJ/dg_1}{dJ/dg_2} = \frac{g_2 g_3 (g_2 + g_3)}{g_1 g_3 (g_1 + g_3)}$$

Therefore, if $g_1 < g_2$, then $dJ/dg_1 > dJ/dg_2$.

This model tells us that differences in structure and function between systems likely derive from differences in the lowest transfer rates. Low transfer rates place bottlenecks in the path of material flow through a system leading to the accumulation of material in the immediately preceding compartment. Therefore, the underlying cause of variation in flux through a system should be apparent in shifts of materials among compartments in the system. For example, if an increase in J is accompanied by a shift of material from compartment 1 to compartments 2 and 3, we may infer that an increase in the function g_1 was responsible. Distinguishing the roles of state variables, internal control feedbacks, and external forcing functions in this change would require additional study of the system, but research efforts are greatly focused by heeding the insights of the systems model.

Relative Transfer

The cycling of nitrogen within a water column by and large follows the path (ammonia $\rightarrow$ nitrate) $\rightarrow$ particulate -N $\rightarrow$ dissolved organic nitrogen (DON) $\rightarrow$ (ammonia $\rightarrow$ nitrate) (algae may assimilate either ammonia or nitrate; nitrite is converted so readily to nitrate that it can be ignored here). In the study of Liao and Lean (1978) on nitrogen transformations in the water column of the Bay of Quinte, Ontario, the sizes of some compartments changed dramatically with season. Between winter and summer, particulate and dissolved organic nitrogen increased while nitrate decreased. This shift implies that the primary difference between the cycling of nitrogen in summer and winter is the rate of uptake of nitrate by phytoplankton. A simple first order systems model will illustrate how we can elaborate this idea.

When the system has achieved a steady state, fluxes into and out of each compartment must be balanced. For example, $J_2 = J_3$, $J_3 = J_4 + J_5$, and so on. Hence, under steady state conditions,

$$g_2X_2 = g_3X_3 = (g_4 + g_5)X_4 = (g_1X_1 + g_4X_4)$$

From the last two quantities, we can show algebraically that $X_1/X_4 = g_5/g_1$, and by making the appropriate substitutions, we obtain the relationship

$$X_1:X_2:X_3:X_4 = \frac{g_5}{g_1(g_4+g_5)}:\frac{1}{g_2}:\frac{1}{g_3}:\frac{1}{(g_4+g_5)}$$

Setting each X_i equal to its proportion of the total nitrogen allows us to estimate relative values for the $g_i s$ during the winter and summer. These indicate quite dramatically that the ratio g_1/g_5 is an order of magnitude lower during the winter than it is during the summer. The sum $g_4 + g_5$ differs little between the seasons, relative to g_2 and g_3, so it is evidently the value of g_1 that decreases so much between

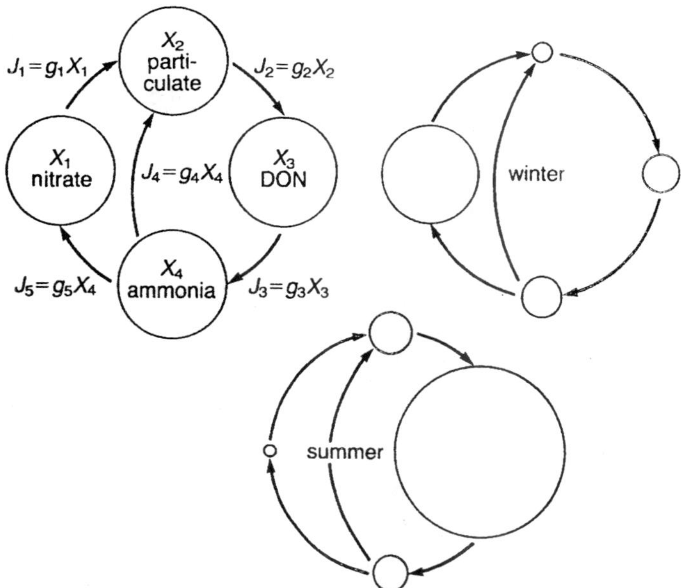

Fig. 18.7. A first-order systems model of nitrogen cycling, showing the changes in compartment sizes between summer and winter.

summer and winter (g_1 describes the rate of assimilation of nitrate-N by plants). During the winter it is clearly much reduced compared to nitrification (g_5) and ammonification (g_3), suggesting that some factor, such as light, might limit production (hence assimilation).

The preceding model may oversimplify or even misrepresent nitrogen transformations in the water column. It was meant to illustrate an approach. Most systems models are much more complicated, often including dozens of compartments and equations of much higher order than one. In our simple model, for example, fluxes J_1 and J_4 almost certainly should have been represented by second order equations involving X_2 (the population of the phytoplankton which accomplish the assimilation; none would be assimilated in the absence of plants). More realistic equations could be developed for particular systems, but simplified models, such as the one that follows, may also serve to direct inquiry into more general comparisons of function between ecosystems.

Soil, Biomass and Detritus

In broad comparisons among terrestrial ecosystems, measurements of net primary productivity (g dry matter m^{-2} yr^{-1}) vary almost 30-fold

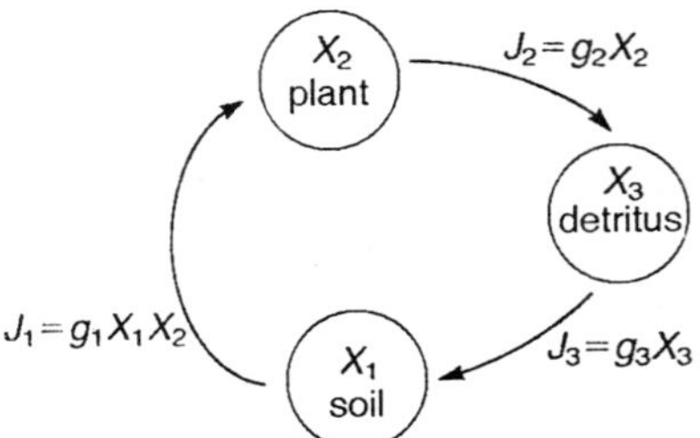

Fig. 18.8. A second order systems model of nitrogen cycling in a forest ecosystem.

between desert shrub (70) and tropical rain forest (2000). This variation clearly is related to climate, yet ecologists have not determined where external forcing variables exert their influence within the system. A systems approach can provide clues to fruitful avenues of study in this case.

Let us consider the cycling of nitrogen again. We shall represent an ecosystem as three compartments: mineral soil (X_1), living plant biomass (X_2), and organic detritus (X_3), with fluxes J_1, J_2, and J_3. (Animals are downgraded in status to that of trivial hangers-on; the activities of microorganisms are implicit in the flux J_3). J_2 (the annual dropping of leaves and other detritus) and J_3 (the mineralization of detritus by microorganisms) can be considered as first order processes. Assimilation must be modeled as a second order process because its rate depends both on nutrient availability and plant abundance. Under steady-state conditions, therefore, we obtain the relationships

$$g_1X_1X_2 = g_2X_2 = g_3X_3$$

and

$$X_1X_2 : X_2 : X_3 = \frac{1}{g_1} : \frac{1}{g_2} : \frac{1}{g_3}$$

This simple model offers some surprises. For example, the level of inorganic nitrogen in the soil, mostly nitrate (X_1), is equal to the ratio of g_2 (rate of detritus production) to g_1 (rate of nitrogen assimilation), and is independent of g_3 (rate of microbial regeneration of inorganic nitrogen). This is not to say that nitrogen flux is independent of g_3. The equation expressing flux as a function of the rates g_i is a bit complex, but with some algebra it can be shown to be

$$J = \left[\frac{g_2g_3(g_1 - g_2)}{g_1(g_2 + g_3)}\right]M$$

Because fluxes of elements are difficult to measure directly, we have to estimate relative values of g_i from the sizes of compartments.

Differences in the relative rates of transfers between compartments in different systems can, however, provide insights into points of feedback control.

D. J. Ovington (1962) tabulated estimates of compartment sizes of various elements in forest systems. For illustration, we shall compare the cycles of nitrogen and phosphorus in a 47-year-old Scots pine (*Pinus sylvestris*) plantation in England and a 50-year-old mixed tropical forest in Ghana. The differences are striking. In England, for both nitrogen and phosphorus, values of g_1, g_2, and g_3 are of the same magnitude. In the nitrogen cycle in Ghana, g_1 and g_2 are similar, but g_3 is almost two orders of magnitude greater. Thus the regeneration of mineral nutrients apparently proceeds much more rapidly in the tropics than in temperate areas, which is confirmed by direct measurement of litter decomposition. Relative transfer rates of phosphorus behave similarly, but reveal another difference between the tropical and temperate forests. In Ghana, the assimilation coefficient (g_1) is relatively much greater than the rate at which vegetation gives up nutrients (g_2), compared to the more nearly equal values in England. These values of g imply that, compared with *Pinus sylvestris*, tropical forest trees assimilate

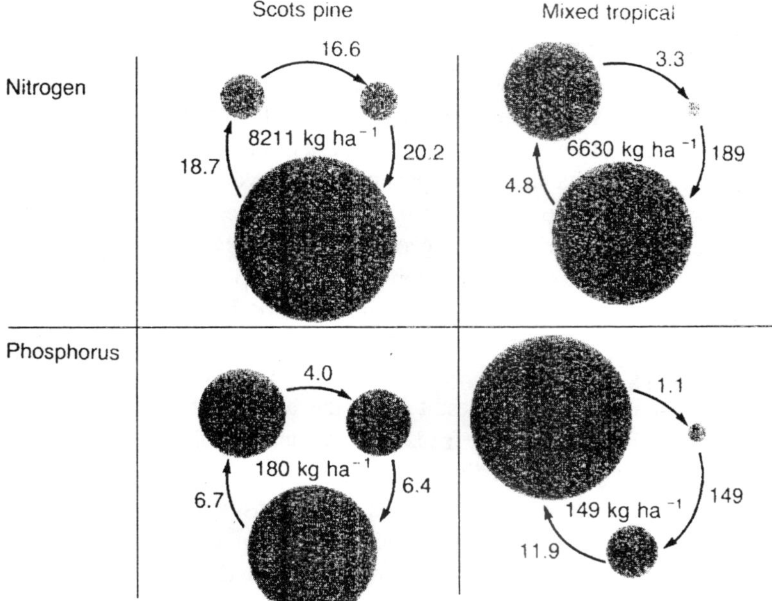

Fig. 18.9. Compartment sizes and values of g_i for nitrogen and phosphorus estimated from compartment sizes in a Scots pine plantation.

phosphorus more efficiently and hold onto it more tightly (for example, by withdrawing phosphorus into the stem before shedding leaves).

The values estimated for *gs* are adjusted so that fluxes (*J*) are equal to 1.0 when compartments are expressed as fractions of the total. Therefore, although the *gs* are internally consistent, they can be compared between localities only when one of the fluxes is known. The flux in either nitrogen or phosphorus in Ghana is probably no more than double that in England and the total amounts of the two elements are similar in the two areas. Therefore, for nitrogen, both g_1 and g_2 are probably lower in the tropical locality than the temperate locality. For phosphorus, g_1 is probably much higher and g_2 somewhat lower. Unquestionably, however, much of the difference between temperate and tropical localities in overall flux is due to the tremendous increase in rate of decomposition of litter.

Regulation of Production

Our understanding of ecosystem processes should be judged by how convincingly we can explain variation in ecosystem structure and function. Attempts to do so for forest production on a global scale suggest that we are far from understanding one of the most intensively studied systems on the earth.

Long-term fertilization and irrigation of forests demonstrate that both biomass (X_2) and production per unit of biomass (g_1X_1) can be increased by adding nitrogen, phosphorus, or water (X_1). Because different treatments in such experiments involve the same species of trees growing on the same soils, g_1 probably does not differ between treatments, and variation in production and biomass results from increased inputs to one or more compartments (X_1) that limit production. Because nutrient addition experiments involve increases in X_1, they can reveal little about differences between forests unless precipitation or weathering inputs contribute importantly to variation. More probably, differences in production result from the effects of external forcing functions or internal control feedbacks on transfer functions (g_i), which cannot be manipulated easily.

Measurements of rates of litter decomposition suggest that the effect of external forcing variables on g_3 can explain part of the variation in production between temperate and tropical forests. This effect may be seen in the forest model by varying transfer rates. For example, when one sets $g_1 = 20$ and $g_2 = 15$, production (*J*) increases rapidly as g_3 increases up to about 20 and then increases much more

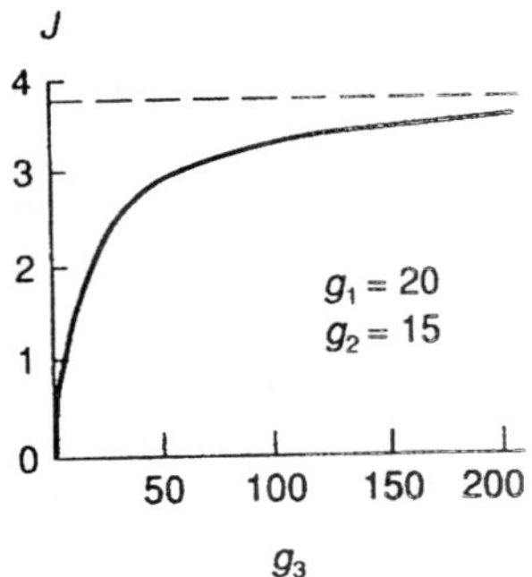

Fig. 18.10. Relationship between flux and the rate constant for decomposition of detritus (g_3) in a second order systems model of a forest.

slowly with further increase in g_3. Production doubles between g_3 = 10 and g_3 = 60, at which point J is 80 per cent of its maximum value, given the values defined for g_1 and g_2.

If, in the tropics, g_3 is uniformly high, production would be relatively insensitive to variation in g_3. How then can we account for variation in production among tropical forests? Vitousek showed that almost half the variation in production is associated with variation in temperature and rainfall. But where do these forcing functions act? When g_3 is very large compared to g_2, which is probably generally true in the lowland tropics, the equation for flux may be simplified to

$$J = \frac{g_2}{g_1}(g_1 - g_2)$$

because $g_3/(g_2 + g_3)$ is approximately 1. With g_3 out of the picture, J varies with g_1 and g_2 only, according to the relationships $dJ/dg_1 = (g_2/g_1)^2$ and $dJ/dg_2 = (g_1 - 2g_2)/g_1$ Note that g_1 depends on physical conditions in the soil, the adaptations of plants for nutrient uptake and carbon fixation, and interactions between roots, bacteria, and fungi; g_2 depends on the life expectancy of plant parts and mechanisms of nutrient retention. Moreover, adaptations of trees that affect g_1 and g_2 may be functionally interrelated in ways that constrain their evolutionary adjustment.

All these factors are beyond our current understanding. Systems models, even the simple ones developed in this chapter, can suggest control points in ecosystem processes, but particular feedback controls can be understood only by additional observation and experiment. And, of course, natural systems are potentially much more complex than such simple models suggest.

Systems Analysis

The models described above represent the compartments and fluxes of materials or energy within the system. Fluxes and compartments are both regulated by transfer functions (the gs in previous equations). The transfer functions depend in part on the conditions of the environment and in part upon the state of other compartments within the system. The terms of the transfer functions (for example, $g_3 = g$ [temperature, moisture, pH, litter chemistry, population sizes of microbes and soil invertebrates, fungus-bacteria interactions, and so on]) are sometimes thought of as the flow of information through the system. In schematic systems models, information and material fluxes are sometimes portrayed differently.

Although "information" is given almost mystical qualities by some systems ecologists, it in fact couples the many material and energy flow circuits within the same system. Consider, for example, the flow of energy and the cycling of nitrogen. These can be represented as separate compartment diagrams, yet they are intimately connected by processes of assimilation and dissimilation. The transfer function governing carbon (energy) fixation must be related to the transfer

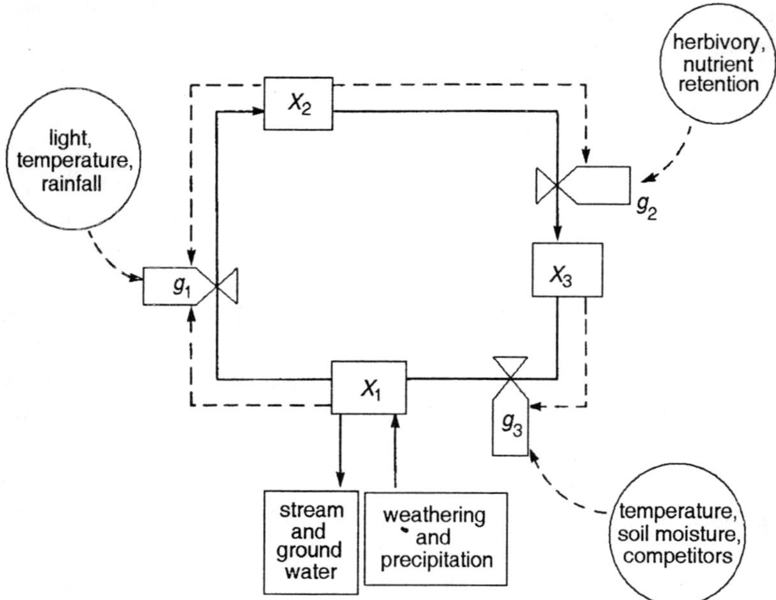

Fig. 18.11. A systems model distinguishing between material fluxes and "information" fluxes.

function governing nitrogen assimilation because both are required by plants in an approximately fixed ratio. What some systems ecologists call information flow is equivalent to the functional integration of the processes of organisms (carbon and nitrogen are assimilated together), the relationship of these processes to conditions in the environment (production increases with soil moisture), and the interactions between organisms (trees compete for soil nutrients; caterpillars eat leaves).

Energy and materials flow through the ecosystem in large part because of the activities of organisms. Understanding the regulation of ecosystem function therefore depends in large part on how organisms respond to physical conditions and how they interact with others of the same and different species. These interactions form the basis for the study of population processes. Although population and ecosystem functions are different expressions of many common processes, the studies of ecosystem and population processes have yet to be fully integrated by ecologists.

19

Molecular Evolution of Society

Humans are the most social of all animals. Societies are sustained by role specialization among members, the interdependence attendant upon specialization, and the cooperation that satisfies the needs of this interdependence. Yet humans also are competitive, to the point of violence, within this mutually supportive structure. Social life balances contrasting tendencies toward mutual help and conflict.

Theories of population biology discussed earlier in this book incorporate as a basic premise competition between individuals of the same species. In population biology, progeny are the measure of success, the prize of the future going to those individuals that outsurvive and outproduce others—those that gather more resources than the average and leave fewer for competitors.

In the previous chapter we discussed conditions under which adults might cooperate to rear their offspring. But in many species sociable behaviour extends far beyond mates: to progeny, extended families, and large groups of unrelated individuals. Some animal populations exhibit much of the complexity of human societies. The social insects—ants, bees, termites—are remarkable for their division of labor and behavioural integration of the hive or nest. Similar subtlety of social interaction, including role specialization and altruistic behaviour, is being discovered increasingly among mammals and birds.

Mated pairs of animals seem to get on quite well together. But most behavioural ecologists believe that the decision to stick together, or not, is subject to selfish motives. Is this also true of higher levels of social behaviour, especially where unrelated individuals cooperate to their mutual benefit? Is all social behaviour personally or genetically self-serving? We humans like to think not. In the past, such beliefs have provided pragmatic justification for conflict, from petty squabbles to abhorrent acts of war. If there were no intrinsic value to societal cooperation (which then surfaces only as an iceburg's tip buoyed up by the societal conflict beneath), then lofty goals of peaceful coexistence, whether between cousins or countries, would diminish beside the practical concern of keeping the peace.

Social behaviour includes all types of interactions between individuals, from cooperation to antagonism. Outright conflict can assume the ritualized appearance of males posturing for social rank or access to mates. Sometimes social behaviour provides a means of organizing and making orderly the expression of basic conflicts within groups. Defense of territories and the establishment of dominance hierarchies serve this purpose with a minimum of social strife. How is it that contestants can agree upon "gentlemanly" conduct to resolve their disputes? One also sees instances that appear as true helping of others at personal cost or risk. How are such behaviours to be reconciled with the exigencies of evolutionary fitness?

Genetic relationship divides social interaction into three categories, each with different evolutionary contexts. First, mates generally are unrelated individuals that join together in the production of offspring. Because individuals of each sex have veto power over the mating act, their common interest in progeny requires cooperation at least to the point of fertilization. Second, unrelated individuals have little common genetic interest in the future; conflict is the usual basis of their interaction. Third, behaviour directed toward closely related individuals is modified by the fact of common genetic descent. When individuals share parents, grandparents, or some more distant set of ancestors, they also share portions of their unique genotypes, thereby giving them a common genetic interest and a basis for cooperative involvement, as we shall see shortly.

Because we are the supremely social species, the study of social behaviour stimulates a particular interest in us. But we must be wary of drawing facile analogies between animal societies and our own. Our lives may sometimes seem like the toil of ants in an ant hill, but human and ant societies differ in fundamental aspects of organization, as we shall see. Probably no other animal society has the cultural tradition of behaviour as ours does, and cultural and genetic evolution may offer substantially different opportunities for achieving social integration. Nevertheless, animal societies will be understood for their own sake, and this knowledge will be applied, rightfully or not, to our own circumstance.

Our goal in this chapter is to understand the ecological conditions and attributes of organisms that combine to nurture various types of social behaviour between individuals. As with any aspect of science, understanding comes from the application of observation, theory, and experiment to a common problem. In the study of social behaviour, in particular, one can see the important roles that instances of each of these has played in the subsequent development of the discipline.

Social Interaction

Rarely do individuals of the same species encounter one another closely without arousing some type of behaviour. Each individual has a personal space within which other individuals from the general population are not tolerated. The distance to which this space is defended—called individual distance—varies from species to species and with the circumstances of the individual, as one might expect. In many cases, individual distance extends no farther than one's reach at a particular moment. At one extreme, chickadees and other birds sometimes roost closely huddled together on cold winter nights; at the other, animals sometimes go out of their way to chase others. Individual distance satisfies the practical need of space for unobstructed movement. Birds cannot take off without risk of injury when others are within a wing's length. The path of motion of any animal is restricted by others close at hand.

Defense of territory is the extension of individual distance to space and objects. When such critical resources as food or breeding sites are defendable, organisms may enlarge their individual distance to include them. Any areas or the spaces around objects defended against the intrusion of others may be regarded as territories. These may be transient or more or less permanent depending on the lability of the resource and the individual's need of it. Shorebirds may defend a particularly good feeding area on a beach until the rising tide covers it over; male hummingbirds may defend a bush against others so long as it is in flower; kittiwake gulls defend their nesting ledge through the breeding season until the young are able to fly. The object of a territorial claim may vary from food to nesting site or mate, but defense typically takes the form of highly ritualized displays, usually without physical contact, at perceived boundaries of exclusive areas.

At times, dispersion of individuals on territories may not be practical because of the social pressure of high population density, transience of the critical resource, or overriding benefits of living in groups. In such circumstances, conflict among individuals also is resolved by contest, with social rank rather than space going to the winner. Once individuals order themselves into a hierarchy of social status, subsequent contests between them are resolved quickly in favour of the higher-ranking individual. When a social hierarchy is linearly ordered, the first ranked member of the group dominates all others, the second ranked dominates all but the first, and so on down the line to the last ranked individual, who dominates none.

Territory and dominance hierarchy are alternative expressions of the same social tendencies. We see this most clearly when a population switches from one to the other as circumstances change. For example, the dragonfly *Leucorrhinia rubicunda* switches from a territorial system with strong site fidelity at low density to one of broadly overlapping feeding areas at high density. In one area of Finland, dragonflies in a sparse population were spaced 3 to 7 meters apart along the edge of a pool; in a dense population individuals were a half meter apart on average. Pajunen scored the intensity of interactions between males (only males are territorial) from low values of 1 for no interaction, through intermediate values corresponding to varying degrees of threat and pursuit, to the highest score (5) for threats followed by fighting. Where the dragonflies were sparse, the frequency and intensity of aggressive interactions were high (average score 4.2) compared to densely inhabited areas (3.1) because of frequent territorial defense in the first area. In dense populations, individual dragonflies rarely returned to resting sites, which are usually stems of emergent vegetation. Instead, they flew over larger areas of the ponds and alighted to rest less frequently than did dragonflies in less dense populations. Both the level of territorial defense and the level of site tenacity decreased as the

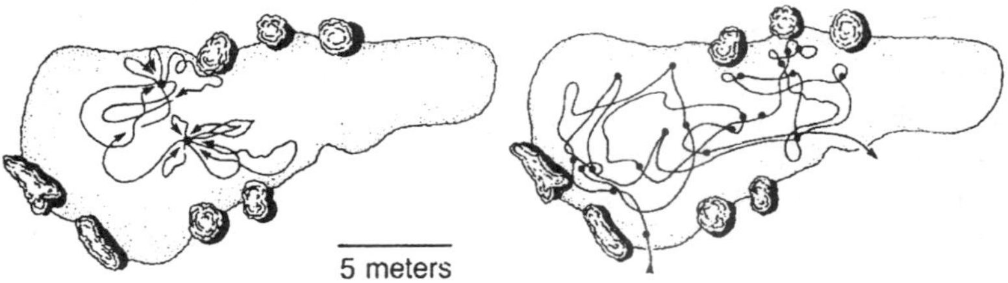

Fig. 19.1. Example of male flight activity in the dragonfly Leucorrhinia rubicunda showing well-developed site attachment (left) and weak site attachment (right.

density of dragonflies in an area increased. Territoriality also is more likely to be manifested as resources become more defendable and valuable. In an experimental study, Anna hummingbirds defended artificial feeders more aggressively as the provision of sugar water was increased to a substantial proportion of the daily energy requirement.

The position of an individual in a dominance hierarchy is sometimes reflected by its spatial position within a group. In large foraging flocks of wood pigeons, individuals low in the dominance hierarchy tend to be at the periphery, where they are more exposed to predators than the dominant individuals occupying the center of the flock. Peripheral birds appeared to be nervous, and because they spend much of their time looking up from feeding, they are often undernourished. Birds in the center of the flock are generally calmer and feed more because they are protected from the surprise attack of a predator by the vigilance of individuals at the periphery.

Territory and dominance hierarchy raise a number of questions concerning the evolutionary origin of such behaviours and their consequences for the population. Behaviourists perceive the decision to defend a territory or not as balancing the costs of defense against the benefits of a secured resource. J. Peterson Myers et al. (1979) showed that the size of territories defended by sanderlings (small shorebirds) feeding on California beaches varied inversely with the density of conspecifics. At high densities, intrusions were frequent and the cost of territorial defense therefore increased. Territory size was smallest in areas of high prey density, primarily because of the larger number of sanderlings that congregated upon such rich sites.

Communication

Most behaviours involved in determining rank or defending territorial boundaries involve ritualized behaviours that rarely lead to risky physical struggle. Certain appearances or behaviours appear to signal higher status than others. Why, then, don't less dominant individuals assume the behaviour of their betters? That is, why don't they cheat the system and deceive other members of the population into thinking that they are more aggressive than they really are? Part of the answer is that some ritualized behaviours allow contestants to judge each other's size, which is difficult to cheat about and which often determines the outcome of earnest combat. Another part is that signaling status and being able to back it up must go together. This has been demonstrated by a remarkable set of experiments by Sievert Rohwer on Harris's sparrows.

Harris's sparrows, which are related to the more familiar white-throated and white-crowned sparrows, breed in the Canadian arctic and winter in small flocks in the central United States. Social status within flocks is highly correlated with the amount of dark colouration in the plumage of the throat and upper breast. Upon first encounters, lighter birds generally avoid darker individuals, and so dark colouration meaningfully signals status. When Rohwer dyed the plumage of light individuals dark, they were mercilessly attacked by others who easily

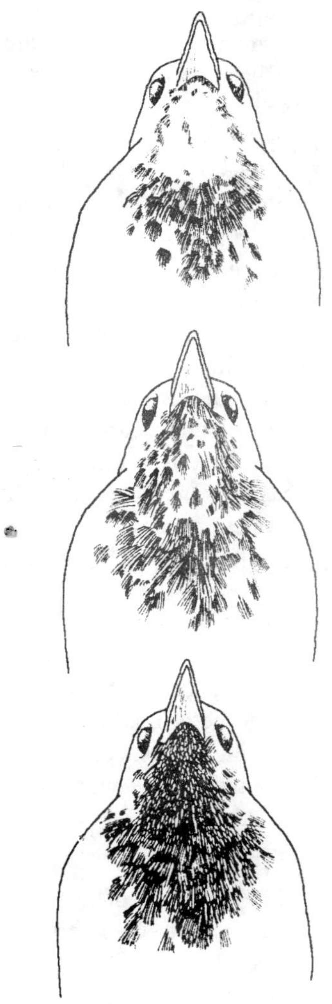

Fig. 19.2. Examples of variation in throat plumage of Harris's sparrows.

saw through the ruse. But dark individuals bleached to a lighter colour found themselves constantly having to attack naturally light birds to regain their status. The system works because plumage and aggression go together—there are no cheaters. When Rohwer implanted testosterone into lighter birds to raise their level of aggression, such individuals that also were dyed darker rose in the dominance hierarchy while those left with their light plumage, but given dark emotions, were persecuted by naturally dark birds and could not rise in status. Behaviour and plumage probably are affected by common physiological conditions associated with hormone levels. When the two do not match, birds get into trouble either because they do not get their due respect or are treated as imposters.

Rohwer's experiments raise a difficult question. If social status depends on the level of a hormone whose production imposes little physiological cost, why don't all members of the population have the highest level of aggression? If a sparrow's social rank can be elevated with a nickel's worth of hormone and dye, what's to stop any individual from assuming the trappings of high rank? Similar hormone implants have turned monogamous song sparrows into bigamists and nonterritorial red grouse into landowners. At this point, we can only presume that status either is linked to age and experience, according to which hormone-mediated levels of behaviour are adjusted to serve other purposes, or that variation in rank is a mixed Evolutionarily Stable Strategy in which dominance has its costs and all members of the hierarchy have roughly equivalent fitness.

Given that status truly represents an individual's capabilities in behavioural encounters, one might also ask: Why do low-ranking individuals, which are excluded from food and mates, and which are exposed to higher risks, remain with the flock? The answer must simply be that it is better to be a low-ranking individual in a flock, perhaps rising in rank with age, than to be a loner. Both the establishment of flocks and the size of flocks must balance costs and benefits to the members. One presumes that individuals won't associate with others unless it is to their personal advantage.

Group Living

Animals get together for a variety of reasons. Sometimes they are independently attracted to suitable habitat or resources and fortuitously form aggregations, like those of vultures around a carcass or dungflies on a cowpat. Within such groups, individuals may interact, usually to contest space, resources, or mates. In other cases, progeny remain with their parents to form family groups. In this case, aggregation results from a failure to disperse. True social groups arise through the attraction of unrelated individuals to each other—that is, through joining together. The evolutionary motivation for such behaviour presumably resides in increased individual fitness either by facilitating feeding, protecting group members from predators, or providing increased access to mates.

Feeding in groups may be facilitated by reducing the time required to look out for predators, as we have seen in the wood pigeon, providing information concerning the location of food, and cooperation in obtaining food. Peter Ward and Amotz Zahavi (1973) suggested that breeding colonies, roosting congregations, and other group formations may serve as "information centers" for food-finding, where the food supply is unpredictable in time and space. Ward and Zahavi envisioned individuals spreading out over vast areas in search of food, conveying information about feeding conditions when they return to the assemblage and, as they leave to feed, following others that have located good foraging areas. Certainly animals are capable of learning from the behaviour of others. But although field observations (Krebs 1974, Brown 1986) and laboratory experiments suggest that individuals can utilize the behaviour of others to locate resources, the fitness components of such behaviour deriving from group living are not understood.

Cooperation involves the coordination of individual behaviour toward a common goal: defense of the herd by male musk oxen, pack-hunting by wolves and killer whales, and other mammals, and by some birds; the coordinated behaviour of aquatic birds that corral fish into a small area where they can be fed upon easily. Such instances of cooperation are not common, but where they do occur social behaviour serves more to foster mutualism between individuals than to organize competition between them: individual behaviour becomes subservient to the group, and groups may assume a characteristic behaviour of their own.

Social Organization

During the last two decades, studies of social behaviour, particularly of birds and mammals, have revealed near-human levels of individual recognition, cooperation, and social subservience. I could pick any of dozens of well-known species, each with its own peculiarities, to illustrate the subtleties of social behaviour in animals. One of my favourites is the pinyon jay (*Gymnorhinus cyanocephalus*), a native of the southwestern United States. As a taxonomic group, the

Fig. 19.3. The pinon jay (Gymnorhinus cyanocephalus).

jays (family Corvidae) exhibit a wide range of social organization, from simple territoriality in the Steller's jay to single family groups on territories in the Florida scrub jay, larger social groups of several families with group territories in the Mexican jay, communal breeding, with several females laying in the same nest, in the Dickey jay, and large, socially integrated flocks in the pinyon jay.

Flock size in the pinyon jay varies up to a maximum of about 250 individuals in the area surrounding Flagstaff, Arizona. The flocks have a complicated organization: year-old birds are excluded from breeding, some cooperative feeding of young exists, sentinels watch for predators while the rest of the flock feeds, individuals show strong behavioural cohesion and tolerance of others, and pinyon nuts are stored in large communal caches.

So long as the nut crop of the pinyon pine is good, the flock remains within a small area, typically about 20 square kilometers, which is not intruded upon by other flocks. During years of nut crop failure, the flock may join others in wandering over hundreds of kilometers in search of food, but always returning eventually to its home base. The flock habitat includes ponderosa pine forest for nesting and pinyon-juniper woodland for gathering pine nuts. Foraging for insect larvae occurs widely during certain times of the year.

In autumn, roughly October to January, loosely organized flocks of 200 to 250 individuals of all age groups are spread over about I hectare while feeding and 2 to 3 hectares during inactive periods. There is virtually no aggression within the flock, even though individuals feeding on snow-free patches of ground or at a salt lick may be crowded shoulder to shoulder. The flock usually moves 2 to 3 kilometers per hour while feeding, but long-distance group movements also occur after considerable mutual stimulation within the group. At first, a few birds fly off a short distance and return to the flock, apparently encouraging other birds to do the same. Eventually this leaving-and-returning behaviour overwhelms the flock as a whole, and all the birds fly off together.

In autumn, foraging individuals concentrate on insect grubs in the soil and on the few pinyon nuts left over in cones from the summer crop. Seeds not eaten are buried on the south sides of tree trunks, where the winter sun first melts the snow and thaws the ground. The jays also choose their mates during the autumn, sealing the bond with a ceremony in which the male feeds seeds to his prospective mate. These ceremonies become more intense as the year wears on, and by winter, mated pairs stay together within the flock.

An unusual feature of the jay flock are the sentries, numbering 4 to 12, which remain at the periphery of the flock perched on high vantage points. Individuals take turns being sentries. At the appearance of hawks, foxes, and other predators, the sentries give warning calls that send the flock into trees and to protective cover; occasionally the sentries mob the intruder while the rest of the flock moves off to a new area.

Breeding begins in earnest in February, when courtship becomes more frequent, often involving chases during which the pair flies rapidly through and over trees, performing sharp turns and steep dives, and mated pairs begin to leave the flock to feed. By this time, the flock consists mostly of first-year birds during the day, but is joined by mated pairs in the early evening after a calling ceremony. Young birds begin the ceremony by uttering soft calls, which then become louder and culminate in a long flight, during which the courting birds rejoin the flock.

Nest-building commences in March over an area of 50 hectares, within which nests are spaced 15 to 150 meters apart. First-year birds stay in a small flock that remains within I kilometer of the nesting area. While females incubate their eggs, males form a separate feeding flock, which individuals occasionally leave in order to feed their mates on the nest.

Nesting is synchronized, with the first eggs of the entire flock being laid within a 3- or 4-day period. As a result, most of the young are of the same age and leave their nests at the same time. During most of the nest period, young are fed by their own parents, but 4 to 5 days before fledging, adults begin feeding chicks of other pairs, as well as their own, with as many as 7 adults providing food at any one nest.

Foster feeding continues for 3 weeks after fledging, during which time the adults and young from nearby nests form an aggregation. At any one time, a few of the adults act as sentinels for the aggregation, others remain hidden with the young in bushes, and the rest forage in a group. Within the whole flock, the individual aggregations remain distinct, even after the young begin to forage for themselves, and do not mix with others.

During August, when the seed crop of the pinyon trees has ripened, all the aggregations converge in the pinyon woodland to gather the nuts and store them in the breeding area for future use. The harvest lasts 2 to 3 weeks and marks the reformation of the large group flock and the completion of the annual cycle.

The particular circumstances of habitat, food, and evolutionary history that promoted the remarkably complex social organization of the pinyon jay are a mystery. Such extreme sociality—by no means unique—illustrates the extent to which the individual can become integrated into, and even subservient to, a larger social unit. Absence of aggression suggests that resources and mates are not strongly contested. When quietly crowded around a bird feeder, the jays seem almost self-domesticated. Sentinel duty, feeding the offspring of other pairs, and storing nuts in communal caches altruistically promote the fitness of other individuals.

Evolutionary Modification

Any social interaction other than mutual display can be dissected into a series of behavioural acts by one individual (the donor of the behaviour) directed toward the other (the recipient). One individual delivers food, the other receives it; one threatens, the other is threatened. When one individual attacks another, it may be thought of as the donor of a behaviour. The attacked individual usually responds by standing its ground or fleeing; in either case, it becomes the donor of a behaviour. The donor-recipient distinction is useful when one considers that each behavioural act results in a change in the fitness of both the donor and receiver of the behaviour. These increments may be positive or negative, depending on the interaction. Four combinations of cost and benefit for donor and recipient organize the outcomes of social interactions into four categories:

Change in fitness of		
Donor	Recipient	Category of behaviour
+	+	Cooperation, mutualism
+	–	Selfishness
–	+	Altruism
–	–	Disruption, spite

Isolated acts of behaviour probably rarely benefit both donor and recipient. By their nature, most social behaviours are asymmetric, instant by instant, although a donor may be a receiver of the same behaviour from the recipient individual at some other time. If the consequences of behaviour for fitness could be judged from the outcome of isolated acts, cooperation would be strongly selected; it probably occurs rarely, however. Spiteful behaviour—reducing another's fitness despite the cost to your own—cannot be favoured under any circumstance. Selfish and altruistic behaviours remain the most likely possibilities; selfishness should normally be selected to the exclusion

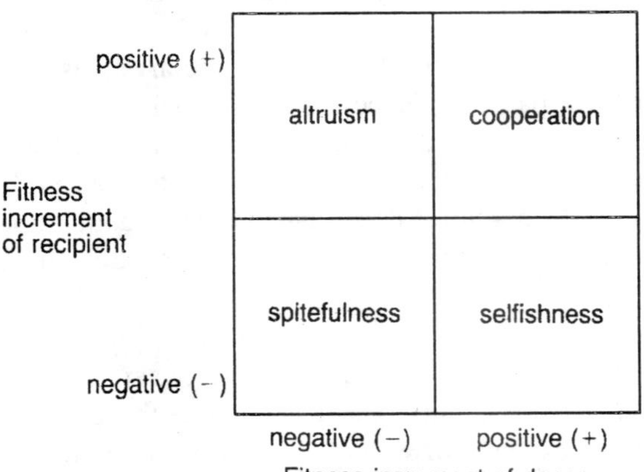

Fig. 19.4. Four types of behaviour classified according to the fitness increments of the donors and recipients of the actions.

of altruism because it increases the fitness of the donor. As we shall see below, however, altruism may be selected, in theory, when selfless behaviour is reciprocated, when interactions occur between close relatives, or when altruism enhances the fitness of discrete social groups.

An important caveat in our discussion of the evolution of behaviour: selection results from the fitness contributions of all aspects of the phenotype influenced by genetic variation; isolated acts of behaviour rarely will have independent genetic determination. Consistent patterns of behaviour are likely to be the simplest expression of genetic variation within a population. But because each act of behaviour results from a long chain of events extending through the nervous and endocrine systems from stimulus receptors to motor effectors, modulated by physiological consequences of experience and by internal rhythms and spontaneous activities, no genetic change is likely to have a simple expression in behaviour, with a simple attendant fitness contribution.

Game Theory

A remarkable aspect of the behaviour of pinyon jays is that individuals appear to share resources without conflict. The usual situation in populations lacking such social integration is for individuals to contest disputes over resources. Contests rarely result in bloodshed because conflict is organized by territorial boundaries or by dominance hierarchies. Yet neither of these systems can become established without some reliable method of evaluating relative rank. Normally this is accomplished by learning to associate certain attributes of other individuals with the probability that they will prevail in a dispute. Individuals may continually change their social status as they age, grow, or acquire particular experiences. In such a system, variation among individuals would be determined environmentally and developmentally rather than be under direct genetic control. The results of evolution must, therefore, be systems of behavioural exploration of social status and agreement over status signals within the population as a whole.

An approach to understanding the evolution of social interaction is to analyze the outcomes of interactions between individuals with different behaviour programs. This can be done through the theory of games, a simple example of which will be considered here by way of illustration. Suppose that individuals are either aggressive or submissive in contests. These opposing behaviour programs, or strategies, may be referred to as hawk and dove. Furthermore, the object of each contest potentially provides a benefit (*B*) to the winner, measured in terms of increments of fitness. The aggressive behaviour of hawks has a cost (*C*) measured in the same units, but doves avoid the cost by avoiding confrontation. Now, when two hawks contest a unit of resource, they split the cost and benefit equally; when two doves meet over a resource, they agreeably split the benefit. When a hawk and a dove meet, the dove always demurs, leaving the hawk the benefit without the cost of confrontation. Hence the hawk-dove interaction is an asymmetric one of selfishness and altruism.

In a pure population of hawks, the fitness contribution (*W*) to each individual resulting from playing the hawk-dove game, *W*(*H*) is 0.5(B -C). In a pure population of doves, *W*(*D*) is equal to 0.5 B. Clearly the doves have the better arrangement because resources are shared without the costs of confrontation. Nevertheless, to determine the evolutionary potential of each behaviour, their interaction must be compared within the same population.

		Strategy of opponent	
		Hawk	Dove
Strategy of player	Hawk	1/2 ($V - C$)	V
	Dove	0	1/2 V

Fig. 19.5. Payoffs in the hawk-dove game.

When hawks and doves encounter each other at random in a mixed population, the payoff of the hawk-dove game to each depends on their relative frequencies. Suppose the frequency of hawks in the population is p (that of doves is $1 - p$); of encounters over potential resources by both types of individuals, proportion p will be with hawks and proportion $(1 - p)$ with doves. Therefore, payoffs of the hawk-dove game in a mixed population are

$$W(H) = \frac{p}{2}(B - C) + (1 - p)B$$

and

$$W(D) = \frac{(1 - p)}{2}B$$

A little algebra shows that, all other things being equal, the proportion of hawks increases—that is, $W(H) > W(D)$—when $B > C$, regardless of the frequency p. It is a simple matter to show that H is an Evolutionarily Stable Strategy, and that D is not, whenever the benefits of contesting a resource outweigh the costs. A dove can invade a pure population of hawks ($p = 1$) only when $0 > 0.5(B - C)$; hence H is an ESS when $B > C$. Moreover, a hawk can invade a pure population of doves ($1 - p = 1$) whenever $B > 0.5B$, which is always. A mixed strategy can result only when costs of conflict outweigh the benefits ($C > B$); as the proportion of hawks increases, they encounter each other more frequently, and therefore more often suffer the costs of confrontation. A mixed ESS occurs when $p = B/C$, at which point the payoffs to both hawks and doves are the same.

The lesson of the hawk-dove game is that altruism is difficult to maintain as an evolutionarily viable strategy whenever altruists can be taken advantage of by selfish genotypes. Does this mean that altruism does not exist? Certainly a direct answer to this question depends upon whether altruism has been observed in nature, or not. Although many behaviours have been interpreted as being selfless, critics have argued each case, finding selfish motivation behind the most saintly of acts. Individuals that utter alarm calls, for example, which many think of as accepting personal risk by making themselves conspicuous in order to warn others of danger, have also been thought of as provoking panic among others in the population so as to divert the attention of a potential predator. Such differences of opinion have not been resolved because the fitness consequences of individual acts of behaviour are difficult to isolate.

A second issue raised by the altruism question concerns the mechanisms by which altruistic genotypes can be maintained in populations, provided they exist.

Reciprocal Altruism

The evolutionary problem posed by altruism has been appreciated for a long time. In *On the Origin of Species* Darwin recognized

> one special difficulty, which at first appeared to me insuperable, and actually fatal to the whole theory of evolution by natural selection. I allude to the neuters or sterile females in insect-communities; for these neuters often differ widely in instinct and in structure from both the males and fertile females, and yet, from being sterile, they cannot propagate their kind.

The distinctive instinct of sterile castes of insects is that they forgo personal reproduction and devote themselves to the fertility of the colony as a whole. Although Darwin was more interested in the evolution of divergent morphology between sterile and fertile individuals, and between different sterile castes, his explanation applies equally well to the evolution of sterility itself:

> This difficulty, though appearing insuperable, is lessened, or, as I believe, disappears, when it is remembered that selection may be applied to the family, as well as to the individual, and may thus gain the desired end. Breeders of cattle wish the flesh and fat to be well marbled together: an animal thus characterized has been slaughtered, but the breeder has gone with confidence to the same stock and has succeeded.

With the concept of selection among families as his point of view, Darwin explained the evolutionary potential of sterile castes of social insects accordingly:

> by the survival of the communities with females which produced most neuters having the advantageous modifications, all the neuters ultimately came to be thus characterized.

Darwin thus considered insect colonies as family groups whose fitness was measured in competition with other colonies. Adaptations of the sterile castes—by extension the existence of sterile castes at all—were selected according to their contribution to the productivity of the colony as a whole.

This concept of intergroup selection, originally applied to insect societies, eventually was adopted more widely to explain the evolution of social behaviour in all kinds of animal groups. The idea was accepted with little interest and more or less uncritically for many years. Then, in the early 1960s, the Scottish zoologist V. C. Wynne-Edwards (1962, 1963) advocated group selection so forcefully, and saw its magnificent consequences for social behaviour so universally, that evolutionary ecologists were forced to consider the argument more carefully.

The hornet's nest of controversy aroused by Wynne-Edwards, and those who differed with him, was discussed earlier in connection with the regulation of population size. Wynne-Edwards believed that individuals restrained themselves from reproduction in order for the population to avoid overtaxing its resources. Restraint from reproduction is altruistic toward selfish individuals, which attempt to monopolize resources and enhance their personal fitnesses within the population. Wynne-Edwards argued that selection of groups that altruistically nurtured their resources predominated over selection of selfish individuals within groups, which led to overexploitation. Opponents of Wynne-Edwards's views offered two counterarguments: first, few populations have the kind of group structure required of intergroup selection and, second, even where such selection could exist, it was bound to be weaker than individual selection because groups displace each other within populations more slowly than individuals displace each other within groups. By the mid-1960s, individual-selectionists had clearly won the day and group-selection theory faded from view. But without group selection, how was one to explain cases of altruistic behaviour? One solution was to argue that such behaviour did not exist; another was an idea proposed by John Maynard Smith (1964) and William D. Hamilton (1964), which they called *kin selection*.

KIN SELECTION

Kin selection results from differences in the fitnesses of genes based upon interactions between related individuals, which have a finite probability of inheriting copies of the same gene from an ancestor. For example, when an individual directs an altruistic act toward a sibling it enhances the personal fitness of an individual with which it shares a substantial part of its genotype. If the altruistic act occurred because of a single-gene mutation inherited from one parent, a brother or sister would have a copy of the same gene with a probability of one-half. Therefore, the fitness of the altruistic gene would exceed that of its selfish alternative so long as the cost to the altruist was less than one-half the benefit to the recipient. In general, a single altruistic act between individuals with genetic relationship r increases the fitness of the altruist gene so long as the cost (C) is less than the benefit (B) times the coefficient of relationship; that is, $C < Br$.

Hamilton (1964) generalized this cost-benefit relationship into a concept of inclusive fitness of a gene. Suppose that the fitness contributions of a single act of behaviour were w_s to the donor (self) and w_i to the recipient. The net fitness contribution of the act would be $w_s + r_i w_i$; this is inclusive fitness because it includes components (which may be gains or losses) realized through the personal fitness of both the donor and the recipient. Summed over all the interactions of the individual, the total fitness contribution is

$$W = w_s + \sum f_i r_i w_i$$

where f_i is the proportion of interactions with individuals having coefficient of genetic relationship r_i. (This equation applies strictly only when genes for the behaviour in question are rare. Otherwise, distantly related individuals may bear the same gene with high probability. It is therefore most useful in evaluating behaviours as Evolutionarily Stable Strategies.)

For behaviour directed toward individuals having coefficient of relationship i, $W > 0$ when $w_s > -r_i w_i$, or $w_i > -w_s/r_i$. Notice that the concept of inclusive fitness applies to both altruistic and selfish behaviour; there are limits to both. When siblings interact, the evolutionarily sound limits of generosity are $-w_s < 0.5w_i$ and the evolutionarily tolerable limits of selfishness are $w_s < -2w_i$. As the coefficient of relationship decreases below 0.5, altruistic behaviour becomes less likely to increase inclusive fitness ($-w_s < r_i w_i$) and selfish behaviour more likely ($w_s < -w_i/r_i$).

The concepts of inclusive fitness and kin selection are based on the coefficient of relationship between the donors and recipients of behaviour. This coefficient may be defined as the probability that one individual has an exact copy (identical by descent from a common ancestor) of a gene carried by another. When one individual has inherited a particular gene, what is the probability that a relative has inherited a copy of the same gene? By this criterion, the relationship between an individual and one of its parents is 0.5, since it must have received the gene in question from one parent or the other. Reciprocally, because each parent contributes half of its genes to each of its offspring the probability that a son or daughter will inherit a particular gene is also 0.5. The genetic relationship between full siblings is 0.5 because the gene possessed by one individual was inherited from either its mother or its father which, in either case, would pass the gene onto a sibling with a probability of 0.5.

The maintenance of altruistic behaviour by kin selection requires that such behaviour be restricted to close relatives within larger groups. Certainly individuals of many species tend to associate in family groups. When dispersal is limited, nearby individuals with which interactions are most frequent are likely to be close relatives. Moreover, individuals seem to be sensitive to their degree of relationship to others even when they have had no family experience with which to learn who are related directly. The cues that indicate relationship must involve some subtle matching of chemical, acoustic, or visual signals that are under genotypic control and are extremely variable within populations.

ALTRUISTIC BEHAVIOUR

Because group selection had been discredited and kin selection could produce altruistic behaviour directly only toward close relatives, an opening remained for a mechanism that could account for cooperation distributed widely throughout a population. One candidate for this position— reciprocal altruism—was proposed by Robert Trivers in 1971.

Trivers (1985) defines reciprocal altruism as the exchange of altruistic acts between individuals. As long as the cost to the donor of each act is less than the benefit to the recipient, and two individuals give and take more or less equally, the arrangement of reciprocal altruism benefits the participants. To avoid bestowing altruistic behaviour on individuals who do not respond in kind, organisms keep track of who reciprocates and who doesn't, and confine their altruism accordingly. Under this system, nonreciprocators are at a disadvantage and may be selected against, leaving a population of highly cooperative individuals. As Trivers (1971, 1985) points out, however, reciprocal altruism can be expected to arise only when individuals have a long association with each other and can discriminate between those that reciprocate and those that don't.

Several studies, such as G. S. Wilkinson's (1984) on food sharing in vampire bats, claim to present evidence for reciprocal altruism, but support is still relatively weak. And although organisms certainly distinguish individuals within a population, it remains to be shown that altruistic behaviour is distributed among individuals according to their past history of interaction.

A further difficulty with reciprocal altruism is that because it requires a reciprocating partner in order to work, it can confer little fitness when reciprocators are rare. In fact, an altruistic gene cannot easily invade a nonreciprocating population. Reciprocal altruism can be an Evolutionarily Stable Strategy. R. Axelrod and Hamilton (1981), Axelrod and Dion (1988), and Smith (1982) used game theory analyses to show that reciprocation can be a successful strategy for a player. For example, suppose that a population consists of three genotypes: selfish individuals, which always take what is offered, with a gain of two units of fitness (the accept strategy. A); altruistic individuals, which always donate help at a cost of one unit (offer, 0); and reciprocators (R), which always do what an individual last did to

		Strategy of recipient		
		Accept	Offer	Reciprocate
Strategy of donor	Accept	0	2	0
	Offer	−1	1	1
	Reciprocate	0	1	1

Fig. 19.6. Payoffs in a reciprocal altruism game.

them, except that, being ever hopeful, their first act is to offer. In a population consisting of all three types, fitnesses are ranked in order $W_A > W_R > W_O$. As the altruistic offerers are eliminated from the population, the reduced payoff matrix in the absence of offerers favours reciprocators over acceptors, and reciprocal altruism prevails. But one can also see that in a pure population of acceptors, the reciprocator has no net fitness advantage. Hence the evolution of altruism through reciprocity requires a mechanism to increase the proportion of reciprocators initially to the point that they can do each other some good. This may be particularly difficult because reciprocation is a complicated strategy involving responses conditioned on prior experience.

Operating Kin Selection

As ecologists turned their attention to the evolution of social behaviour in the 1960s, several phenomena provided foci for discussion. One of these was the evolution of alarm calling. Although many dismissed this behaviour as being selfish, Paul Sherman's (1977) observations on the behaviour of Belding's ground squirrels in the Sierra Nevada of California offered strong support for the hypothesis that alarm trills given in the presence of terrestrial predators are maintained by kin selection.

Because ground squirrels are social creatures and exposed to view in their montane meadow habitat, the threat of predation is great. Reducing this threat with alarm calls warning of danger is an important factor in their lives. Sherman found that alarm callers to terrestrial predators were adult and first-year females; males and yearling females did not exhibit the behaviour. He also found that males disperse widely from their place of birth, whereas females are sedentary. Moreover, females called more frequently when they had occupied the same territory for many years, hence female relatives were likely to be nearby, or when relatives were known to be present in the population.

Sherman also determined that alarm calling to terrestrial predators has a cost. When predators approached groups of ground squirrels, at least one of whom gave an alarm call, 14 of 107 calling individuals were attacked (13 per cent) compared with 8 of 168 silent individuals (5 per cent). These observations are consistent with kin selection playing a role in the evolution of alarm calling.

Kin selection arguments also have been applied to explain the phenomenon of helping at the nest, observed in many species of birds. Initial reports of the behaviour, by Alexander Skutch (1935, 1961) and C. Hilary Fry (1972), indicated that in several species, particularly in the tropics, the young in a single nest were fed by more than a single pair of adults. Subsequent observations on populations in which birds were colourbanded for individual recognition revealed that the helpers were usually the pair's offspring from previous nestings. Hence "helping" usually involves individuals assisting their parents to raise younger siblings, or assisting their siblings to raise their nieces and nephews. Frequently, helpers are birds raised the previous year, hence they are at least a year old and physiologically capable of breeding on their own.

Helping is viewed as altruistic in the sense that helpers forgo producing their own offspring for the sake of enhancing the personal fitness of their parents. But helping may either benefit or hinder the parents, and it may benefit or cost the helping individual. Because helpers are related to both their parents and their siblings by coefficient 0.5, the combinations of cost and benefit to donor and recipient favoured by kin selection are fairly straightforward. Most studies indicate that helpers increase the productivity of their parents substantially, but rarely to the extent of individuals breeding on their own. As Glen Woolfenden and John Fitzpatrick (1984) point out, individuals may also enhance their own personal fitness by remaining with their parents: they are more likely to survive on familiar territory, and they may

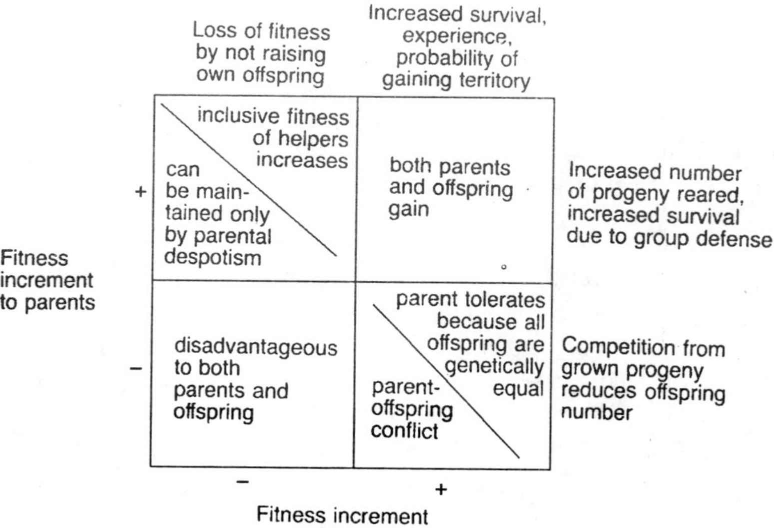

Fig. 19.7. Explanation of fitness increments to parents and offspring arising through helping at the nest (offspring helping their parents to near siblings).

inherit the family territory when their father dies. Helping also may substitute for investment of the true parents, thereby increasing their survival.

Why is helping more frequent in the tropics? High adult survival rates in the tropics allow family groups to remain intact from year to year with high probability; low frequency of death and replacement of one mate also means that helpers assist in rearing full siblings more frequently than half siblings. In addition, when populations are regulated by density-dependent factors, and when recruitment of one-year-old birds exceeds death of adults, young birds would have to compete with more experienced older birds for breeding territories. Under such conditions, which seem common in the tropics, the inclusive fitness of genes which dictate that a young bird remain with its parents may exceed the inclusive fitness of genetic tendencies to set out on one's own. Although helping could evolve under conditions in which such cooperation enhanced the personal fitness of both parent and offspring, inclusive fitness considerably widens the range of costs and benefits to each under which the behaviour is favoured.

Warning Colouration

Warning colouration signals predators that potential prey items are distasteful, poisonous, or both. One can see straight off that conspicuous advertising of this quality benefits the vulnerable prey if it dissuades would-be captors; recognizing the meaning of the display also benefits a predator, who avoids the regret of unwise eating. Adopted by all members of both prey and predator populations, these conventions of advertisement and appropriate response are unquestionably Evolutionarily Stable Strategies. But how does the system get started? Can a genetic mutation leading to warning colouration invade a population of distasteful, cryptic prey hunted mercilessly by visually oriented predators which cannot distinguish them from other palatable species? A single individual with warning colouration would be rendered conspicuous and vulnerable in a population of otherwise cryptic organisms, and would likely be nailed by an unsuspecting predator. Noxiousness and warning colouration can confer fitness only if potential predators have opportunities to learn their meaning—by eating someone else.

In the case of a novel mutation, the most likely other individual in a population to have the same rare gene (identical by descent) is a parent or sibling. Half of the individuals in a single brood are likely to have the same unique gene inherited from one parent. If the

Appearance	Dispersion of larvae: Aggregated	Solitary
Aposematic	9 species	11
Cryptic	0	44

Fig. 19.8. Relationship between appearance of larvae of various species of lepidoptera and their degree of aggregation.

conditioning stimulus is strong enough, one attempt to eat a noxious, aposematic individual may save the lives of several of its brothers and sisters. With its inclusive fitness thus enhanced, the mutation could spread rapidly through the population.

This scenario requires that members of a brood remain in close proximity so that the experience of a single predator is translated into the rejection of a sibling. Among butterflies and moths, warning colouration and aggregation of broods of larvae are strongly associated, which is consistent with the idea that kin selection is important to the evolution of warning colouration. Among species that disperse their eggs widely over many food plants, warning colouration is less frequent.

Parental Investment

Parental investment represents an arrangement between parents and their offspring. Rather than passively accepting whatever their parents offer, most offspring actively solicit care. Young animals beg for food and solicit brooding; eggs actively take up yolk from the ovarian tissues or bloodstream of the mother. For the most part, the fitness interests of parent and offspring are compatible; when progeny thrive, so do their parents' genes. But when the selfish accumulation of resources by one offspring reduces the overall fecundity of the parents, parent and offspring can come into conflict. Trivers (1974) was the first to give parent-offspring conflict a theoretical basis. Each act of parental care benefits the offspring by enhancing its survival, but costs the parent by decreasing the number of other contemporary or future offspring. Resources allocated to one child cannot be delivered to others; prolonged care delays the birth of subsequent children; risks of caring for today's children decrease the probability that parents will survive to rear tomorrow's.

From the standpoint of the individual offspring and relative to the benefit bestowed upon its own genotype, the cost of parental care—measured by a reduction in the number of its siblings—must be discounted by one-half; a sibling shares only one-half of an individual's genes identically by descent. Therefore, when an individual possesses a genetically determined trait that increases the delivery of parental care to it, that trait is favoured so long as the cost, in terms of number of siblings, is less than twice the benefit to the individual.

As offspring develop and become more self-sufficient, the ratio of benefit to cost of continuing to care for them decreases. The cost of a particular parental act may change little over the age of the offspring but, as the young mature and are better able to care for themselves, the benefits of parental care dwindle. When the benefit/cost ratio drops below 1, the parent should cease to provide care to that offspring in favour of producing additional ones. Suppose, however, that a child has a gene that increases its solicitation of parental care. Because the inclusive fitness of the child discounts the cost of care to the parent by one-half, from the standpoint of the child, parental acts should be continued until the benefit/cost ratio is one-half. Thus the period of time between the ages at which B/C is 1 and 0.5 is one of conflict between parent and offspring. With respect to the timing of weaning, Trivers stated:

> Weaning conflict is usually assumed to occur either because transitions in nature are assumed always to be imperfect or because such conflict is assumed to serve the interests of both parent and offspring by informing each of the needs of the other. In either case, the marked inefficiency of weaning conflict seems the clearest argument in favour of the view that such conflict results from an underlying conflict in the way in which the inclusive fitness of mother and offspring are maximized. Weaning conflict in baboons, for example, may last for weeks or months, involving daily competitive interactions and loud cries from the infant in a species otherwise strongly selected for silence. Interactions that inefficient within a multicellular organism would be cause for some surprise, since, unlike mother and offspring, the somatic cells within an organism are identically related.

ESS reasoning has been applied to the problem of parent-offspring conflict in a series of difficult papers by G. A. Parker and M. R. MacNair (1978, 1979) and MacNair and Parker (1978, 1979). The major perplexity of the evolutionary resolution of parent-offspring conflict is the fact that conflicting offspring themselves grow up to be parents of conflicting offspring, presumably themselves suffering the same way they made their own parents suffer. This can be shown not to be a difficulty by a simple model of the fate of a conflictor gene passed down through progeny to grandprogeny. The examples clearly show that such a gene will invade a population of nonconflictors as long as the cost/benefit ratio is less than 2. When conflict affects the number of half-siblings (r = 1/4), rather than full siblings (r = 1/2), the cost/benefit ratio must be less than 4, and both the intensity and duration of parent-offspring conflict will be greater.

Given that conflict exists, to what extent can parents control how much care they deliver to their offspring? Parents certainly are physically in control in the sense of behavioural dominance. But parents also must rely on solicitation signals from their offspring to determine how much care should be delivered. Young can sense their state of nutrition better than their parents can, and so parents have to rely on communication from the offspring to adjust feeding rates. Offspring might increase their rate of solicitation of care to support increased growth rate, greater activity and perhaps behavioural dominance over siblings, or storage of nutrients for use after weaning. If they changed any of these things, increased demand would accurately reflect their greater requirements, which could be provided only at the expense of other, more reticent young. This could be countered from the parent's standpoint only through reduced responsiveness to solicitation, which would in turn favour offspring with lower requirements.

Regardless of whether parents are in control or not, Amotz Zahavi (1977) has pointed out that, by punishing or withholding resources from selfish offspring, parents reduce their own fitness. Ousting an unrelated competitor from one's territory or killing a microbe that has invaded the body is one thing, but kicking a child out of the house or destroying a growing ovum would be self-defeating. Marcus Feldman and I. Eshel (1982) took a population genetics approach to this intriguing problem. They showed, first, that an altruist cannot invade a brood of selfish offspring when the decrease in personal fitness of the altruist compared to its selfish siblings is more than twice the benefit of increased fecundity of the parent. Hence conflicting up to a B/C ratio of 2 is an ESS. Then they asked whether a gene for parental interference could invade if it depressed the fitness of the selfish individual and redistributed that fitness to the rest of the brood. They found that when the overall fitness of a brood is lowered by parental interference, and when there is free recombination between the conflictor and interferor gene loci, fixation of selfish offspring and noninterfering parents is stable. When the loci are tightly linked genetically, that combination may be unstable and parental interference may be able to prevent the invasion of the conflictor gene. These models greatly oversimplify the genetic basis of parental care, and their ambivalence in spite of this simplicity suggests that the issue will be difficult to resolve.

Evidence bearing on conflict between parent and offspring is summarized by Trivers (1984) and Shaanker et al. (1988). There is little doubt that such conflict occurs and that both the level of care and its duration are worked out with some difficulty. The issue of conflict over weaning time is clouded by the development of offspring, whose behavioural capabilities (hence ability to solicit actively) continually increase with age. Perhaps the apparently rising "conflict" over weaning observed in many species merely reflects growth and maturation of the young. Furthermore, the costs and benefits of "conflicting" behaviour have yet to be measured to determine whether the cutoff line is $B/C = 1$ or 1/2. At present, therefore, the concept of parent-offspring conflict should provide a point of reference for discussing behaviour and physiology governing parental investment in offspring.

Social Insects

The complex societies of the termites, ants, bees, and wasps have presentee a formidable challenge to evolutionary ecologists, primarily because of the existence of nonreproductive castes. Darwin recognized that the evolutionary future of such castes was assured by their contribution to the fitness of the whole colony, and suggested that evolution proceeds in such species by a process similar to family selection in artificial breeding programs. George C. Williams and Doris C. Williams (1957) placed the evolution of sterile castes of insects in the context of competition between families. Subsequent discussion of the problem has become more complicated as arguments based upon kin selection, cooperation, and parental manipulation have been raised. Thus insect societies have caught the attention of all camps concerned with the evolution of social and family behaviour more generally.

Before considering the evolutionary issues raised by insect societies, we shall discuss a bit of their natural history. The best reference remains E. 0. Wilson's (1971) book *The Insect Societies*. Wilson defines several grades of sociality, the highest of which is eusociality. This grade is characterized by several adults living together in groups, overlapping generations—that is, parents and offspring together in the same colony—cooperation in nest building and brood care, and reproductive dominance, including the presence of sterile castes. Defined thusly, eusociality is limited among insects to the termites (Isoptera) and the ants, bees, and wasps (Hymenoptera), although the elements of eusociality are present in one mammal, the African naked mole rat.

The complex organization of the insect society is dominated by one or a few egg-laying queens. Nonreproductive progeny of the queen gather food and care for developing brothers and sisters, some of which become sexually mature, leave the colony to mate, and establish new colonies. Most insect societies are huge extended families.

Bee societies are simply organized; females are divided among a sterile worker caste and a reproductive caste that is produced seasonally. Whether an individual will become a sterile worker or a fertile reproductive is controlled by the quality of nutrition given the developing larvae. In general, differentiation of sterile castes is stimulated by environmental, usually nutritional, factors. The development of sexual forms can be inhibited by substances produced by the queen and fed to the larvae. In bees, the worker caste represents an arrested stage in the development of the reproductive female, stopped short of sexual maturity.

Ant and termite colonies often have a continuous gradation of worker castes, ranging from very small individuals that are primarily responsible for the nutrition of the colony to larger individuals that are specialized morphologically to defend against intruders. The so-called soldiers are often equipped with formidable mandibles and stings; those of some species of termites can produce noxious gases (*Rhinotermes*) or direct a stream of noxious fluid at an intruder (*Nasutotermes*). In termite colonies, workers are both male and female; in hymenopteran societies, workers are all females produced from fertilized eggs. Males, which develop from unfertilized eggs, appear in colonies only as reproductives (drones) which leave to seek mates.

The queen in many insect societies is highly specialized as an egg-laying machine. The queens of some termites may lay as many as 6000 to 7000 eggs per day for many years. Termite queens do not store sperm, so sexually active males must accompany them continuously. By contrast, most bee and wasp queens can store millions of sperm for many years and produce an entire colony from a single mating. Some ant queens have been known to be sexually active for as long as 15 years, laying more than a million eggs.

The extreme specialization of the queen is a problem at colony-founding time because there are no progeny to take care of the queen's

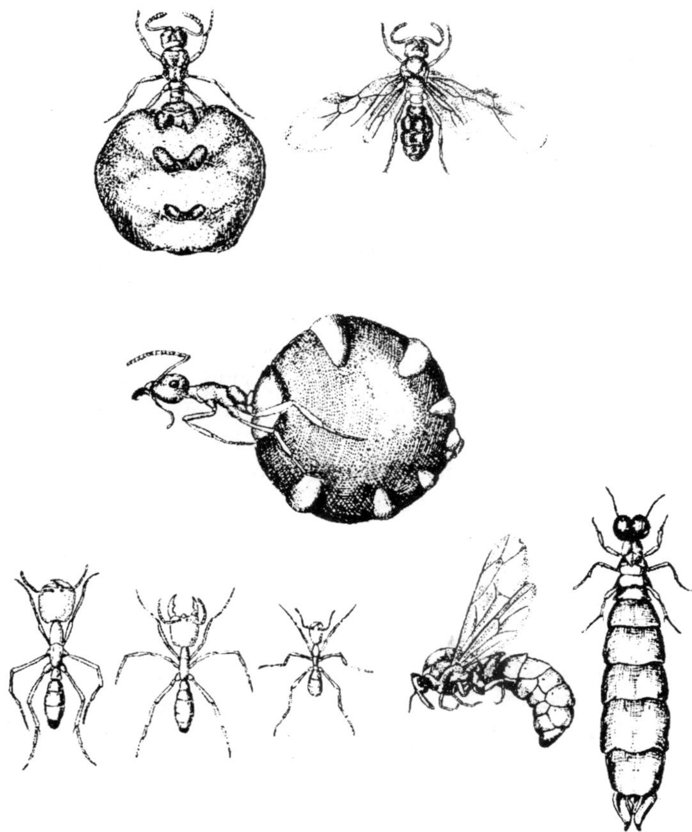

Fig. 19.9. Castes of several species of ants: a virgin queen (top right) and an old egg-laying queen (top left) of the workerless social parasite Anergates atralulus of Europe; a replate of the hone ant Myrmecosystus melliger from Mexico (center); three sizes of blind workers, a winged male, and a queen, blind and wingless, of the African visiting ant Dorylus nigricans (from left at bottom).

first brood. The female must initially provide food for her young. Honeybee and army ant queens have overcome this problem by taking part of the parental colony in which they were born to help found their own colonies and to help care for the first broods. In most other species of ants and in termites, new queens feed their first larvae with secretions from their body, and then wait until their brood are large enough to gather food before continuing to reproduce. In some of the more primitive ponerine ants of Australia, the queen herself, when founding a new colony, builds the first shelter and gathers food for her brood. Colony growth in these ants is slow at first because the queen must first perform most of the tasks eventually provided by workers before she can develop into an efficient egg-laying machine.

Evolution of Eusociality

From its distribution across taxonomic groups, it is clear that eusociality has evolved independently many times in the bees, wasps, and ants. It is less clear what route was followed. The most widely accepted sequence of evolutionary steps in the development of insect societies is that envisioned by William M. Wheeler (1928). The steps must have included a lengthened period of parental care of the developing brood, either guarding the nest or continual provisioning of the larve in a manner similar to birds feeding their young. If a parent lived and continued to produce eggs after its first progeny emerged as adults, then the offspring would be in a position to help raise the subsequent brood. This overlapping of generations, which is not common among insects, and extensive parental care are necessary ingredients in the recipe for eusociality. Once progeny remain with their mother after they attain adulthood, the way is open to relinquishing their own reproductive function solely to support hers.

Only Charles D. Michener (1958, 1969) and N. Linn and Michener (1972) have proposed a serious alternative to Wheeler's scenario. They suggested that insect societies originated in the aggregation of unrelated adult individuals, most of which eventually became subordinated to a single reproductive female. Some possible early stages of this sequence are known from bees and wasps, and while most cases of several females occupying the same nest are parasitic or mutually aggressive, benefits may accrue to each participant through joint defense of the nest against predators and parasites. Kenneth M. Noonan (1981) has shown, however, that foundresses may be close relatives, actively chosen through kin recognition and contesting nest sites with unrelated individuals.

Regardless of their origins, insect societies are not necessarily peacefully maintained. There is considerable conflict between colonies, reinforced by acute senses of kin recognition, and between queens within colonies. Workers and queens also dispute egg-laying; one encounters a spectrum from species in which workers regularly produce eggs, to those in which workers become fecund only after a queen dies, and those in which the workers are permanently sterile. Indeed, the workers of slave-making ants establish dominance hierarchies among themselves.

Most ideas about the evolution of eusociality have centered on the sterile caste as an alternative strategy of reproduction that maximizes the sterile individual's contribution to the inclusive fitness of its genes. This could occur where there is strong local resource competition among close relatives for some resource essential to reproduction, and this competition can be minimized by specialization of function, as between egg-laying and food gathering/brood tending.

Inclusive fitness arguments have dominated thinking about the evolution of eusociality since William D. Hamilton (1964, 1972) pointed out the curious asymmetry of genetic relationship within colonies of social hymenoptera resulting from their haplodiploid breeding system. Remember that males are haploid and themselves have no father; females are diploid offspring of two parents. Therefore, of the genotype of a female wasp, a male sibling contains only one-quarter of the genes identically by descent (none of the male parent, and one-half of the female parent), whereas a female sibling contains three-quarters of the genes identically by descent (the entire set of the father and half the set of the mother). Because the genetic relationship between females and either their sons or daughters is one-half, Hamilton reasoned that a female wasp should prefer to raise sisters than to raise its own offspring. This should greatly favour the evolution of eusociality and worker sterility.

Richard D. Alexander (1974) argued against Hamilton's idea on the grounds that female workers in bee and ant colonies produced roughly equal numbers of both male and female reproductives, whose average degree of genetic relatedness is one-half, and that termites have highly developed eusociality but the more usual sex-determining mechanism in which both males and females develop from fertilized eggs and are interrelated with genetic coefficients of correlation of one-half. Alexander preferred the idea that insect societies evolved through parental manipulation of broods to raise additional offspring rather than to lay their own eggs or go off on their own to found new colonies. Quite simply, any worker that attempted to mature sexually is killed by the queen. Only when a queen dies can workers become reproductives, and then there may be vicious fighting to determine which ascends the throne. As colonies increase in size, and the odds that an individual worker can become a queen following the death of its parent diminish, the best strategy of offspring may be to increase inclusive fitness by working, so long as the queen lives. In colonies of millions of individuals, the queen signals her presence by producing a distinctive and reassuring odor, or pheromone.

INDEX